ALSO BY *Frank R. Wilson*

Tone Deaf and All Thumbs?

CO-EDITOR WITH FRANZ ROEHMANN

The Biology of Music Making

Music and Child Development

The Hand

The Hand

How Its Use Shapes the Brain, Language, and Human Culture

FRANK R. WILSON

Vintage Books

A Division of Random House, Inc.

New York

FIRST VINTAGE BOOKS EDITION, SEPTEMBER 1999

 Published in the United States by Vintage Books, a division of Random House, Inc., New York, and simultaneously in Canada by Random House of Canada Limited, Toronto. Originally published in hardcover in the United States by Pantheon Books, New York, in 1998.

Permissions acknowledgments appear on pages 379–80.

The Library of Congress has cataloged the Pantheon edition as follows:
Wilson, Frank R.
The hand / Frank R. Wilson.
p. cm.
Includes bibliographical references and index.
ISBN 0-679-41249-2
1. Hand—Physiology. I. Title.
QP334.W53 1998
612'.91—dc21 97-46427
CIP

Vintage ISBN: 0-679-74047-3

www.vintagebooks.com

Printed in the United States of America
10 9

FOR PAT, SUZANNA, AND JEFF,

without whom there

would be no point

to anything

I am quite a wise old bird, but I am no desert hermit who can only prophesy when his guts are knotted with hunger. I am deep in the old man's puzzle, trying to link the wisdom of the body with the wisdom of the spirit until the two are one. At my age you cannot divide spirit from body without anguish and destruction, from which you will speak nothing but crazy lies.

—Padre Ignacio Blazon
in *Fifth Business* by Robertson Davies

Contents

Acknowledgments

Work on this book began in the summer of 1989, as I was starting a sabbatical year in the department of neurology at the University of Düsseldorf, in what was then West Germany. Nineteen eighty-nine turned out to be a good year for this trip: the Berlin Wall came down, the oppressive governments of Eastern Europe fell, and the world began to seek a new understanding of itself. No one knew, or could guess, what was coming next.

As you will see, *little did we know what was coming next* is a classic evolutionary and developmental theme, visited and revisited in this book. It is also a specific description of the journey my wife and I began as we headed out from New York toward Southampton aboard the *QEII.* More than once since that bright and hopeful departure I have had reason to recall the prophetic words of Lauren Sosniak (about whose work with the University of Chicago's Development of Talent Research Project you will read later in the book): "No one had any idea what they were getting involved in at the start; no idea how long it would take, no idea where it would lead."

Writing this book has been like that—*exactly* like that—and now I face the serious but happy task of thanking the many people who contributed over an extended period to the unexpected achievement I believe our joint efforts represent. No one could have planned the book that finally emerged from this cooperative venture, nor can anyone know what comes next—how these ideas and experiences will be received. But I think we can be proud of what we have attempted to say, and to ask.

First and foremost, I wish to express my profound gratitude to Dan Frank, my editor at Pantheon, who lavished his attention and vision on this project throughout its long gestation, patiently guiding the development of its theme and the definition of its boundaries. As publication of the book neared, I discovered and came to depend upon several others at Pantheon whose affectionate attention to this project made me feel as if I had become part of their family: Kristen Bearse, Ed Cohen, Sharon Dougherty, Kathy Grasso, Jay Gress, Altie Karper, Grace McVeigh, Claudine O'Hearn, and Meryl Zegarek. Equally, my deepest thanks to my agent Rosalie Siegel, who never lost her patience as two years turned into four, and four into eight. Rosalie never lost her sense of humor, never let a dark cloud linger on our horizon.

Shortly after Pat and I moved to Düsseldorf, I met and then formally interviewed Anton Bachleitner, the director of the Düsseldorf Marionette Theater. This surprising encounter triggered more than a score of similar interviews whose powerful collective message reminded me of the famous dictum of Wilder Penfield, one of the great patriarchs of neurology and neurosurgery: *La neurologie cherche à comprendre l'homme lui-même*—it is the task of neurology to understand man himself. To all those who shared their stories with me (most, but not all of whom you will meet in these pages), I thank you, and assure you that each and every one of you lives equally in the heart of this book: Robert Albo, Anton Bachleitner, Jeanne Bamberger, Anat Baniel, Meg Bennett, Jochen Blum, Nancy Byl, Joseph Cleary, Matthew Dickey, Leonard Gordon, David Hall, Barbara Hansen, Reed Hearon, Elizabeth Jones, Harlan Lane, George McLean, Mark Mitton, Richard Moore, Robert Norman, Patrick O'Brien, Loren Pedersen, Serge Percelly, David Ransom, Jack Schafer, Dorothy Taubman, Richard Unger, and Richard Young.

Several individuals were particularly generous with their time, professional knowledge, and enthusiasm in helping me develop and assemble the book's scientific, educational, and professional themes: Jeanne Bamberger, Harlan Lane, Mary Marzke, Mark Mitton, and Pat O'Brien. I am also grateful to Donald Johanson for making available for illustration specimens from the collection of the Institute of Human Origins at Arizona State University. Important information about the history of puppets was provided by two prominent British puppeteers: George Speaight and Ted Beresford. Dr. Anastassios Simonidis, Consul-General of the Republic of Cyprus in San Francisco, kindly and patiently led me through a maze of

Greek terminology to a treasury of crucial information about ancient Greek medicine and puppetry.

Several individuals provided access, counsel, solitude, and/or sanctuary at crucial times: Marit Bakke, John Beebe, Martha Crewe, Susan and Robert Guerguy, Bonnie Helms, Robert Hopcke, Elizabeth Jones, Steve Kaye, Katie Clare Mazzeo, Helmut Möller, Loren Pedersen, Earl Reeve, Lynn Rogers, Barry and Norma Fisher Saipe, David Turner, and Richard Unger (and Alana, and Shannon McCarthy). Peter F. Ostwald—mentor, colleague, and friend—provided inspiration and example over many years of our association at the University of California School of Medicine in San Francisco, and I delight that our friendship continues to gain meaning through the warmth and spirit of Lise Dechamps Ostwald, who has carried Peter's work forward with extraordinary grace and skill.

Reaching back past the formal beginnings of this book, and continuing to the present, many professional colleagues have influenced and encouraged the ideas in this book and have exerted a powerful influence on its character and direction. For all this encouragement and nurture, and for the unparalleled joy of a life in medicine and neurology, my deepest thanks and appreciation: Susan Brown, Ted Boroian, Nancy Byl, Wally Cook, David Cooke, Robert Feldman, Robert Fishman, Robert Freeman, Hans-Joachim and Elsche Freund, Sid Gilman, Mark Hallett, Volker Hömberg, Robert Joynt, Julie Krevans, David Marsden, Mike Merzenich, Sorna and Eva Rajan, Flaviu Romanul, Oliver Sacks, Seymour Sarason, Robert Silverman, Holly Smith, Raoul Tubiana, and Christoph Wagner.

Several friends have steadfastly "attended" in unique and invaluable ways, from the very beginning: George Moore, neurophysiologist, cellist, and provocateur-at-large, who presented me with Sir Charles Bell's *The Hand* at the beginning and thereafter never stopped reminding me of the book I had to write; Jack Meehan, master teacher of music and a master of the art of balanced living, who has kept me in tune with his hands, his sense of humor, and his heart; Clara Toms, my office nurse for many years, who handled every problem as if it were nothing, who anticipated every need, and who watched over me and my family as if we were her own; Franz Roehmann, composer, educator, mountain man, skier, and sax player, who has been the fastest and kindest friend anyone could possibly hope to have;

my younger sister, Julie Massey, whose passion for theater brightens and energizes everyone's day at The Goodman in Chicago, and who will awaken you with wit and zest, as actors do, when you are not watching or listening as closely as you should.

Finally, with deepest love I wish to remember several people whose devoted work and love of life left a deep and lasting imprint—who in a superficial, *earthly* way, have vanished: my mother and father (whose warm and caring hands I can still see and feel); Eloise Ristad, John Blacking, Rosario Mazzeo, and Michael Hendren. Each, I know, continues to incite mirth, love, and a passion for the open road among the unsuspecting but lucky living, as predicted by D. H. Lawrence in *Fantasia of the Unconscious*:

> The living live, and then die. They pass away, as we know, to dust and to oxygen and nitrogen and so on. But what we don't know, and what we might perhaps know a little more, is how they pass away direct to life itself—that is, direct into the living. That is, how many dead souls fly over our untidiness like swallows and build under the eaves of the living? How many dead souls, like swallows, twitter and breed thoughts and instincts under the thatch of my hair and the eaves of my forehead? I don't know, but I believe a good many. And I hope they have a good time. And I hope not too many are bats.[1]

The Hand

Prologue

Early this morning, even before you were out of bed, your hands and arms came to life, goading your weak and helpless body into the new day. Perhaps your day began with a lunge at the snooze bar on the bedside radio, or a roundhouse swing at the alarm clock. As the shock of coming awake subsided, you probably flapped the numb, tingling arm you had been sleeping on, scratched yourself, and maybe even rubbed or hugged someone next to you.

After tugging at the covers and sheets and rolling yourself into a more comfortable position, you realized that you really *did* have to get out of bed. Next came the whole circus routine of noisy bathroom antics: the twisting of faucet handles, opening and closing of cabinet and shower doors, putting the toilet seat back where it belongs. There were slippery things to play with: soap, brushes, tubes, and little jars with caps and lids to twist or flip open. If you shaved, there was a razor to steer around the nose and over the chin; if you put on makeup, there were pencils, brushes, and tubes to bring color to eyelids, cheeks, and lips.

Each morning begins with a ritual dash through our own private obstacle course—objects to be opened or closed, lifted or pushed, twisted or turned, pulled, twiddled, or tied, and some sort of breakfast to be peeled or unwrapped, toasted, brewed, boiled, or fried. The hands move so ably over this terrain that we think nothing of the accomplishment.[1] Whatever your own particular early-morning routine happens to be, it is nothing short of a virtuoso display of highly choreographed manual skill.

Where *would* we be without our hands? Our lives are so full of commonplace experience in which the hands are so skillfully and silently involved that we rarely consider how dependent upon them we actually are.

We notice our hands when we are washing them, when our fingernails need to be trimmed, or when little brown spots and wrinkles crop up and begin to annoy us. We also pay attention to a hand that hurts or has been injured.

The book you are holding is a meditation on the human hand, born of nearly two decades of personal and professional experiences that caused me to want to know more about the hand. Among these, two had the greatest impact: first, as an adult musical novice, I tried to learn how to play the piano; second, as an experienced neurologist, I began to see patients who were having difficulty using their hands. Each experience afforded its own indelible lessons; each spawned its own progeny of questions.

Like most people, I have spent the better part of my life oblivious to the workings of my own hands. My first extended attempt to master a specific manual skill for its own sake took place at the piano. I was in my early forties at the time and in my dual role as parent and neurologist had become enchanted by the pianistic flights of my twelve-year-old daughter, Suzanna. "How does she make her fingers go so fast?" was the question that occurred to me when I interrupted my listening long enough to watch her play. I read everything I could about the subject and finally realized I would never find the answer until I took myself to the piano to find out.[2]

As a beginning student I imagined that music learning would go just as it is depicted by music teachers: begin with simple pieces, learn the names of the notes, practice scales and exercises, memorize, play in student recitals, then move on (shakily or steadily) to more and more difficult music. But over the course of five years of study my personal experience deviated further and further from this itinerary. It was not that I was fast or slow, musical or unmusical; at various times I was each of those. Despite the guidance of a seasoned teacher armed with the highly polished canons of music pedagogy, the whole enterprise was rife with unexpected turns, detours, and diversions. Inside me, it seems, there was *already* a plan for being a musician—a modest one, but a plan nonetheless: the protocols of music had simply set the specific cognitive, motor, emotional, and social terms according to which hand and finger movements that were initially unsure and clumsy would gradually become more accurate and fluent. As I hope to demonstrate—even to the satisfaction of music teachers—I might as easily have been in a woodcarving class, or learning how to arrange flowers or build racing-car engines.

After several years of piano study I began to see musicians as patients. Most came expecting that a doctor with musical training would better understand their physical problems than one without such experience. Later, the "hand cases" also came from restaurants, banks, police stations, dental offices, machine shops, beauty parlors, hospitals, ranches. All came for the same simple reason: they could not do their jobs without a working pair of hands.

A major turning point in my thinking about the hand came as the result of a presentation I made to a group of musicians about a particularly difficult and puzzling problem called musician's cramp. I had brought along a video clip to show during the talk. It was a brief clinical-musical medley of hands that had either been injured or had mysteriously lost their former skill; formerly graceful, lithe, dazzlingly fast hands could barely limp through the notes they sought to draw out of pianos, guitars, flutes, and violins. Just a few minutes after the film began, a guitarist in the audience fainted. I was amazed. This was not the sort of grotesque display one sometimes sees in medical movies; these were just musicians unable to play their instruments. When the same thing happened at subsequent presentations—a second and then a third time—I was genuinely puzzled. I decided I must have missed subtleties or hidden meaning in these films apparent only to very few viewers. It was not until much later that I came to understand the real message these fainting musicians were expressing.

I now understand that I had failed to appreciate how the commitment to a career in music differs from even the most serious amateur interest. Although I had worked very hard as a beginning piano student, took the work seriously and spent a great deal of time at it, it was not my *life.* Consequently I did not anticipate the profound empathy for the injured musicians that would be felt by some viewers of these films. Moreover—and this is a lesson I learned, one person at a time, as I conducted interviews with nonmusicians for this book—when personal desire prompts anyone to learn to do something well with the hands, an extremely complicated process is initiated that endows the work with a powerful emotional charge. People are changed, significantly and irreversibly it seems, when movement, thought, and feeling fuse during the active, long-term pursuit of personal goals.

Serious musicians are emotional about their work not simply because they are committed to it, nor because their work demands the public expression of emotion. The musicians' concern for their hands is a byproduct of the intense striving through which they turn them into the essential physical instrument for realization of their own ideas or the communication of closely held feelings. The same is true of sculptors, woodcarvers, jewelers, jugglers, and surgeons when they are fully immersed in their work. It is more than simple satisfaction or contentedness: musicians, for example, love to work and are miserable when they cannot; they rarely welcome an unscheduled vacation unless it is very brief. How peculiar it is that people who normally permit themselves so little rest from an extreme and, by some standards, unrewarding discipline cannot bear to be disengaged from it. The musician in full flight is an ecstatic creature, and the same person with wings clipped is unexploded dynamite with the fuse lit. The word "passion" describes attachments that are this strong. As I came to learn how such attachments are generated, it became the mission of this book to expose the hidden physical roots of the unique human capacity for passionate and creative work. It is now abundantly clear to me that these roots are more than deep and more than merely ancient. They reach down, and backward in time, *past* the dawn of human history to the beginning of primate life on this planet.

Paleoanthropology—the study of ancient human origins—has until recently been better known to the public through cartoon images than through its serious work. But this seemingly dryasdust discipline is now followed by an enthralled public because of the stunning discoveries and brilliant reporting of its most prominent modern pioneers, including the Leakey family in Kenya, Donald Johanson, and, of course, Stephen Jay Gould. New information harvested from fossilized skeletal fragments millions of years old has both enlivened evolutionary theory and joined it to the developmental and behavioral sciences, linguistics, and even the neurosciences. Charles Darwin's name and his ideas are again as widely discussed and debated as they were in the middle of the last century. Indeed, the explosion of recent publications about Darwin, *neo*-Darwinism, *universal* Darwinism, and even *neural* Darwinism certify his genius; with the passage of time the impact of his insights and his work simply grows and grows.

Reawakened interest in Darwin finds a quiet but highly significant counterpart in a recent growing awareness of the remarkable life and work of Sir Charles Bell, a Scottish surgeon who was not only a contemporary of Darwin but one of the most respected comparative anatomists of his day. As a young boy, Bell had not only studied drawing but assisted his older brother in the teaching of anatomy. In 1806, having moved from Edinburgh to London and having become an anatomy teacher himself, he published *Essays on the Anatomy of Expression in Painting,* a book which was popular with both artists and surgeons and which remained in print for over forty years. Bell's work on comparative anatomy was well known to Darwin, and his *Essays* presaged Darwin's publication, in 1872, of *The Expression of the Emotions in Man and Animals.*

In 1833, with Darwin near the midpoint of his epic five-year voyage on the *Beagle,* Bell completed and published the Fourth Bridgewater Treatise: *The Hand, Its Mechanism and Vital Endowments, as Evincing Design.* In keeping with the terms of the Bridgewater endowment, Bell had intended that his book would help to establish biology as a support for religious faith. But this was not the result. His analyses of the behavioral consequences of variation in anatomic structure, and his insights into the relationship between movement, perception, and learning, were revolutionary and seminal. The book, and Bell's continuing work on the anatomy of the nervous system, had a far greater influence on the development of the science of physiology of the nervous system than on religious thought or polemic.

It is genuinely startling to read Bell's *Hand* now, because its singular message—that no serious account of human life can ignore the central importance of the human hand—remains as trenchant as when it was first published.[3] This message deserves vigorous renewal as an admonition to cognitive science. Indeed, I would go further: I would argue that any theory of human intelligence which ignores the interdependence of hand and brain function, the historic origins of that relationship, or the impact of that history on developmental dynamics in modern humans, is grossly misleading and sterile.

Following Bell, we will begin with a brief review of what is known of the human (and the hand's) evolutionary timetable, and then move to the present—to the "Decade of the Brain"—to consider the most recent efforts by anthropologists and brain scientists to create a comparable timetable, or track, for the evolution of intelligence. This review is an essential preliminary to a later chapter on human language and a discussion of the role some theorists attribute to the hand in the emergence of symbolic thought.

We will continue with a compact overview of the anatomic and physiologic nuts and bolts pertinent to hand function. It is not possible to understand the hand as a dynamic part of the body, or to safely tackle broader issues concerning the hand in relation to brain function or human development, without at least a minimal grasp of the fundamentals of its physical structure and function. But what do we mean by "the hand"? Should we define it on the basis of its visible physical boundaries? From the perspective of classical *surface* anatomy, the hand extends from the wrist to the fingertips. But under the skin this boundary is just an abstraction, a pencil line drawn by mapmakers, giving no clue as to what the hand is or how it actually works.

On both sides of the wrist, under a thin layer of skin and connective tissue, pale white, cordlike tendons and nerves pass from the hand into the forearm. Are the tendons above the wrist—that is, in the forearm—part of the hand? After all, we are able to hammer nails or use a pencil only because of the pull of tendons and muscles near the elbow. From the perspective of *biomechanical* anatomy, the hand is an integral part of the entire arm, in effect a specialized termination of a cranelike structure suspended from the neck and the upper chest. Should we agree that the hand must be conceptualized in biomechanical terms, we invite further complexities of definition. We would know very little about the living actions of the hand except for observations of the effects of injury on its function; such observations are well documented from the time of ancient Greece, when it was known that muscles could be permanently paralyzed by cutting a thin white cord that somehow activates the muscle. Such cords are called nerves, and physicians and anatomists in ancient Alexandria already knew that nerves originated in the spinal cord. What are we to do with this fact? Are the nerves controlling the muscles and tendons that cause the hand to move also part of the hand?

Another set of observations, beginning a little over a century ago, has made it clear that the hand can be rendered useless by damage to the brain from injury (a fall or a gunshot wound) or as the result of a disease process (stroke, multiple sclerosis, or Parkinsonism, for example). Pathologic change associated with specific diseases or injuries, when confined to different parts of the brain, can have quite different and distinctive effects on hand function. Should those parts of the brain that regulate hand function be considered part of the hand? The perspective of *physiological* or *functional* anatomy suggests that the answer is yes. We need go no further than this to realize that a precise definition of the hand may be beyond us. Although we understand what is meant conventionally by the simple anatomic term, we can no longer say with certainty where the hand itself, or its control or influence, begins or ends in the body.

The problem of understanding what the hand *is* becomes infinitely more complicated, and the inquiry far more difficult to contain, if we try to account for differences in the way people use their hands, or if we try to understand how individuals acquire skill in the use of their hands. When we connect the hands to real life, in other words, we confront the open-ended and overlapping worlds of sensorimotor and cognitive function and the endless combinations of speed, strength, and dexterity seen in individual human skill and performance. We also confront the vagaries of human learning. Consider the following sequence of events:

> Two people of the same sex and roughly the same age, physical makeup, and education both begin piano lessons and juggling lessons. At the end of one month, the first student seems to be progressing at the piano faster than at juggling, and the second student seems to be doing the opposite, and each reports that her hands seem to have more of a "knack" for the one skill than for the other. In response to these divergent results, piano and juggling lessons are modified for each student, introducing whatever changes seem most likely to equalize skill levels in the two students. However, as time goes on, and despite all efforts at remediation, the differences

in performance *increase.* No matter what is done, the first student continues to improve faster at the piano than at juggling, and the second still does the opposite.

How can this be? Are there significant but unseen structural differences in the hands of these two individuals? If we knew more about the detailed anatomy and biomechanics of their hands and arms, could we explain the differences in their abilities to refine these special skills? Perhaps. Or should we look to *brain science* to explain the discrepancy? The answer here is also, at best, perhaps. If it is true that the hand does not merely wave from the end of the wrist, it is equally true that the brain is not a solitary command center, floating free in its cozy cranial cabin. Bodily movement and brain activity are functionally interdependent, and their synergy is so powerfully formulated that no single science or discipline can independently explain human skill or behavior. In fact, it is not clear that what we have asked can be called a *scientific* question. The hand is so widely represented in the brain, the hand's neurologic and biomechanical elements are so prone to spontaneous interaction and reorganization, and the motivations and efforts which give rise to individual use of the hand are so deeply and widely rooted, that we must admit we are trying to explain a basic imperative of human life.

Ultimately, this "meditation" seeks to juxtapose and integrate three quite different perspectives on the role of the hand in human life:

1. the anthropological and evolutionary perspective: where the human hand came from and how it acquired the repertoire of movements that have given it a central role in human life and survival;

2. the biomechanical and physiological perspective: the engineer's view of the specialized structure and function of a forelimb no longer used for weight-bearing and whose terminal configuration is adapted for control of external objects;

3. the neurobehavioral and developmental perspective: how the dynamic interactions of hand and brain are developed and refined, and how that process relates to the unique character of human thought, growth, and creativity.

The last of these three perspectives is the one that seems to me most in need of illumination.

Early in 1990 I was living in Germany, having taken my interest in performance problems of musicians to a research laboratory at the University of Düsseldorf. My particular interest, as I have mentioned, was in hand cramps, a problem that suggested to my imagination a marionette whose strings had knotted up of their own accord. Since Düsseldorf happens to be the home of a prominent marionette theater, I sought out its director, Anton Bachleitner, to find out how these puppets actually work. Inevitably, our discussion of puppets led to a discussion of his own interest in them. Bachleitner, then in his thirties and a man who had lived since the age of eight in the world of marionettes, insisted that he knew the first day he set eyes on a puppet that he had found his life's work. A master woodcarver, he has designed, carved, and painted almost all the puppets for the shows produced by his company; he personally trains all his players, chooses and adapts the plays, and performs in them as well. Every bit as much as any musician I have known, he lives through his hands.

How could Bachleitner have known when he was eight years old that this was what he would do for the rest of his life? His description of that realization did not seem to be just a young boy's escapist fantasy; he *knew* what he wanted to do with his life, he acted on that knowledge immediately, and he got what he wanted. What could possibly account for his clear, early vision of a future life and the fortuitous mix of aptitudes he would later need? And how can we possibly explain the directness and ingenuity with which he got himself where he knew he belonged?

As I located and interviewed others whose careers depended on unusually refined hand control I found that most could spell out in five minutes the purely procedural demands of their work. But to understand fully how they had incorporated that knowledge and had turned it into a career was another matter. Each had made a succession of discoveries that had been followed by a strengthening of the desire to learn more and a determination to "get it right," or "find the truth," no matter what the obstacles. This process always resulted in a distinctive personalization of their work, and a growing sense of (and demand for) independence. The process usually turned out to have been not unlike my own experience with piano lessons:

improvisational—nothing like what was in the books. I also found in these stories a hint of inevitability, as though, like Bachleitner, at least some had known all along where they belonged.

Are people genetically predisposed toward a particular life's work based on a biological aptitude? If so, could genetic makeup predispose certain people toward careers in which refined hand skills are indispensable? If that were true, what would be the implications for our cherished notions of intelligence and aptitude? A deeper question about the nature of "innate talent" also arises, since we are a species evolving genetically at a far slower pace than the world in which we as individuals must survive.

The "design" of the modern human brain was completed 100,000 years ago, perhaps even earlier. Does that mean that each living person is locked into a certain kind of mind as tightly as he or she is locked into bone structure, hair and eye color, sex and limited life span? And how could we possibly have in our midst someone "born" to be an astronaut? That is, how can any human manage the physical, mental, and psychological demands of space flight? Far more mundane and commonplace human accomplishments are equally difficult to explain. How do we manage to drive our cars at freeway speed at night, seemingly guided by nothing more than our own headlights? How did we ever learn the nocturnal trick of *inferring* the true state of the road, its obstacles, and the other moving vehicles, from the tiny parabolas of light streaking across our retinas? How do we explain the melodic flights of a great jazz pianist or violinist, or the marksmanship of a golfer?

These questions are neither exotic nor frivolous. If behavioral potential has any significant degree of genetic underpinning, how do we even begin to tolerate the modern world we live in? Where is the fulfillment for a modern office- or factory-worker performing automated or repetitious tasks within a physical and social context that scarcely resembles any environmental ensemble from the formative eons of human prehistory? What are we to do if the human "gene pool" dictates the regeneration of stable percentages of individuals with aptitudes of little or no value to modern society? It is probably not a joke that computer games, spectator sports, television violence fantasies, and weekend hunting and fishing expeditions are the necessary transformations of outmoded but undiminished vestigial drives and skills that humans still carry with them. But is the creation of a menu of imaginative diversions our only recourse to the unremitting sway of an obsolete "hunter-gatherer" heritage?

In order to examine these questions, we will look at several examples of the cultural transformation of the human career. Here, using as guides people whose work is not only based in the hand but steeped in the oldest traditions of every human culture, we will see how powerfully a personal motive can invest and lend great meaning even to modern endeavors when they are oriented toward the satisfaction of basic human need. We will consider the celebration of food, the rituals of medicine and magic, and the affirmations of music. We will also consider what has been called "the permanent immaturity" of the human brain, and whether human culture may have become our own ("virtual") Galápagos, changing the direction and the timetable of human evolution.

Finally, and inevitably, we shall consider the impossible job we have given teachers, and the equally impossible job our children face, trying to absorb all that we insistently thrust upon them in the name of the future we would like them to have. If sports and video games rejuvenate the psychic connections to a primitive past, it is the schools that bear us into the future.

Since the Industrial Revolution, parents have expected that organized educational systems will tame and modernize their children and "prepare them for life." Such is the theory. But education—ritualized, formal education, at least—is not an all-purpose solution to the problem of inexperience and mental immaturity among the young. I was completely unprepared for the frequency with which I heard the people whom I interviewed either dismiss or actively *denounce* the time they had spent in school. Most of my interview subjects, although I never asked them directly, said quite forcefully that they had clarified their own thinking and their lives as a result of what they were doing with their hands. Not only were most of them essentially self-taught, but a few had engineered their personally unique repertoire of skills and expertise in open retreat from painful experiences in a school system that had dictated the form and content of their education in order to prepare them for a life modeled on conventional norms of success.

Apart from a grudging deference to what might be called the "right-brain lobby," what is there in our theories of education that respects the *bio-*

logic principles governing cognitive processing in the brain and behavioral change in the individual? How does, or should, the education system accommodate the fact that the hand is not merely a metaphor or an icon for humanness, but often the real-life focal point—the lever or the launching pad—of a successful and genuinely fulfilling life?

We cannot escape the fleeting character of our lives: each of us moves within a single frame of a *very* long movie. But we are not passive recipients of the particular model of the brain that ended up inside our own personal skull. We know beyond any doubt that education and experience alter the way the brain functions, but we cannot agree how best to apply that principle to the benefit of our children and ourselves. We devour the latest pronouncements of educational psychologists and cognitive neuroscientists, but do not know what the term "learning" means with respect to the brain itself, apart from the rather dry notion of altered probabilities of "synaptic strength" or "neural net" function. There is a *lot* we don't know.

When I began work on this book, I believed both in the basic human desire for autonomy and in people's resourcefulness. Time and again the people I interviewed reaffirmed that belief, enlarged and enriched it. These people also made it clear that self-definition, even when it seems to have strong behavioral presets, is not a passive process. Both literally and figuratively, it must be a hands-on and hands-in affair. Sometimes it begins with the realization that the assumptions and demands of formal education must be ignored or actively resisted. Once launched, the process of self-education and development never really stops. People are *born* resourceful and they *become* skillful and "thoughtful" when they genuinely care about what they are doing. One begins to understand the origins—and learns to appreciate the interdependence—of human skill, intelligence, and vitality by looking at the details, one piece and one person at a time. That is the real story I hope readers will find in these pages.

1
Dawn

OUR TEXTBOOKS LIKE TO ILLUSTRATE EVOLUTION WITH EXAMPLES OF OPTIMAL DESIGN—NEARLY PERFECT MIMICRY OF A DEAD LEAF BY A BUTTERFLY OR OF A POISONOUS SPECIES BY A PALATABLE RELATIVE. BUT IDEAL DESIGN IS A LOUSY ARGUMENT FOR EVOLUTION, FOR IT MIMICS THE POSTULATED ACTION OF AN OMNIPOTENT CREATOR. ODD ARRANGEMENTS AND FUNNY SOLUTIONS ARE THE PROOF OF EVOLUTION—PATHS THAT A SENSIBLE GOD WOULD NEVER TREAD BUT THAT A NATURAL PROCESS, CONSTRAINED BY HISTORY, FOLLOWS PERFORCE.

—Stephen Jay Gould[1]

THE EARLIEST DIRECT HUMAN ANCESTORS were the australopithecines, "southern apes" of Africa who walked upright. The best known is Lucy, who lived some 3.2 million years ago in Hadar (in eastern Africa) and whose discovery created an enormous sensation not only in the anthropological world but with the public at large. Twenty years of careful research on her species have solidified her claim to primacy.* She was the first anatomically bipedal human ancestor to be discovered, and she had an unapelike hand and a chimpanzee-size brain.[2] At the time of her discovery there had already been a few tentative suggestions that the modern human brain might have evolved as a consequence of the increase in tool use among Lucy's descendants. This specific assertion was made by anthropologist Sherwood Washburn, writing in *Scientific American*, just as the first

* From the beginning, Lucy has had to fight for her title as a human ancestor; see this chapter's endnote 13 to learn about the latest challenge.

reports of a tool-using hominid* (*Homo habilis*) in East Africa were circulating.

> Now it appears that man-apes—creatures able to run but not yet walk on two legs, and with brains no larger than those of apes now living—had already learned to make and use tools. It follows that the structure of modern man must be the result of the change in the terms of natural selection that came with the tool-using way of life. . . . From the short-term point of view, human structure makes human behavior possible. From the evolutionary point of view, behavior and structure form an interacting complex, with each change in one affecting the other. Man began when populations of apes, about a million years ago, started the bipedal, tool-using way of life.[3]

Washburn's thesis actually contains three distinct assertions:

> 1. The brain and the musculoskeletal systems, *as organs,* evolve just as organisms themselves do, by modification of structure and function over time. Consequently, the behavior of any living members of any species, at any given time, reflects the operating characteristics of separate parts of the body in general, and (for Washburn), of the brain and musculoskeletal system in particular.
>
> 2. Two critical modifications in the musculoskeletal system contributed to the launching of the hominid line. The first—as Darwin himself asserted—was the adoption of a bipedal gait. Subsequent changes in the upper limb, altering the repertoire of hand movements in ways that favored tool use, were the final catalyst for the subsequent split of humans from the same primate line that had produced the great and the lesser apes much earlier.
>
> 3. The driving force behind hominid brain evolution (which Washburn, respecting Sir Arthur Keith's influential opinion, equated with

* *Hominidae* (or, the hominids) is the name given to the family that includes *Australopithecus* and *Homo.*

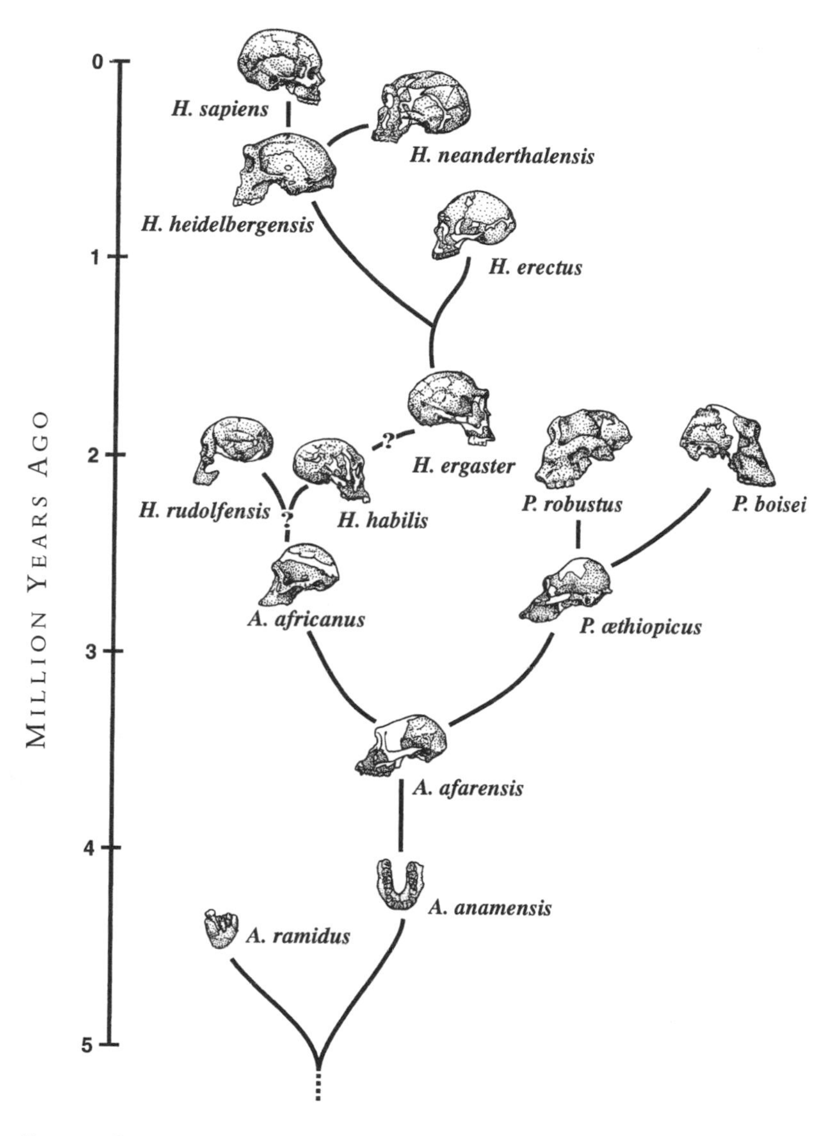

FIG. 1.1 A recently constructed time line showing one possible scheme of relationships among the various species of the human family, beginning with the emergence of primate bipedalism. Although hand and wrist fossils recovered from *A. afarensis* are complete enough to permit detailed inferences about hand function, the fossil record thereafter is very sketchy until *H. erectus,* and specimens of the ulnar side of the wrist are not complete until *H. neanderthalensis.* (Courtesy of Ian Tattersall.)

> increased brain volume) was not simply "selection pressure" created by external environmental change. The *brain itself,* and then *society,* ultimately overwhelmed the jungle (metaphorically speaking) as a destabilizing element and an organizing force in this process, with increasingly powerful effects on a host of adaptive anatomic and behavioral changes as hominid lines expanded their range into new habitats.*

Washburn quite specifically insisted that the modern human brain came into being *after* the hominid hand became "handier" with tools, maintaining that the brain was the last organ to evolve. It is a daring idea, one which requires us to look very closely at the evolutionary background of this hand, and at the changes that brought it its present anatomic configuration and functional capabilities.[4]

That we know anything at all about the earliest human origins is due in large measure to the epic lives and work of Louis and Mary Leakey, whose critically important African discoveries beginning in the 1950s gave us not only our first vision of early hominid life but some very specific ideas about the antiquity of the human brain itself.[5] Until recently, the search for a "founder" human brain has rested on the premise that distinctive human behavior, particularly language and tool use and the phenomenology of the mind, would be found to spring from brain size.[6] Under that particular formulation, four reference dates assume special importance for anthropologists. The first marks the appearance of Lucy's clan, the australopithecines, beginning with *Australopithecus anamemsis* between 3.9 and 4.2 million years ago (mya). The second is the date of the first member of the *Homo* family, *Homo habilis,* who appeared about 2 mya. Next, sometime earlier than 1 mya, *Homo erectus,* also appeared. The fourth date is that of the appearance of modern *Homo sapiens,* established by fossils whose age has been dated at 100,000 years.[7] During this time the brain grew from 400–500 cc (the australopithecines) to 600–700 cc (*Homo habilis*), then to 900–1,100 cc (*Homo*

* This proposition is at the heart of what has been called "universal Darwinism," by which is meant the operation of the three cardinal elements of the theory—variation, selection by fitness, and transmission—in *any* dynamic and protracted biologic, social, or even intellectual process.

erectus), and finally to our own approximately 1,350 cc capacity.* These dates give us a finite period to examine for evidence about the state of and ongoing changes in the *hand* that might be associated with both brain and behavioral evolution.[8]

In point of fact, fossil specimens from arms and hands are even rarer than those of skulls. But the importance of finding and dating them can hardly be overstated: if we *could* learn something about the sequence and timing of modifications that prepared the primate upper extremity for refined tool use, we might be in a position to associate those changes with the emergence of the modern human brain. Enough is now known about the evolution of the primate upper limb at least to make clear what we need to look for next.[9]

The earliest primates, so far as we know, were Paleocene† mammals, mouse- to cat-size creatures who began to make the complex adaptations that an above-ground food source demanded by way of hunting and gathering skill. Taken as a whole, primates attacked the problem of successful tree life by accomplishing the following generic physical changes:

1. orbits and eyes moved to a forward position in the head, permitting binocular vision, certainly an advantage for navigating in three-dimensional space and for finding and catching small prey at close distances;

---

* The near doubling of cranial capacity from *H. habilis* to *H. erectus* was easier to explain when it appeared that their emergence was separated by nearly one million years. If, indeed, they appeared virtually contemporaneously, a hypothetical direct-line increase in brain capacity becomes extremely difficult to defend. Based on present evidence, one would have to postulate a far greater antiquity for *H. habilis* than has been documented, or that *H. habilis* was possibly (like the robust australopithecines) a hominid form which became an extinct offshoot in the hominid line leading to *H. sapiens.*

† Our time scale encompasses the Cenozoic *era* (the age of mammals), which began 65 mya, and continues to the present. The oldest *epoch,* a subdivision of the era, is the Paleocene (65–58 mya); next came the Eocene (58–35 mya); then the Oligocene (35–24 mya); then the Miocene (24–5 mya); then the Pliocene (5–1.6 mya); then the Pleistocene (1.6–.01 mya to 10,000 ya); and finally the Holocene epoch, extending from 10,000 ya to the present time. See Ian Tattersall, *The Fossil Trail,* pp. 3–16.

2. forearm and collarbone structure (a gift from insectivore ancestors) were modified to permit greater flexibility and perhaps greater safety in arboreal travel and dining;

3. paws retained the archaic but extremely useful five-ray (pentadactyl) pattern, permitting the animal to grasp with individual digits; toes and thumbs acquired the ability to close the gap between the thumb and first digit (i.e., they became convergent, though not yet opposable); nails replaced claws on the dorsal surface of terminal digits, while palmar surfaces acquired sensitive, ridged pulps—all these changes permitted improved climbing and locomotion along trunks and branches, and better grasping and holding of fruits, leaves, and insects;

4. the snout shortened, vision began to supersede smell as the dominant sense, and jaws, skull, and teeth changed, consistent with dietary change;

5. the brain changed in size and configuration, probably to accommodate the geometrically more complex (and physically riskier) living and hunting environment.

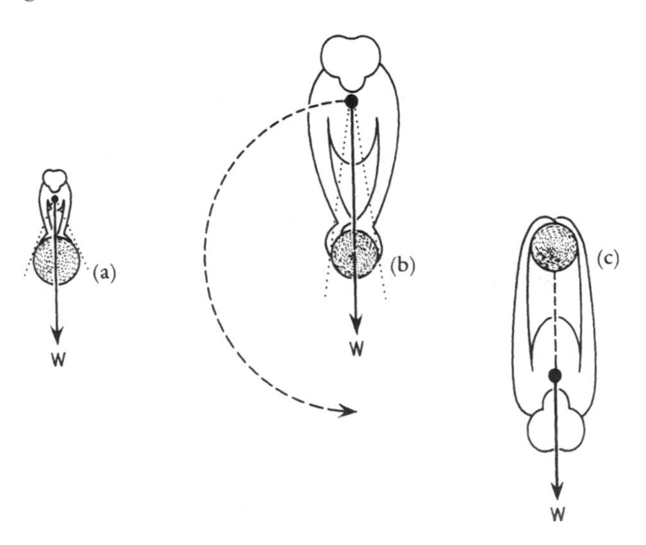

FIG. 1.2 Napier illustrates his proposed explanation for the emergence of brachiation. As arboreal primates increased in body size, moving along branches became more stable from a suspended position. (From John Napier, *Hands,* 1980.)

These changes were just beginning during the Paleocene epoch (65–58 mya) and had become well established by the end of the Eocene epoch (35 mya). It was also probably near the end of the Eocene epoch that the anthropoids (or "simians") appeared for the first time. The early anthropoids lived in the trees and were exclusively quadrupedal: none had yet made the shoulder or other upper-extremity modifications that permit suspended locomotion in trees. Those changes, and more advanced changes to the chest wall and the bony structure of the shoulder girdle permitting full brachiation, are believed to have taken place in the mid or late Oligocene.*

By the beginning of the Miocene (24 mya), the anthropoids had branched into monkeys and apes. It is not clear why the monkeys and apes should have diverged, but the reason may be found in simple physics: apes are bigger than monkeys, and their size must have become a problem at the tops of the trees, where the supply of food is rich but the branches are small. Napier has suggested that a quadrupedal method of movement along the branches becomes unsatisfactory when body size and weight create an unfavorable center of gravity.[10]

By the end of the Miocene epoch the African apes were moving out of the trees and back down onto the ground; the great apes were large animals, and their dietary habits had changed somewhat. Gorillas, the largest by far, had almost become a sort of land-based whale, large enough to be safe from attack by most predators, and able to exist on a very simple vegetarian diet. Chimpanzees were in a state of locomotor and dietary transition, with a highly mixed diet (including small animals) and very limited bipedal walking, usually assisted by weight-bearing on the knuckles.

At the beginning of the Pliocene (when the hominid story begins) the hand of monkeys and apes existed in several forms. First, in all these animals the four fingers had essentially the same form and functional capacity. The finger bones (phalanges) were relatively flat and slightly curved in the direction of flexion. In apes, and a few monkeys, there were strong attach-

---

* Brachiation means "moving (swinging) suspended from the arms"—later we will consider the remarkable mechanics of this trick in greater detail. In general, although monkeys have the ability to move while suspending themselves from their arms (and tails), most lack the limb structure for true or "full" brachiation.

ments on the underside to keep the flexor tendons tightly held next to the bone. Opposition to the degree possible in the human hand was uncommon, although gorillas and some old-world monkeys could come close.* Isolated digital movements were possible to permit scratching, picking, and digging movements, or for stripping, and small objects could be pinched between the thumb and the index finger. (Grooming is a favorite leisure activity with this hand among all its present owners, just as it is with us.) Objects could be held and carried in any of these hands exactly the way we hold a suitcase.

The greatest variability found in the prehominid hand was in the thumb, which tended to be short compared to the fingers: shortest in the chimpanzee and orangutan, more like hominids in the gorilla, sometimes even absent in monkeys. Interestingly, although the gorilla thumb most closely resembles the human thumb among these precursors, the gorilla has never been observed spontaneously to attempt tool use. Chimpanzees, with a much smaller thumb, stand out among pongids as prolific spontaneous tool users and as avid learners and improvisers in environments where the animals can be influenced by human artifact and teaching.[11]

Another development that appeared with apes was a freeing of the attachment of the far end of the ulna (the major forearm bone of the elbow

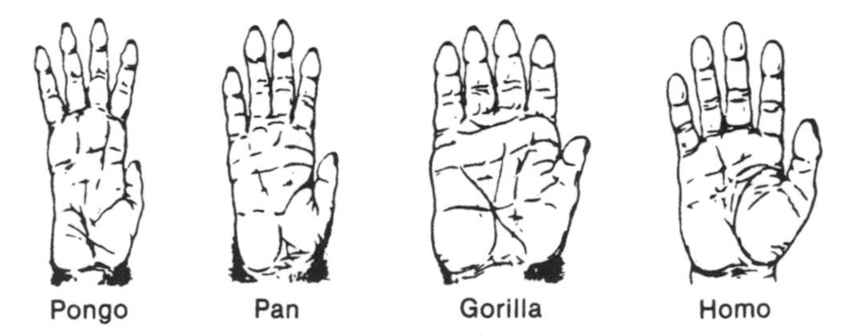

FIG. 1.3 Human hand compared with that of apes sharing a common ancestry. The ape hands have been scaled to human size. *Pongo* is the orangutan, *Pan* the chimpanzee. (From A. Schulz, *The Life of Primates*, 1969.)

* Full opposition of the thumb in humans is due not only to rotation of the thumb and to its relative length, but also to rotational movements of the index and other fingers. These do not occur in any prehominid hand.

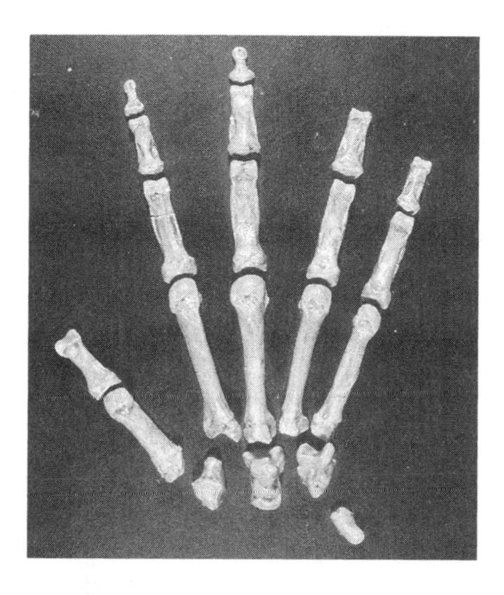

FIG. 1.4 Lucy's left hand (composite example from *A. afarensis* fossils) seen from the palmar side. The tips (distal phalanges) of the thumb, fourth and fifth fingers, and four of the wrist bones are missing. (Specimen courtesy the Institute of Human Origins, Arizona State University, photographed by Diane Hawkey.)

that meets the wrist on its small-finger side). This change must be considered critical to brachiation, since it increases the twisting range of the arm below the elbow needed to swing the body forward under the arm. It also allows the hand to tilt at the wrist away from the thumb.[12]

We are now staring at a maddeningly blank page in our hand story, because the best example we have of a prehuman hominid hand is that of Lucy, who had acquired several humanlike features in the hand and also managed a major overhaul in pelvic and leg structure. In fact, it is her lower extremities that place her (and the other australopithecines) at, or very close to, the head of the line leading from the apes directly to man.[13] The pelvis is short, and the configuration of the femoral joint at the hip and at the knee, like nothing ever seen before, is for all practical purposes the same structure humans now possess in an elongated version. *A. anamemsis* is thought to have walked upright on the basis of the structure of the tibia; in Lucy, the tibia, femur, and pelvis have been recovered, and there is no question that she was bipedal.

Mary Marzke, a physical anthropologist at Arizona State University, has spent a great deal of time looking at the hand and wrist bones of *Australopithecus afarensis.* Professor Marzke points out that although Lucy does have an opposable thumb, other primates also have this feature. Chimps and monkeys, in fact, are quite good at bringing the thumb to the side of the index finger. What they *don't* do well (as Lucy herself could not do) is bring the thumb tip all the way across the hand to the fourth and fifth fingers. Also, neither the apes nor Lucy flex the fingers on the ulnar side of the hand (the side with the little finger) toward the base of the thumb in the movement known as "ulnar opposition." We humans do this without

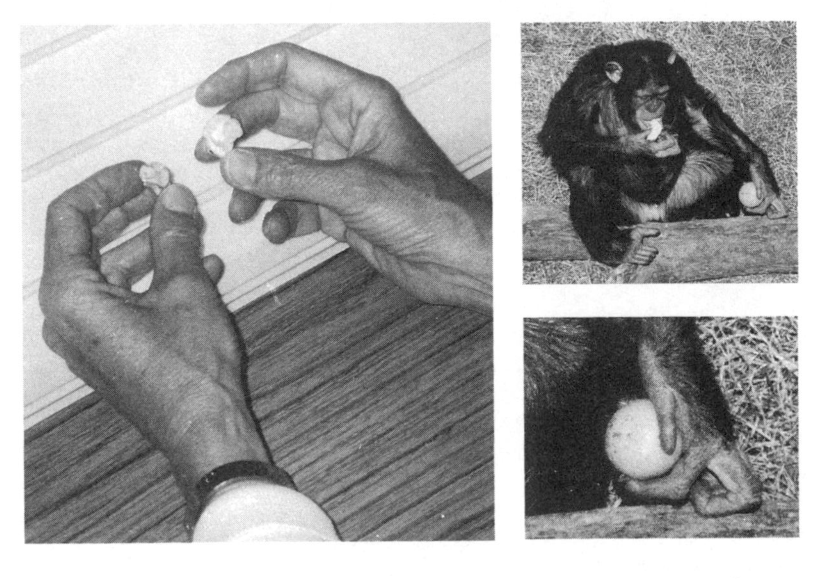

FIG. 1.5 Anthropologist Mary Marzke compares the hamate wrist bones from *A. afarensis* (Lucy in Marzke's left hand) and *Pan troglodytes* (chimpanzee) in her laboratory at Arizona State University. *Above, right:* Allie, a chimpanzee Mary has studied, relaxes at home in Bastrop, Texas. *Below, right:* enlarged view of Allie's left hand. While the differences between these hamate bones are not so easy to appreciate, the differences between chimp and human hands are strikingly apparent. In particular, note the size of the thumb and how its position in relation to the fingers affects the capacity for manipulation of objects held within the hand. (Photograph of Allie with permission of the Department of Veterinary Sciences, University of Texas, M. D. Anderson Cancer Center, Bastrop, Texas.)

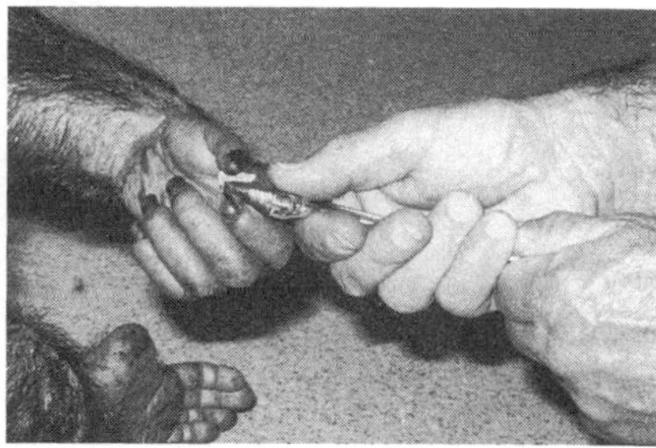

FIG. 1.6 Allie's avid interest in tools illustrates critical similarities and differences between human and chimp grips. *In the top photo,* she holds a screwdriver in a power grip almost the way a human user would, closely approximating the oblique positioning that humans use to turn the screwdriver by rotating the forearm. A more natural hold for her is shown *bottom left,* in which she flexes the wrist to turn the instrument. Because of the small size of her thumb and her inability to move the fourth and fifth fingers toward the base of the thumb, her oblique squeeze grip is comparatively weak in this task. *Bottom, right:* The structural basis of two major differences between chimp and human grip is clearly seen as Allie's handler offers her a wrench: (1) all of her fingers fold straight into the hand, whereas his fingers curve toward the base of the thumb; (2) although chimp and human finger size is comparable, the human thumb is proportionately much larger. (Photographs by Mary Marzke, with permission of the Department of Veterinary Sciences, University of Texas, M. D. Anderson Cancer Center, Bastrop, Texas.)

the slightest sense of marvel whenever we grasp the handle of a hammer, a golf club, or a tennis racket and prepare to take a swing.

The advances in Lucy's hand are nearly impossible to appreciate without studying her wrist bones—fitting them together, moving them around, and comparing them to the wrist bones of other anthropoids. Having spent some years doing exactly this, Marzke concludes that Lucy's hand is at least partially "modern." The most impressive evidence for its modified design resides in the joint surfaces at the base of the thumb, index, and middle fingers, and in the changes in the size and orientation of the joint surfaces of the wrist bones closest to those digits. The thumb is long in comparison with the fingers, the ratio approaching that found in modern humans and in few other primates. Taken together, these changes move the *radial* (or thumb) side of Lucy's hand dramatically toward the twentieth century.[14] The apparent functional advantages of the changes are:

- the thumb, index, and middle fingers can form a "three-jaw chuck," which means the hand can conform to, grasp, and firmly retain irregular solid shapes (such as stones);
- finer control can be exerted over objects held between the thumb and the tips of the index and middle fingers;
- rocks can be held within the hand to pound repeatedly on other hard objects (nuts, for example), or to dig for roots, because the new wrist structure is able to absorb (dissipate) the shock of repeated hard strikes more effectively than in the ape hand.

These changes would have given *A. afarensis* the capacity to conform the thumb and first two fingers to a very wide range of object sizes and shapes, allowing them to be held and manipulated easily on the thumb side of the hand. Changes in the ligaments suggest that Lucy had the capacity for prolonged periods of percussion using small stones.[15]

The area in Ethiopia where Lucy and her close relatives were found is strewn with plain stones, so it is not fanciful to postulate that pounding and digging with stones (or wood, or parts of animal skeletons) was normal behavior for *A. afarensis.* But there is more. As we have noted, Lucy also had a pelvis and the legs of a primate capable of habitual bipedal locomotion. That means that she did not simply happen to get up on her feet occasionally; this was her natural standing posture. Yet her legs were not particularly long, and she probably didn't run any faster than her own ape neighbors or ancestors. What could possibly have been the survival value of this change in body structure? Marzke has examined the pelvic specimens carefully to determine sizes of muscles and their attachments and has come to a surprising conclusion: the bony and muscular structure (especially of gluteus maximus) strongly suggest that Lucy was equipped both to pound stones and to *throw* them, accurately and with speed. Lucy, in other words, might have been at home on a pitcher's mound.[16]

Chimps can and do throw stones, but mainly to display alarm or aggression. They almost always throw underarm because they cannot use hip rotation to accelerate the torso during an arm swing. Lucy was not restricted in this way. She could have thrown overarm because her shoulder had the capacity for full brachiation (including forearm supination), her hand was capable of a "three-jaw chuck" grip, and her pelvis and its musculature permitted a whipsaw movement of the full axis of the body during the windup and throwing motion.[17]

Throwing may well have become more an attack skill in Lucy than it ever could be in chimps, but a major improvement in clubbing had to await the changes in the configuration of the ulnar side of the wrist and hand, which came after the time of *A. afarensis.* The trick of ulnar opposition is unique to modern humans, and as the final piece in a mosaic of changes involving the entire extent of the upper extremity from shoulder to fingertips, it may well have unleashed the final stage of a unique mammalian strategy for long-term species survival. Opposition of the fourth and fifth fingers combined with ulnar deviation of the wrist permits a stick to be seated tightly in the hand and oriented along the axis of the arm, so that the swinging radius of arm-plus-stick (and, therefore, the force of a blow) increases.* Having the ability to telescope the arm outward would convey a lethal advantage in close, hostile encounters, and once this obliquely oriented "squeeze" grip was introduced into the hominid hand, no adversary or prey in the same weight class was safe in its presence without being unusually fleet-footed, hard-headed, or thick-skinned.[18]

A second effect of ulnar opposition can be seen in an improved precision

---

* See figures 6.2, 6.3, and 7.1 for two important examples and uses of the oblique squeeze grip.

grip, in which small objects are manipulated between the fingers without contacting the palm.* The hand with this biomechanical ability would have been able not only to wield a large club and to manipulate and throw stones but also to use all five digits in the fine control of small objects. This one "small" modification, in other words, would have greatly enlarged the functional potential of the hand at both ends of the existing behavioral repertoire, opening the possibility for both a more combative and a more digitally dexterous individual.

Darwin is credited with the first formulation of the potential impact of an upright walking posture: a hand freed of the obligation to support body weight can take on other tasks. But the paw-to-hand conversion began in tree-dwelling primates, and to evolve into what we now know as the human hand it had to be extensively reworked. The reworking involved far more than the paw itself: structural refinement was needed throughout the entire upper extremity, including the shoulder, before the hand could even begin to capitalize on the freedom it had been given. And during this long process, the owner of the hand had grown heavy, powerful, and aggressive.

In the earliest tree-dwelling primates, the need to cling to trunks and small branches favored a tactilely refined hand, and one whose movement and *control* repertoire had impressive range: from delicate, dissecting movements of individual digits to powerful grasping movements used both in sustained holding (prolonged suspension) and in the sequences of quick hold and quick release movements used in swinging and jumping. As experience with locomotion in the trees was gained, visual-motor and tactile-motor neurologic connections had to have undergone enormous change: the visual and kinesthetic support of successful arboreal acrobatics and

---

* John Napier defined the two major grips of the hand as follows: "Stability may be achieved in the normal hand in one of two ways: (1) The object may be held in a clamp formed by the partly flexed fingers and the palm, counter pressure being applied by the thumb lying more or less in the plane of the palm. This is referred to as the power grip. (2) The object may be pinched between the flexor aspects of the fingers and the opposing thumb. This is called the precision grip." John R. Napier, "The Prehensile Movements of the Human Hand," *Journal of Bone and Joint Surgery* 38 B, no. 4 (November 1956), p. 508. We will consider the issue of precision grips and precision handling in greater detail in chapter 7.

predation demanded a bigger brain with very specialized control characteristics.

Brachiation itself may have begun simply as a biomechanical accommodation to increasing size and weight or to the advantages of suspended feeding (being able to reach for a meal not just from a sitting position but while hanging from a branch as well), or to both, as Napier proposed. But the resulting increased mobility in the upper arm effectively permitted the animal to locate either hand virtually anywhere within a sphere whose center was at the pivot point of the shoulder. The brain must have responded to this purely biomechanical change by increasing the complexity of its representation of the arm and hand in space, and there is no question that we have benefited from the overall change in very specific ways: without this freedom of movement, together with the neurologic ability to monitor and refine a host of new movements, ball and racquet sports would be nonexistent and car mechanics and plumbers would be useless working in cramped quarters with impaired visibility. In sum, we must infer that brachiation placed an enormous burden on the brain's kinesthetic monitoring and spatial computing power, since there were so many new places the hands could actually *be* while they were doing their job. Eventually there must also have been countless new tasks for the arms and hands that required differentiated use of the left and right hands, as well as the capacity for refined coordination of the two hands during bimanual tasks.

The importance of bipedalism itself must not be overlooked in assessing the impact of changes in the upper limb. The great apes and the larger monkeys have retained powerful musculature to support the head, which itself is a primary attack and defense structure for these animals. Cats, dogs, chimpanzees, and baboons are all equipped to use their jaws and teeth in mortal combat. But the survival of hominids, whose vertically oriented and delicately balanced head is a weapon only in *verbal* warfare, would either have depended on stealth as a defense or have come from transforming the upper extremity into a platform for offensive weaponry. As Marzke suggests, even Lucy would have been a remarkably dangerous predator had she been able to throw stones accurately. Once her descendants had the ability to prosthetically lengthen the arm, or to fashion more exotic projectiles and launching devices, size and build per se would have become essentially immaterial to the hominids, and the freedom to expand into new environments much greater.*

Both the brain of the chimpanzee and that of the australopithecines weigh approximately 400 grams. No one knows whether brain enlargement is specifically related to increased tool use, but it is known that tools did not become complicated in their structure, nor were they kept and transported for long periods by their users, until quite recently. It is a virtual certainty that complex social structure—and language—developed gradually in association with the spread of more highly elaborated tool design, manufacture, and use; in the next chapter important new theories about the interaction of these major human advances will be reviewed.

We know that sometime after Lucy, a more mobile joint developed at the base of the *small* finger, but the origin of this modification in hand structure is unknown. To date, no fossil specimens from this region of the hand of *H. erectus* have been recovered, so it cannot be said whether the increase in brain weight from 400 grams to 1,100 grams occurred before or after this change in hand structure. This particular unanswered question is surely one of the great remaining buried treasures of human anthropology (see Fig. 1.7).

Given the also-ran status of the arms and hands of apes and monkeys, and the disappearance of the australopithecines, it is probably no exaggeration to say that the final biomechanical change at the base of the pinkie may have conferred an advantage to the hominid hand comparable to supplying its owner not just with gunpowder but with the biomechanical and computational infrastructure for an entire ballistics technology. Considering everything else that was already present, there was *nothing* this hand could

---

* There would of necessity be major neurophysiologic ramifications to the linkage of bipedality and throwing, as is proposed in *A. afarensis.* Although the monkey's visuomotor control of upper limb function is very precise, maintaining visual fixation on a relatively stationary target (a branch) while swinging toward it is a very different problem from tracking a *moving* target at a distance and using visual fixation to calculate the control of the arm as a missile launcher. Study of control mechanisms, or confirmation of control models, in the case of such complex movements is well beyond our technologic capabilities at present, and likely to remain so for some time.

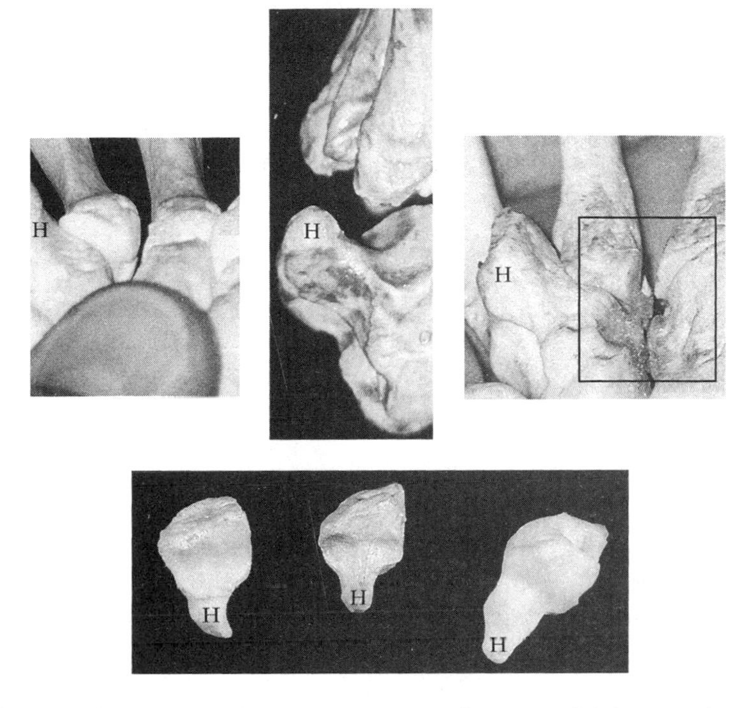

Fig. 1.7 The ulnar side of the wrist in transition, illustrating subtle but critical structural modifications in the hamate and the complex articulations between it and the hand bones (*metacarpals*) of the middle, ring, and small fingers. The anatomy of this part of the hand and wrist (which includes the arrangements of ligaments, tendons, and muscles acting on the joints) sets limits on the movement of the ring and small fingers during closing movements of the hand (flexion). In humans, the hamate-metacarpal articulations function together as a modified saddle joint, comparable to the joint at the base of the thumb. *From left to right, top and bottom rows:* the specimens are human, *A. afarensis* (Lucy), and chimpanzee. The letter *H* marks a palm-side projection (the *hook*) of each hamate. *Top left:* view of the contact surface between the base of the third and fourth metacarpals. In this specimen, the curvature between the base of the third and fourth metacarpals would favor rotation of the ulnar side of the hand toward the thumb (supination and ulnar opposition). Compare these joint surfaces with the same area in the chimpanzee, shown inside the box, *top right.* The articulation in this specimen is flat, preventing rotation of the fourth metacarpal during flexion. Also note in the chimpanzee that the hook of the hamate is long, thereby restricting the movement of the fourth and fifth metacarpals. *Top center:* Lucy's hamate, seen from the side, has separate hollowed-out (concave) surfaces for the fourth and fifth metacarpals, which would tend to limit ulnar opposition. The hook of the hamate in Lucy is short compared with that of the chimpanzee. (Specimens courtesy the Department of Anthropology, Arizona State University. Lower photograph by Diane Hawkey.)

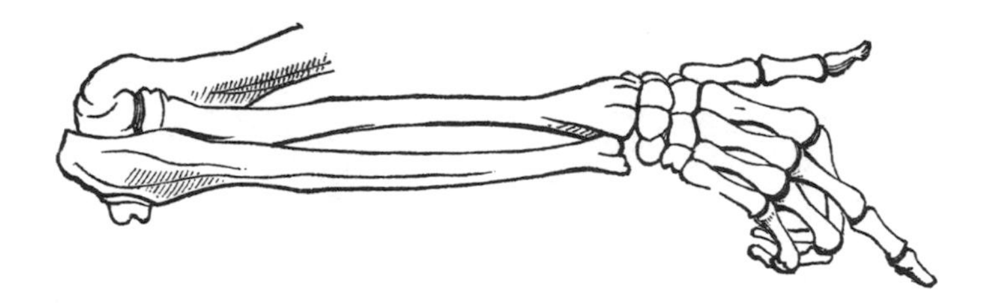

FIG. 1.8 Charles Bell's drawing of the human arm . . . pronation and supination (Charles Bell, *The Hand,* p. 69). This simple and graceful rendering depicts a unique and fateful suite of "minor" modifications to elbow, wrist, and hand structure which permitted humans to become, at once, both the most delicate and the most dangerous of the primates.

not do if it could learn how to do it. And apparently with time, and with other opportunities and challenges encountered in new environments, the brain *did* rise to the challenge, and greatly modify itself in the process,* since there is very little now that this hand, which makes tools that make machines that make computers that make machines *and* tools (and so on), cannot do.

The sequence from *Australopithecus anamemsis* through Lucy to *Homo habilis, Homo erectus,* and finally *Homo sapiens* has been well worked out and exhaustively documented. Woven into that sequence, certain behavioral traits acquired a dominant role in the struggle for species survival. In retrospect, we have come to view these as distinctively and exclusively human: tool use, language, reason, and self-consciousness. But we still search in vain for defining moments or events in this behavioral evolution; they were not so much implanted as emergent. Quadrupedal apes with occasional bipedality gave rise to habitually bipedal australopithecines; increasing use of the hand in the employment and modification of found objects conveyed a survival benefit to the next descendants. Early tool use and manufacture were associated with a modest increase in brain size in *Homo habilis.* The increasing refinement and perhaps specialization of manipulative, hunting, and offensive skills, as well as the ramification of social interactions enabled

* The implications of this process of self-modification by the brain are addressed in the next chapter and in the Epilogue.

by more structured communication (and migration) by *Homo erectus,* had a further "kindling" effect on brain operations and structure. Finally, intraspecies cooperation and competition greatly increased the need for an elaborated social structure and communication and for coordinated industry, all of which demanded a more powerful and more versatile brain. When he finally had enough brain to be able to guess what the brain itself was doing, *Homo* pronounced himself *sapiens.*[19]

Unless we are prepared to argue that language and reason appeared *de novo* in the cortex of the brain, we must grant Professor Washburn his point: the whole list of recently acquired and uniquely human behavioral attributes must have arisen during the long process of brain enlargement that began with the expansion of novel and inventive tool use by *Homo habilis* and with the myriad new experiences and environments that successful employment of such behavior inevitably provided.[20]

Recently, anthropologist Peter C. Reynolds has pointed out that although stone tool manufacture is usually regarded in his profession as a solitary endeavor, it need not have been. Reynolds suggests that complex tools, such as axes and knives, may in fact have been customarily manufactured by small groups of people working together, each performing some part of the task. The possible importance of this alternative transcends the mere pragmatics of shared labor. Any such cooperative efforts would have required a means of communicating, which would probably have taken the form of hand signals and other bodily gestures or vocalizations, or both. In other words, cooperative tool manufacture could have provided a crucial precondition for the evolution of language. An emerging language based in the growth of cooperative tool manufacture would have fostered the evolution not only of a more sophisticated tool manufacture but also of a more complex social culture and a more refined language. As has already been pointed out, these two behaviors, mutually interdependent and mutually reinforcing, would also have been capable of gaining enhanced representation in the brain. To whatever extent language and tool use became specific, heritable traits (even when they demanded priming through cultural exposure), they would have profoundly altered survival prospects for all individuals with this genetic endowment.[21]

A related and more practical question for us concerns the neurologic and behavioral implications of an obligatory hand-brain marriage on a collapsed time scale. What might be the connection between tool use, language, and thinking during the span of a single human life? Recall that Washburn said (referring to what is fixed in our own genetically determined anatomy), "From the short-term point of view, human structure makes human behavior possible."

If language and the employment of the hands for tool manufacture and tool use co-evolved—effectively forging a new domain of hominid brain operations and mental potentials that we collectively refer to as "human cognition"—then we *should* find analogous links, or reinforcing effects, between purposive hand use, language, and cognition in the individual histories of living people. Think what this means. "Intelligent" hand use might not be merely an incidental bequest of our hominid heritage, but—along with the language instinct—an elemental force in the genesis of what we refer to as the "mind," activated at the time of birth.

2

The Hand-Thought-Language Nexus

IT IS NOT A WORD THAT IS DIFFICULT TO COMPREHEND, BUT THE CONCEPT BEHIND THE WORD WHICH THE CHILD DOES NOT UNDERSTAND. THE RELATION OF WORD TO THOUGHT, AND THE CREATION OF NEW CONCEPTS, IS A COMPLEX, DELICATE, AND MYSTERIOUS PROCESS.

—Leo Tolstoy[1]

PHILOSOPHERS, PSYCHOLOGISTS (especially cognitive scientists and psycholinguists), anthropologists, archaeologists, and plain, ordinary neurologists share a singular fascination—*mystification* is perhaps more accurate—over intractable problems raised by the attempt to link together a small number of basic and widely accepted ideas concerning the origins of human intelligence.

What exactly about the brain explains our unique dominion over the rest of the natural world? We assert that the brain accounts for what we call "intelligence." What do we mean by intelligence? In general terms, we mean the capacity to discover, weigh, and relate facts in order to solve problems. We are, of course, not the only animals with intelligence, according to this or any other serviceable definition. However, humans are, for all practical purposes, unique in the animal kingdom in our elaborate and essentially obligatory employment of two problem-solving strategies.

First, we design and manufacture an indescribably large, diverse, and specialized inventory of tools to help us. Some of these tools are quite simple, and in fact work on the same principles as those commonly used by a variety of animals (to some of whom we undoubtedly owe our start in this direction). Humans did not originate the use of rocks and sticks as weapons, hammers, probes, or levers. But over the last tens of thousands of years human toolmaking has taken a very different path from that of the most advanced living primates, including our most recent ancestors. What exactly have we done that is so different? We have made technology the centerpiece of our survival strategy.[2]

Second, we have a trick that we call language. Actually, we have many kinds of language, each of which is based on a formal system of codes and/or symbols through which we represent states of the world. "Represent" means, of course, to "stand for," but in a way that leaves a great of deal of latitude in ordinary usage. When referring to a postulated in-the-brain version of anything commonly known as a real-world object or process, the word is a sort of placeholder, implying but by no means proving that there are or even can be neural structures, operations, or states directly corresponding to any object or process outside the brain itself.

With respect to language, as with tools, we owe a great deal to other animals who communicate in a variety of ways for a wide range of purposes relating to their social lives and to their survival. There can be no doubt that human ancestors were influenced by the rich diversity of examples available to them. Like spoken human language, animal calls are effective and widely used means of signaling between members of the same species.[3] Animals (including insects) also use "body language" to express emotions and intentions over basic survival issues: food, mating, and territory.[4] Animals communicate by means of objects and tokens, just as humans do. And, in common with all the diverse forms of nonverbal animal language—most transparently exemplified by the universal language of facial expressions and by the sort of "emblematic gestures" (to borrow linguist David McNeill's term) we use to gain the attention of waiters in restaurants—human language includes a large assortment of social signals that can be read by anyone anywhere in the world. Here, however, the similarities end.

Human language is not simply an elaborated improvement on animal language; it involves the use of codes and symbols whose correspondence to objects and events in the real world (that is, whose *meaning*) is established by agreement between people. Its underpinnings in the brain are so uniquely localized *and* generalized, and its developmental history in the young is so specific and so powerfully prescribed, that it demands its own explanation.

No one disputes the basic propositions that humans are behaviorally defined by their uniquely elaborated and refined use of tools and language, lacking either of which human society (and for all practical purposes, individual human life) would not continue. Human language presupposes expressly cooperative relationships between people with a common encoding-decoding plan. In other words, language is prima facie evidence of an information-sharing life between people. When people created formal languages, they created mechanisms for sharing knowledge, and in so doing authenticated the existence of mutual awareness and cohesive purpose in their lives. The word we use for *that* arrangement is "culture."

The partnership of language and culture is so deeply woven into human history, and so compelling a force in our own personal development and acculturation, that we quite naturally come to regard language as the trait that both explains and defines our intelligence. But we must be extremely careful not to equate specific behavioral strategies with a more general inclination to solve problems in the interest of our well-being or survival. What is needed is a better understanding of the various relations among tool use, language, and intelligence, or perhaps (for now) just better theories concerning those relations. Fortunately, a few theories of this kind have recently been offered. None accommodates all of the known facts; none is completely persuasive. But they are serious attempts at explaining how modifications in size and structure brought the hominid brain to its present state, and at relating those modifications to the evolution of human behavior—and thus they demand our attention. The following three are of particular importance to the thesis of this book.[5]

Robin Dunbar's Theory of Brain Growth, Language, and Intelligence

Robin Dunbar, who is professor of biology at the University of Liverpool, has recently published a book titled *Grooming, Gossip, and the Evolution of Language.* Succinctly stated, his basic and intuitively appealing proposition is that brain growth, the evolution of language, and the need for intelligent behavior were necessary concomitants of the increasing social complexity of hominid community life. As early hominids expanded their territory and faced greater environmental challenges, their survival became increasingly dependent upon group cooperation. Because large, cooperative groups depend upon working relationships, which in turn depend upon the degree to which members not only know each other as individuals but understand the social fabric and dynamics of the entire group, the minimal qualification for membership in the group is what we might call social savvy. The more populous the group, the greater the demands for a high level of social intelligence, which seems to depend on having a big brain. Some critical benefit of this kind *must* accrue to the ownership of a large brain, because, as pointed out by psychologist Harry Jerison, the metabolic cost of this privilege is quite high.*

Of course, there are other possible circumstances in primate history that could account for the investment in a large brain. As Dunbar notes, some have suspected it to be a consequence of dietary change. The primate arboreal, fruit-eating life requires color vision and a large feeding range, and would almost certainly demand a bigger brain to accommodate the neural controls for the visual upgrade and for mapping a widely dispersed harvesting territory. But an examination of the actual numbers—primate brain size in relation to the fruit-eating life—showed no evidence of such a correlation. Dunbar then guessed that the analysis should consider not the entire brain but only *neocortex,* which is the most recently evolved part of the primate brain. Plotting neocortex size alone against *stable group size* yielded a nearly perfect progression: the larger the "tribe," the larger the neocortex. This correlation held not only for seventy primate species examined, but for

* Jerison's name is strongly associated with what is called the "encephalization hypothesis," which is based on calculations of the ratio between actual brain size and brain size predicted by weight. See Harry Jerison, *Evolution of Brain and Intelligence* (New York: Academic Press, 1973). Using the Jerison equation, anthropologist Ralph Holloway calculates an "encephalization quotient" (EQ) for humans of 2.87; chimpanzee and gorilla values, respectively, are 1.14 and 0.75 (Holloway, 1996, pp. 89–90). The human brain accounts for 2 percent of body weight but consumes 20 percent of the energy supplied by the diet; in computer slang one might say the human brain is an extravagantly expensive operating system to own and operate.

the two important nonprimate mammalians who were studied. Vampire bats, which are highly social creatures, have a large neocortex; carnivores who are solitary hunters have a very small neocortex.[6]

The search for insight into the pragmatics of this relationship led to the observation that monkeys spend a great deal of time grooming each other in order to maintain what are called coalitions. Longtime grooming partners are dependable allies, and in complex societies one needs allies. In Dunbar's research, grooming time emerged as a universal, indexable, quantifiable means for tracking the interpersonal servicing necessary for individual survival in large groups. He calculated from neocortex size that humans should ideally be living in cohorts of up to 150 individuals, then found a number of examples confirming this estimate. Take, for example: the size of a company in the military; the size of Hutterite communities, which are required to break up when their membership exceeds 150; and even the "ideal" size of small businesses, which tend to abandon personal contact in favor of written procedures and memoranda once their size exceeds about 120 individuals.* Neocortex size is a reliable predictor of group size because intelligence is mainly *social* intelligence; the more people there are to keep track of, the greater the complexity of relationships to be kept in mind and orchestrated, and the more time which must be spent maintaining coalitions.†

What does language specifically have to do with all this? The expansion of the brain to meet the computational requirements of a calculating social life could not continue unchecked, nor could grooming time expand indefinitely. Language, an alternative mode for the exercise of social intelligence, proved an effective way of ending both the upward spiral of brain size and the intolerable drain on time demanded by one-on-one grooming. A single

---

* See Dunbar, 1996, chapter 4: "Of Brains and Groups and Evolution."

† There is an interesting reverse argument for the primacy of social intelligence. Psychologist Henry Plotkin, whose work we consider in some detail later in this chapter, notes that people "think adaptively rather than logically." See Henry Ploktin, *Darwin Machines and the Nature of Knowledge* (Cambridge, Mass.: Harvard University Press, 1993), pp. 190–198. He cites examples from the research of Wason and Cosmides, who have shown that logical puzzles which cannot be solved by people when presented abstractly are readily solved when they are presented in the guise of common social problems.

human being, asserts Dunbar, tends to maintain a circle of friends (people from whom he or she would willingly accept a dinner invitation, for example) of a maximum of about 150 individuals, and does so by means of intimate and cozy verbal communications with those individuals; gossip, in effect, is a *verbal* surrogate for grooming. According to Dunbar there are competent surveys showing that most of what people talk about amounts to pure gossip—even (he claims) in universities.

Convincing or not, this hypothesis hardly explains where modern human language itself actually *came* from. Dunbar approaches that problem by suggesting that there was a sort of proto-language that was social long before it had to be technical. The language became technical of necessity when tool use and the worldwide migration of *Homo erectus* required substantive information exchange between individuals and groups of individuals.

What Dunbar accomplishes with this formulation is important: he makes a strong case for the linkage of brain enlargement to the mental and social complexities of group life in the higher primates, and he backs his thesis with a quantitative (but highly selective) arrangement of biologic and ethnographic data. What he plainly does *not* do is help us understand how human language, *as neurologists and psycholinguists characterize it,* fits into this picture. For that, we are obliged to look elsewhere.

MERLIN DONALD'S THEORY OF CULTURAL AND COGNITIVE EVOLUTION

Merlin Donald is professor of psychology at Queen's University, Kingston, Ontario. His book *Origins of the Modern Mind: Three Stages in the Evolution of Culture and Cognition* was published in 1991.[7] Donald proposes that the human brain, complete with language, arrived at its present state through a three-stage process whose internal two boundaries coincide with periods of comparatively abrupt transformations in the physical structure and behavior of apes and hominids. Each of the three stages is defined by a specific profile of motor and social behavior, and each corresponds to well-documented species descriptions and to the established chronology of human ancestry. The behavioral profiles form the basis for inferences about the "cognitive architecture"* of the brain at each stage, and the central postulate is that each stage was marked by a fundamental change in the way the brain had come to respond to the world around it:

> The essence of my hypothesis is that the modern human mind evolved from the primate mind through a series of major adaptations, each of which led to the emergence of a new representational system. Each successive representational system has remained intact within our current mental architecture, so that the modern mind is a mosaic structure of cognitive vestiges from earlier stages of human emergence. . . . The key word here is *representation.* Humans did not simply evolve a larger brain, an expanded memory, a lexicon, or a special speech apparatus; we evolved new systems for representing reality. (*Origins,* pp. 2–3)

What Donald has really proposed is that, beginning with the apes, knowledge itself began undergoing an essential change. What does that mean? Perhaps the clearest explanation can be found in the form of an illustration. Consider the following: a goat, walking through a densely forested area, smells and then sees a discarded Mars candy bar wrapper. It sniffs the wrapper, which still has some chocolate on it, eats the wrapper, and moves on. If we can verbalize for the goat, the meaning of this discovery was "food," and the motivational part of the story ends with the ingestion of the wrapper.

Let us now place in this same situation a chimpanzee who has previously been in contact with humans. The behavior of the chimp goes like this: spot the wrapper, pick it up, turn it over, wrinkle it, shake it, sniff it, then lick it, throw it away, and scamper off. This being a backward, *nonverbal* chimp, we must supply words for the "meaning" of his discovery: "Tasty paper . . . maybe there's more where this came from." The motivational (and the nutritional) story—the chimp's quest for human snack food—may

* The term "cognitive architecture" refers to the organizational rules and operating characteristics of the mind inferred from observations of its output (expressed ideas and manifest behavior), completely without reference to its real anatomy or to any physiologic processes by which it does its "minding."

FIG. 2.1 Ape and man in mutual contemplation. The commonalities of physicality, instinct, and intelligence bespeak a kinship undiminished by the differences in respective behavior and cultures. (*Le Singe et l'Homme A* by Pierre-Yves Trémois. Copyright © 1998 Artists Rights Society [ARS], N.Y./ADAGP, Paris.)

or may not continue, depending upon how intent he is on finding more chocolate. Next along the forest path comes a human who stops, picks up the wrapper, looks around, scratches his head, suddenly goes red in the face and gasps, drops the paper, turns around and starts running in the direction he just came from. What kind of story is this?

In each of these vignettes, the candy wrapper is "information," or, if you prefer, "data." From the goat's behavior we infer that it had the sensory equipment and the experience to classify the wrapper as food, and to act on its own judgment by eating it. The knowledge the goat brought to this encounter was compact, focused, and pragmatic, and its response to the wrapper was predictable.

The chimp's handling of the "wrapper as data" is more complicated. We can make the same default inferences about sensory equipment and categorizing skills: the chimp, like the goat, has a *food/nonfood* schema for partitioning objects in the world. But the chimp's behavior is more discriminating, more selective, than that of the goat. To him, only *part* of this wrapper is food. We can't be sure if the chimp avoids eating candy wrapping because he lacks the enzymes to digest this material,* or because his manners are more delicate than those of the goat, or for some other reason. But clearly the chimp's decision to lick the chocolate from the wrapper and then throw the paper away was influenced not only by the raw visual and olfactory data present in the paper, but by additional knowledge that he alone possessed, or by supplementary information about the wrapper that, *because* of something unique to his general store of knowledge, he sought and obtained through inspection. The chimp's behavior has some stray elements in it, but is *almost* as predictable as that of the goat.

As for the odd behavior of the human after he had picked up the wrapper, it can only be called totally unpredictable. However, since the human has *language,* he can speak for himself:

> "I'd been out looking for Jack all afternoon; bloke disappeared after lunch and now we're stuck. The truck needs a new axle, the radio's on the blink, and he's the only one can fix it, and so I go after him. About a half mile out I see this Mars bar wrapper on the trail—fresh!—and I figure I'm going to catch up with him. Then suddenly I realized Jack *hates* Mars bars, and in fact the only person I know or even heard of who can stand 'em is Blakesly, who's out here somewhere and swore to shoot me on sight if he ever found me. Boy, was *that* ever a close call!"

I have invented this small problem-story to illustrate what Merlin Donald has set out to do in his book. Somehow we have to account for the consistent and striking differences between what humans and other animals know about the world and how the *way* each knows the world informs their

---

* The chimp, of course, could not know this. However, he could have an instinctive distaste for paper, or for anything with the smell of ink on it (even when masked by the delicious aroma of chocolate); or he might have seen and remembered his favorite Uncle Creaky, doubled over and screaming after eating just such a paper.

behavior or sets boundaries on the behavioral repertoire which they exhibit. The story as presented invites us to notice how the goat, chimp, and man differ in their approach to the wrapper in both cognitive and motor behavior. The striking differences, in my view, are between the motor behavior of goat and chimp, and between the cognitive behavior of chimp and man.

First, the motoric issue. The goat chews the wrapper and swallows it. The chimp, however, goes through a procedure of the kind you might expect from a member of your own family who has found leftover frosting on the paper doily under your birthday cake. He picks up and handles the paper, looks at it, smells and tastes it; following these tantalizing preliminaries, he licks off the chocolate and throws the paper away.*

There may be a cognitive difference between the goat and the chimp in this example but, strictly speaking, we cannot describe the behavior of the chimp as "more intelligent" than that of the goat, because the goat almost certainly does not see the separation of paper and chocolate as a problem requiring a solution. Nevertheless, something—empathy, perhaps—leads us to infer that the chimp's behavior is more thoughtful than that of the goat. But make no mistake: the goat is not stupid. Mountain goats are among the most successful of the large mammals and flourish under environmental conditions so extremely harsh as to exclude humans except in very limited numbers and then only with very special survival provisions. Mountain goats are "adapted" to this life on a genetic basis; they are so good at it that they do not have to agonize over the details, or even think about what they are doing at all.

When we look at the difference between the behaviors of the chimp and the human, we encounter a major *mental* discontinuity, indicated by the fact that the human does not treat either the wrapper or the adhering chocolate as food at all. Thus, based on the story he tells us, we must describe his

---

* Note that there is no direct gain to the chimp (in the harsh world of Darwinian realities) from having a free upper limb, and fractionation movements of his digits, in this situation; the goat consumes as much chocolate as the chimp does. Were he able to comment on the situation, the goat might say that this proves that the vaunted upper limb is not such an advance after all. The chimp, however, might reply that the preliminaries that have been opened to him through the dexterity of the hand have become *meaningful* to him, and have added pleasure to the experience.

behavior as vastly more "intelligent" than that of the chimp—for even if we do not consider him to be a particularly bright person, there is no doubt that he is operating in a problem-solving mode, and there is nothing simple about his problems.

But there is more. Notice how the man's tale confronts us with the real magnitude of the cognitive gulf between chimps and humans. The man can *read;* he associates the markings on the paper with trivial (but now critically important) habits of two other persons known to him. In an instant, so rapidly that we hardly notice he is doing it, he generates a complex hypothesis concerning the search he is conducting ("Jack must be nearby"), tests it through a series of mental operations, rejects it, generates another hypothesis (unsolicited and immediately alarming), accepts *that* in favor of the first—almost certainly by doing a crude, fast estimation of probabilities and the important outcomes associated with those probabilities—and, finally, in a flashy show of inductive reasoning, runs for his life.

And there is *still* more. In his incidental comments our human narrator makes fleeting but telltale references to human science and technology: the wheel, the internal-combustion engine, and construction by subassemblies, electricity, telecommunications, and gunpowder plus ballistics. The importance of specialized human cognitive and manual skill is also acknowledged: the radio (a device for sending coded or voice messages) doesn't work and *Jack is the only one who knows how to fix it.* As if all this were not enough, the human in our story displays a *theory of mind.* This means that he can generate and attribute to others (in this case, Blakesly) specific thoughts that are not his own, and can modify his own behavior as a response to the mental state he has fabricated and attached to another person.* Finally, of course, he can relate all of this to his companions (one of whom, we would discover later, speaks only French). How did we get from the chimp to all of *this?*

Henry Plotkin, to whom we shall turn momentarily, would point out that the placement of this story in a book for you to read illustrates the operation of a *nested hierarchy.* It is like the path of a ball balanced on a stick

---

* His parents noticed that he had become quite adept at this by the time he was four years old, as do almost all children.

standing on the end of a spoon clutched in the teeth of an acrobat doing a handstand on a swinging trapeze. I, the author of the book you are reading, having actually invented the story as a map to another story (which we are in the middle of), have told it to you, the reader, *as if it were real;* and you, in turn, having deciphered its message, discover yourself as its true object, nested inside, now fairly bursting with wonder because you have been made conscious of the acrobatics of your own "representational system."

But I have left an important piece of the story out, and it is, in fact, the heart of Merlin Donald's remarkable work. It is the part of the story *he* invented to get us from chimps to ourselves. The background to Donald's invention is this: true, improvisational problem-solving, and with it what we can begin to identify as intelligence, begins with monkeys, increases in apes, and increases further still with the australopithecines. Bipedality created unprecedented opportunities *and* dangers for the hominids moving away from the protection of the disappearing forest. But by Lucy's time, nearly 4 million years ago (as we discussed in chapter 1), there was already a long history of increasingly varied use of the forelimb and hand, and an unrestrained tendency in apes—chimpanzees, in particular—to exploit the hand's manipulative capabilities. Moreover, by the time of Lucy, the radial (thumb) side of the hand and its articulations with the wrist were beginning to allow an extended range of grasping movements.

Donald attaches great importance to this history:

> The chimp uses its hands for a wide variety of functions, and its remarkable tool-using abilities have been known since the time of Darwin. Although gorillas have some manual skills, theirs are not as developed as the chimpanzee's, particularly with regard to exploration. This suggests that chimps have a more sophisticated self-representation system, at least as it relates to manual exploration.*
>
> Note that hand control involves, for the first time in evolution, a com-

---

* One of the most distinctive elicited behaviors of the chimp, which does not occur in the monkey, is its use of its hand to explore its own image in a mirror; it uses the mirror to guide its own self-examination when marks are placed on the face; this reaction is taken by primatologists as strong confirmation of self-awareness.

ing together of visual, tactile, and proprioceptive* feedback on the same action system. Hand control may be regarded as the crossing of a biological Rubicon in that a dominant distal sense—vision—comes to control and modulate actions directly. (*Origins,* p. 147)

Donald proposes that prehominid apes and hominids at this level of development can best be characterized as having an "episodic culture."

Their behavior, complex as it is, seems unreflective, concrete, and situation-bound. Even their use of signing and their social behavior are immediate, short-term responses to the environment. . . . Their lives are lived entirely in the present, as a series of concrete episodes, and the highest element in their system of memory representation seems to be at the level of event representation. (*Origins,* p. 149)

Donald's central argument concerns what happened between the beginning of the appearance of *Homo erectus* and 200,000 years ago—approximately when the transition between *H. erectus* and *H. sapiens* occurred. Anatomically, cranial capacity averaged 1,000 cc (with a trend toward enlargement) during this period, compared to approximately 400–600 cc for chimpanzees and the australopithecines. As noted in chapter 1, there are no published reports of fossil remains from the ulnar side of the hand from this period, so it is still not known when the change occurred that allowed ulnar opposition, the most recent functional advance of the modern human hand.

What is known is that by the end of this period *H. erectus* was behaviorally extremely advanced:

Erectus developed a variety of sophisticated manufactured tools and spread over the entire Eurasian landmass, adapting to a wide variety of climates and living in a society where cooperation and social coordination of action were central to the species' survival strategy. . . . Some features of the cul-

* "Proprioception" refers to awareness of position and arises in the body through the action of special sensors in muscles, joints, tendons, ligaments, and skin.

> ture of *erectus* suggest qualitative changes in cognition, rather than more of the same. Widespread tool manufacture required both an elaborate mechanism for inventing and remembering complex sets of procedures and the social skills to teach and coordinate these procedures. *Erectus* also cooperated in seasonal hunting, migrated over long distances, used fire, cooked food, and evolved a brain that reached 80% of the volume of the modern human brain. (*Origins,* pp. 163–64)

Donald proposes that *erectus* evolved a new system of knowledge in the course of its worldwide migration, which not only supported the geographic and population expansion but (as Dunbar also proposes) was demanded by it. By the end of an era of approximately 1.5 million years of honing of the *erectus* survival strategy, evolution had prepared the stage for a new species: a small boost in *erectus* brain capacity and the acquisition of speech was the formula for *H. sapiens.*

What *erectus* really achieved during this period—its winning strategy—Donald calls a *mimetic* culture.[8] The foundation of *erectus* culture was mimetic skill:

> Mimetic skill or mimesis rests on the ability to produce conscious, self-initiated, representational acts that are intentional, but not linguistic. . . . Mimesis is fundamentally different from imitation and mimicry in that it involves *invention* of intentional representations. . . . Mimetic skill results in the sharing of knowledge without every member of the group having to reinvent that knowledge. . . . The primary form of mimetic expression was, and continues to be, visuomotor. The mimetic skills basic to child-rearing, toolmaking, cooperative gathering and hunting, the sharing of food and other resources, finding, constructing, and sharing shelter, and expressing social hierarchies and custom would have involved visuomotor behavior. (*Origins,* pp. 169–77)

In view of another comprehensive theory that places great emphasis on the role of toolmaking in creating a cognitive base for language, Donald's position on this point is of special interest:

> The example of toolmaking shows how these general and special skills might have combined to enable an important new mimetic skill. . . . Toolmaking is primarily a visual-manual skill, but it also involves obtaining the necessary materials, fashioning the appropriate tools at the right time, apportioning responsibility, and so on. . . . Innovative tool use could have occurred countless thousands of times without resulting in an established toolmaking industry, unless the individual who "invented" the tool could remember and re-enact or reproduce the operations involved and then communicate them to others. (*Origins,* p. 179)

Donald suggests that during this period control of facial musculature was increasing, permitting an increased range of emotion that could be registered in facial expression. But the anatomy and function of the respiratory tract (especially the position and function of the larynx and upper airway) had not changed significantly, and the neurologic mechanisms necessary for control of airflow and articulation could not have reached their present state, so speech as we understand it now could not have existed. Donald rejects the possibility that a visual language based in the hands—sign language, which in its modern form is equivalent to speech in its linguistic range and power—was in use.

> The recency of formal sign language, and its resemblance to some forms of writing, argues in favor of its classification with ideographic writing and other modern visuosymbolic inventions, rather than with oral narrative skills and speech. (*Origins,* p. 308)

Gestural language, however, was almost certainly employed in communication, and Donald allows this as a likely form of mimetic communication and perhaps even "the precursor of the more advanced semiotic inventions underlying speech" (*Origins,* p. 220). However one finally chooses to construe the historical interaction of speech and gesture, there is no doubt they overlap and reinforce each other in human language. The association of gesture with human thought is a slightly different but closely related question. Here, developmental observations may hold the answer:

cognitive and developmental psychologists regard the appearance in children of pointing as a "gesture of intentionality" (at about fourteen months) to be an important milestone in their mental development and consider its exclusive use by humans a demarcation from chimpanzee cognitive potential. Chimpanzees neither spontaneously produce this gesture nor acquire it through training.[9]

Donald's position on the centrality of speech as the primary enabler—if not the original *agent provocateur*—of language is clear: formal language, with fully developed and heritable neurologic control for its expression, is grounded in speech. Speech, in turn, is the outcome of convergence in evolution of (1) widespread advances in cognitive capacity (representational and mimetic skill); (2) the articulators of speech; and (3) recently evolved neural structures—concentrated in but not confined to the "speech areas" of the left hemisphere in the vast majority of people—dedicated to the control of speech production and the processing of speech sound.*

While Dunbar's and Donald's accounts differ in emphasis and specificity, each identifies the appearance of language as the outgrowth of highly coordinated interactions between evolving *Homo* and the environment. They also make clear that in the case of humans, the organism-environment interaction is a novel evolutionary story, in that the classic environmental selection pressures came to include cooperative (and internally competitive) organizations of higher social primates. This is a rather remarkable two-for-one feat: it shows us where language may have come from, and how it is related to intelligence; it also shows us how society itself—culture—became part of the dynamic process of selection.

Henry Plotkin's Theory of Intelligence as a "Secondary Heuristic"

Henry Plotkin is professor of psychobiology in the Department of Psychology at University College in London. In 1993 he published *Darwin Machines and the Nature of Knowledge.*[10] Plotkin's book and the develop-

* This "subthesis" is presented in detail in Donald's *Origins*, pp. 236–56.

ment of his thesis complement Donald's so seamlessly, that together—for me, at least—they stand as a single, definitive (and unparalleled) work.

Plotkin regards the term "knowledge" as signifying more than words or facts in the head. Knowledge is *any* state in an organism that bears a relationship to the world. A bullfighter who has been gored by a bull has the words *el toro* in his head, and a memory and powerful emotions connected with that incident blended into (or "informing") his own internal "representation" of the animal, and he has a scar on his chest wall as well. The scar also "represents" the bull—it is a registration *on and in the body* of the bullfighter, a permanent reference to and symbol of the bull and of the encounter with it, and (in Plotkin's extremely interesting way of looking at things) the scar qualifies as part of this bullfighter's *knowledge* of the bull.

And the bullfighter's knowledge of the bull goes even deeper: it is the sum of multiple integrations of joint angles and muscular attachments at his hip, in the spine, and in the knees, ankles, and feet, transformed into synergies of poised, fluid, and gracefully explosive movement. This knowledge also includes a thirst for adrenaline, a capacity for total concentration, and—just in case the bull is unimpressed—the flawless chemistry of clotting factors and antibodies in the blood.

The convergence of all these elements in a single individual is the achievement of evolution, which, through the inexorable process of genetic variation, selection, and transmission now associated with Darwin's name, captured the bullfighter mix, stored it as a heritable genetic code in a few unsuspecting adults, and (magically one night) brought one absolutely unique version of it to life in the person of our young friend with piercing eyes and one buckled rib.

Plotkin points out that the evolutionary process is indifferent to this and all other bullfighters as individuals but is unswerving in its devotion to the stability and longevity of their genes.* It does this by the simple device of

---

* That is, genes look out for themselves, period. By universal accord, Richard Dawkins, inaugural holder of the Charles Simonyi Chair of Public Understanding of Science at Oxford University, is the official spokesman for this Darwinian axiom. Three important and highly readable books by Dawkins present the argument in full: *The Blind Watchmaker* (1986), *The Selfish Gene* (1989), and *Climbing Mount Improbable* (1996).

adding to the genes something that, when it unwinds in the brain of every living creature, becomes a powerful urge to stay alive long enough to mate. Enough creatures live long enough not only to experience this urge, but to yield to it, that the genetic blueprints tend to survive. Whenever such a blueprint (or genome) does survive, *even though it may be an imperfect copy of the donated genes of the parents,* it will attempt to inject itself (literally) into yet another generation.

Each new combination of genes *tries* to be a faithful copy; however, a very slight tendency for errors to occur has crept into this biologic formula. Sometimes, these errors prove fatal: a pregnancy is aborted, a child fails to reach reproductive age, or an adult is infertile. Occasionally, however, the "mistake" proves providential. Almost always when that happens—when a genetic mutation provides a survival advantage—it is because the world changed enough to disprove the prediction implicit in the original code, but in a way that, through sheer chance, validates the prediction implicit in the mutation. Once in a while—comparatively seldom but with satisfying regularity over the long haul—surprising and quite wonderful things can happen; a cold, machine-like, statistical process transforms itself into a genetic Cinderella story.*

Plotkin calls this process—the genetic device whose operation produces longevity in families of genes through the generation, testing, and regeneration of genetic codes—"*the primary heuristic.*" This is why, and this is what it means:

> Heuristic ordinarily has the meaning of "that which leads to discovery and invention." . . . The primary heuristic has two features that are worth noting. One is that it takes the logical form of induction, generalizing into

---

* For those who would like to remember the serendipity of Darwinism in a catchy way, I donate the following bit of doggerel as a mnemonic, asking you to recall that "bugs" in the genetic program produce phenotypic "jitter."

Darwin's theory speaks of hits
And misses, and of wobbles.
If jitterbugs arouse a prince,
Might slippers dance on cobbles?

> the future what worked in the past. That is, the successful variants are fed back into the gene pool, where they will be available for sampling by future generations. This is the conservative, pragmatic part of the heuristic. The other is the generation of novel variants by chance processes. This is the radical, inventive component of the heuristic. It is nature's way of injecting new components into the system in order, possibly, to make up for the deficiencies that may occur if what worked in the past no longer does so because the world has changed. (*Darwin Machines,* pp. 84, 139)

Every living organism carries genes that contain, because of the operation of the primary heuristic, the plan for a life that can succeed even in the face of known adversity. Built into the body by the genetic code are structures (hard head, strong muscles, good eyes) and functions (being able to duck without thinking, for example). Structure, function, special adaptations, instincts—*all* of these presets in the organism provided for by the genes—are "knowledge." This kind of knowledge exists in every animal. When the knowledge inherent in the primary heuristic was upgraded in large, mobile, long-living, social primates—remember, these animals were expanding into unfamiliar environments containing unpredictable dangers and so needed the mental equipment for unprecedented freedom of action—Merlin Donald's "episodic culture" came into being. As Plotkin accounts for it, intelligence is one of the true Cinderella stories of evolution: it is the principal adaptation that evolved in primates to meet the "Uncertain Futures Problem."

> The Uncertain Futures Problem concerns an organism going through life, equipped only with instructions given at conception (and hence perhaps correct only at that time) on how to survive, and having to interact with a world that may be different from that in which its life began. . . . There are several ways by which the problem of change can be confronted. . . . [One] is by giving rise to change *within* phenotypes; that is, by producing phenotypes that change in response to changes in the world. I will call this the tracking option. (*Darwin Machines,* pp. 144, 147)

With respect to the issue of the knowledge and intelligence required for truly unpredictable change (the location of a water hole, or one's own nest, or who is dangerous), the following is necessary:

> Knowledge-gaining devices whose internal states match those features of the world that we are calling short-term stabilities. Such tracking devices would be set in place by the usual evolutionary processes of the primary heuristic and hence would operate within certain limits. But the exact values of those limits that these devices will settle to, and for how long, are not within the power of genes to decide. So devices such as these have a degree of autonomy in their functioning that makes them partially independent of both genes and development. (*Darwin Machines,* p. 149)

Intelligence is what Plotkin is talking about here, and he defines it in the context of his previous conceptualization as a *secondary* heuristic. Returning to the Donald and Dunbar narratives, we can now state that *social* intelligence was almost certainly the first iteration of Plotkin's secondary heuristic. This is not saying anything new, but it is a way of defining the fit between these theories.

The real payoff for us in this coupled formulation is the directness with which it tells us that we have not explained *what* exactly (other than the vicissitudes of migration and political intrigue) had the power to drive *erectus* all the way to *sapiens.* Can we really accept that this magnificent creature, *Homo erectus,* spent 1.5 million years just sharpening small talk, learning to cook and sew, building lean-tos, and inventing improvisational theater? This is not simply too pale a story; it is a story that ignores major implications of evolution's gift of an exploded repertoire of hand movements harnessed to the cognitive power of the brain.

At the conclusion of his book, Plotkin returns us to the Gordian knot, the nexus of knowledge, language, and intelligence:

> Given that language is unique to our species, that must mean that some part of that portion of our genetic makeup that is unique to us as a species is a part-determiner of the human ability to function in the realm of language. In other words, we must be genetically predisposed to learn, think

> in and communicate by language. This uniqueness of language as a human form of knowing about the world, as well as its apparent multiple roles in human psychology, means that an understanding of language is an essential part of a science of human knowledge.
>
> This nexus of roles for language means that the relationship between internal organization and external features of the world that characterizes all forms of knowledge takes, in the case of language, a dynamic double form: a word, or a sentence, as external event is matched to internal representations of the meaning of the utterance, and meaning itself may take more than one form. It is this symbolic quality of language that makes it so central to our psychology. Language so dominates our lives, especially our interior mental lives, that it is difficult not to think of it as anything other than the central plank of human understanding and knowledge. (*Darwin Machines*, pp. 201–202)

Plotkin lays out in plain language the essential background for what has become the central burning question of linguistic theory: Where did human language come from? Modern linguistics really began with the work of Noam Chomsky, who proved the essential dissimilarity of human language from other forms of animal communication by exposing its unique operational and developmental characteristics. The cornerstone of Chomsky's revolution was his insistence that the uniformly spectacular (and spectacularly uniform) linguistic accomplishments of children cannot be accounted for by their experiences alone. Children do not hear nearly enough in their early years to account for what they *know* about language by the time they begin to speak. This is Chomsky's "poverty of the stimulus" argument, and it is the centerpiece of his ruinous attack on B. F. Skinner and the behaviorist school of psychology, who had until 1959 persuasively contended that language exists in society and that children acquire it entirely on the basis of parental (or other) teaching.* Putting this argument in a nutshell, Plotkin says:

> What is extraordinary is the contrast between the casual immersion into a language environment which the children experience and the highly spe-

* See Noam Chomsky, "Review of B. F. Skinner's *Verbal Behavior*," in *Language* 35: 26–58, 1959.

cific and intricate language structures that emerge. Language learning in the human child is well described by the seeming paradox of the child knowing what it has to learn. (*Darwin Machines,* p. 203)

Buttressing this argument (that you can prevent children from learning to read or ride bicycles but you can't stop them from learning to talk), Chomsky had pointed to two other universals in human language: that its emergence in children follows a very precise timetable of development, no matter where they live or which particular language is the first they learn; and that language itself has an innate structure. Chomsky has recently reminded audiences that the origins of the *structure* of language—how semantics and syntax interact—remain as "arcane" as do its behavioral and neurologic roots.[11] Chomsky himself finds nothing in classical Darwinism to account for human language.* And for that reason, says Plotkin, linguistics is left with a major theoretical dilemma. If human language is a heritable trait but one that represents a complete discontinuity from animal communicative behavior, *where did it come from?*

Chomsky's views threaten central assumptions widely shared by anthropologists, cognitive scientists, and linguists by insisting that the following three statements are incontrovertibly true with respect to human language:

- It is a heritable, species-level trait.
- It is not a special function of general intelligence.
- It has no analogues in animal communication.

Within the past few years a record number of books and scientific articles concerning the origins of human language have appeared.† The most

* Eventually, Chomsky may well be seen less as a spoiler in the evolutionary account of human language than as the uncompromising critic of an evolutionary theory that itself fails to address the uncontested facts of human language. That is, it might be more accurate to describe his position as that of someone exhorting the cognitive and evolutionary theorists to go back to the drawing boards until they come up with something more plausible, more *unifying,* in a theory purporting to explain the origin of human language.

† The best known of these, Steven Pinker's *The Language Instinct,* will be considered in detail in chapter 10, "The Articulate Hand."

recent of these, Terrence Deacon's *The Symbolic Species,* takes the position that human speech is the "natural" organ of human language, and that the underlying structure of language can be explained on the basis of adaptations of the vocal tract, which permitted the mapping of internal symbolic representations onto control mechanisms for vocalization. While siding with the eminent linguist Philip Lieberman concerning the essential ties between speech and linguistic ability, Deacon emphasizes that it would be an oversimplification to state that language capacity was in effect silently accumulating throughout mammalian and primate evolution, and then somehow was *released* once the complex human vocal tract was finally ready to produce articulate speech sounds.* What chimpanzees lack, Deacon says, is not so much a capacity for vocal production as a capacity for vocal *learning.*[12]

In *Darwin Machines,* Plotkin had concluded his discussion of language by stating a problem:

> What I cannot yet understand, and neither can anyone else, is exactly what the functional origins of language are. (*Darwin Machines,* p. 206)

Has Deacon finally solved the "Chomsky problem"? In *Evolution in Mind,* Plotkin suggests that Deacon is at least on the right track:

> For Deacon, symbolic reference is what lies at the heart of human language. He takes it as read that the mind is the product of evolution and hence is innately constrained in terms of basic sensory processing, perception, attention, rates of information-processing, memory, and other functionally general cognitive processes that provide the processing power upon which symbolic reference is built. . . . [Thus, there] is a convergence in every language on certain universal structures of language. But this is not because universal grammar is built into our brains by our genes and development, but rather because "language structures at all levels are the products of powerful multilevel evolutionary processes, to which innate

---

* See Philip Lieberman, *Uniquely Human: The Evolution of Speech, Thought, and Selfless Behavior* (Boston: Harvard University Press, 1991).

mental tendencies contribute only one subtle source of the Darwinian selection biases." (*Evolution in Mind,* pp. 158–59)

While strongly agreeing with Deacon's proposal that language must be understood as an effect of a complex co-evolutionary process, I cannot accept that Deacon's model tells us where language *syntax* came from. That is, if we accept that the movement of muscles within the human vocal tract explains how *sound* units (phonemes) are synthesized into longer sound bursts that can be detected and sorted by the human auditory system into *meaning* units (words), we are still not quite home. In fact, we may still be a very long way from home. What this process leaves entirely unexplained is why and how the human brain organizes the transformation of *words* into *sentences* as it does. As we shall consider in chapter 10, "The Articulate Hand," there *is* a way to explain where sentence structure comes from, and that explanation makes it clear why neither acoustics, nor the physiology of voice production, nor auditory perception offers us the most parsimonious—or the most credible—explanation for origins of sentence structure.

The work reported in this book, while not undertaken with either Chomsky's or Plotkin's concerns in mind, could nevertheless assist in the development of an alternative, evolutionarily informed account of the relation between human thought and language. What stands out in that regard is the central role of the hand in human thought and action, backed by a growing understanding of the specializations and enhanced survival value of the primate upper limb as it had evolved even before the time of the australopithecines, long before the advent of language.

What I am suggesting is that perhaps we *do* know what the functional origins of language are. It is likely that sometime during its stewardship of the genetic lineage of *Homo, erectus* completed the final revisions to evolution's remodeling of the hand, opening the door to an enormously augmented range of movements and the possibility of an unprecedented extension of manual activities. As a collateral event, the brain was laying the foundations of cognitive and communicative capacity. I would not suggest that a tiny modification of this ancient pentadactyl structure by itself closed, or even catalyzed the closure of, the narrowing gap between Merlin Donald's mimetic culture and its successor, the *mythic* culture.* Rather, this new hand reflected a modification of the primary heuristic, and brought with it the opportunity for a new class of situational knowledge based on as yet unexplored and undefined use of the hand. This change by itself was nothing but a mutation until its utility gave it the status of an adaptation. Absent what preceded it, surrounded it, and was still to come, it would have been neither burr nor spur. But, with the advantage of hindsight, we can guess that events following this anatomic change conspired to produce a second iteration of Plotkin's secondary heuristic: "manual intelligence," or just plain "hand smarts."

The task undertaken by the present book from this point will be to explore and critically examine this proposal. The handyman's hand was more than just an explorer and discoverer of things in the objective world; it was a divider, a joiner, an enumerator, dissector, and an assembler. The handyman's hand could be loving, aggressive, or playful. Eventually, it found in the intimate touch of grooming the secret to the power of healing. It may also have been the instigator of human language.

There is growing evidence that *H. sapiens* acquired in its new hand not simply the *mechanical* capacity for refined manipulative and tool-using skills but, as time passed and events unfolded, an impetus to the redesign, or reallocation, of the brain's circuitry. The new way of mapping the world was an extension of ancient neural representations that satisfy the brain's need for gravitational and inertial control of locomotion. Elementary physics, of course, was written into the brain and spinal cord of mammals a very, very long time ago, and had endowed the monkey's limbs with assured, acrobatic

---

* That's us (and the Neanderthals), beginning roughly between 150,000 and 200,000 years ago. Merlin Donald describes its central theme:

> Mythical thought, in our terms, might be regarded as a unified, collectively held belief system of explanatory and regulatory metaphors. The mind had expanded its reach beyond the episodic perception of events, beyond the mimetic reconstruction of episodes, to a comprehensive modeling of the entire human universe. Causal explanation, prediction, control—myth constitutes an attempt at all three, and every aspect of life is permeated by myth. (*Origins,* p. 214)

The second half of Donald's book is devoted to a discussion of the mythic culture, and its (present) successor, the modern theoretic culture.

genius and hands like computerized magnets. But a new physics would eventually have to come into this brain, a new way of registering and representing the behavior of objects moving and changing under the control of the hand. It is *precisely* such a representational system—a syntax of cause and effect, of stories and of experiments, each having a beginning, a middle, and an end—that one finds at the deepest levels of the organization of human language.

Robertson Davies, the distinguished Canadian novelist, made an extremely provocative suggestion in *What's Bred in the Bone* when he observed that "the hand speaks to the brain as surely as the brain speaks to the hand."[13] I think he wasn't just telling a story about painters and painting. I think he was telling us (again!) that the time has come to repair our prevailing, perversely one-sided—shall I call them cephalocentric?—theories of brain, mind, language, and action.

3
The Arm We Brought Down from the Trees

LOOK AT THE SHOULDERS ON THAT GUY!

—*overheard on the boardwalk, Venice Beach, California*

SO SLIGHT ARE THE ANATOMIC DIFFERENCES between the ape and the human hand that anthropologists discounted their importance in the *functional* capabilities of the two hands until quite recently. But focusing on the hand alone misses an extremely important piece of evolutionary history: the arm was such a sophisticated instrument when its owner moved to the ground that very little further modification was needed at the far end—in the hand—to transform the entire limb into what it has now become. However worthy the shoulder and arm may be of their high aesthetic ranking in a society obsessed with the appearance of the body, the engineering beneath the skin is where the real glamour is.

Consider for a moment the winning formula for quadrupedal locomotion in tree branches. Better yet, just step outside and watch a squirrel racing through the branches of a tree. Monkeys added a few tricks to this very successful method of above-ground travel, mainly by giving greater responsibility for safe landings to the arms, and adding the further security of a prehensile tail. The simian's hand-arm apparatus proved perfectly satisfactory for locomotion (which is why the playground rig has always been called "monkey bars") while permitting feeding by hanging from one arm and plucking with the other. Chimps have the same basic equipment, minus the tail, and have a beefed-up arm and shoulder; movement in suspension (brachiation) was improved by an increase in the supination range of the forearm.

The chimps brought this state of shoulder-arm engineering down with them to the ground, but their travel on the ground was a kind of goofy quadrupedal gait in which the body weight is supported by the soles of the feet and the knuckles of the hands. When the australopithecines started walking upright, the upper limb had not changed much, and it is unlikely that the control mechanisms were different, either. Had there been any major changes, you might suspect that the arm was destined to be downgraded—either to a structure with merely ornamental function, or to a shrunken version, possibly like the forelimbs of the kangaroo. After all, why *wouldn't* the shoulder and arm just recede into the background once their primary responsibilities in locomotion had been eliminated?

But the hominid arm did not wither, either anatomically or functionally. Why? One possible answer is that the brachiating arm (unlike the forelimb of mammals lacking its unique experience and capabilities) had already secured a major "presence" for itself in the brain—that is, its complex functions had come to be widely represented and "networked" into expanded sensorimotor systems within the central nervous system. Given the enormous risks of terrestrial life, Darwin's selection engine almost certainly would have favored the propagation of those hominid bipeds who could exploit *on the ground* the agility of this arm and the considerable dexterity of this hand. There was more than ample reason in the new life not to allow this amazing structure to slip into retirement. (Perhaps there *were* families of new hominids whose arms just withered. Fortune, it would seem, did not smile on them.)

So, as an essential prelude to the main story of the hand, we will consider just what the prehuman *arm* was and what the chimps and monkeys were doing with it. Perhaps the most instructive way to begin is with a visit to the world of construction, where mechanical versions of the arm abound and where its truly remarkable abilities are easier to appreciate. We are going to learn just a little about coupling, guidance, and propulsion.

In 1992, the *Wall Street Journal* published a story about the use of the backhoe as a competition vehicle. The headline in the *Journal* read: TO WIN THIS RODEO, MAN AND BACKHOE MUST BECOME ONE.[1] The story had been prompted by the first North American Backhoe Rodeo Championship in Phoenix, Arizona, and it was spiced with a rich assortment of quotes from the operators competing in the rodeo. The comments ranged from pithy winning tips to more thoughtful reflections on mastery of the backhoe; taken together, they left no doubt that the backhoe operator is a movement specialist of the highest order.

Jim Hart, a Bangor, Maine, water district employee, practiced for the finals by trying to strike kitchen matches with the bucket of his hoe. Another contestant practiced by "plucking an egg from a bucket of sand using a tablespoon taped to a tooth of his backhoe."

But mere practice is not enough:

> "You must empty your mind and think of nothing so that the backhoe becomes an extension of your arm." (Louis Forget, a construction worker from Quebec who rode his first backhoe at age nine)

> "You're part of the machine. It's part of you." (Harvey Neigum, the eventual winner of the contest)

The idea of "becoming one" with a backhoe is no more exotic than the idea of a rider becoming one with a horse or a carpenter becoming one with a hammer, and this phenomenon itself may take its origin from countless monkeys who spent countless eons becoming one with tree branches. The mystical feel comes from the combination of a good mechanical marriage and something in the nervous system that can make an object external to the body feel as if it had sprouted from the hand, foot, or (rarely) some other place on the body where your skin makes contact with it.*

Shortly after it appeared, I sent the backhoe story to Richard Young, a young Californian whose profession is training crane operators.[2] I told him I was interested in cranes because they reminded me of the arm, which has to position the hand, support it, and move it, in order for it to do its job. Richard's perspective on cranes was more practical than mine: what really

* The contexts in which this bonding occurs are so varied that there is no single word that adequately conveys either the process or the many variants of its final form. One term that might qualify is "incorporation"—bringing something into, or making it part of, the body. It is a commonplace experience, familiar to anyone who has ever played a musical instrument, eaten with a fork or chopsticks, ridden a bicycle, or driven a car.

matters with cranes, he said, is preventing them from falling over. Cranes are now lifting bigger loads higher, and operating between buildings that are closer together. There are more kinds of mistakes to be made, and when trouble *does* come, there may not be sufficient time for even a highly experienced human operator to take corrective action.

> When something happens, something the operator can't see, he'll never be able to correct it if he hasn't foreseen it as a possibility. Maybe he's operating the hoist, he's got his boom out at an angle, load lines down, it's hooked up to a heavy weight and he's picking it up. All of a sudden, for some reason, maybe the boom tips, or a wire rope slips off. He has a hundred-ton load sitting there, and it drops just a little. That's enough to shock-load the crane and make his load start bouncing up and down. Then a gust of wind comes up and swings the load. So this guy's sitting in the cab, and he has to know how to stop what's happening. *But he has to know it's coming in the first place.* Otherwise he's lost. That's how most accidents happen. Really, it doesn't take much.

Richard's emphasis on the constant hazard of a crane tip-over raised the hidden issue of balance. How *do* upright structures stay upright when they are moving loads of varying weights at varying distances from their center of gravity? Cranes are thrown off balance and fall all the time; how could they not? The dynamic management of weight distribution is critical; unexpected weight shifts can occur suddenly and without warning; the corrective measures open to the operator are relatively crude; and the time during which corrective measures can be effectively applied is exceedingly brief.

If the risk of falling complicates the control of cranes, it must be far more a problem for us. Whether we are carrying bags of groceries into the house, dashing to appointments carrying a briefcase in one hand and coffee in the other, or playing in a ball game, getting from point A to point B without falling is *far* more difficult if we are carrying something. What does the arm have to do with balance?[3]

Consider a laborer carrying a bag of cement on one shoulder up a flight of stairs, then up a ladder, and then across a beam, before dropping it into a wheelbarrow or a mixer. Each time he takes a step he is actually throwing himself off balance by shifting support of the load from one leg to the other. He does the same if he leans slightly to one side or the other, or shifts the bag from one shoulder to the other, or repositions the bag vertically or laterally anywhere along his shoulder or arm. If he steps on a wet or dusty surface his foot may slide, and he must instantaneously restabilize the load or risk falling. At the same time, he must also be aware of internal sensations arising from his musculoskeletal system: if he exceeds the load- and tension-bearing capacities of his own muscles, bones, joints, and ligaments he may seriously injure himself.

This problem is so complicated for the neuromuscular system that the perceptual and control side of the job is handled almost entirely outside conscious awareness. There is far too much information about joint angles and weight distribution to be gathered and analyzed, and there are far too many inertial and force equations for the brain to solve rapidly and simultaneously. Richard had pointed out with considerable pride certain advances in crane instrumentation: load-moment, boom-angle and level

FIG. 3.1 Steelworkers during the construction of New York City's famed Waldorf-Astoria Hotel in 1930 take their lunch break—a scene that reminds us of an important but insufficiently appreciated gift from human arboreal ancestry. (© 1930, Keystone/Sygma, reprinted with permission)

indicators. The biologic analogues of this instrumentation exists in each of us, too. But the engineers have installed all this instrumentation in larger and faster cranes that *stand still on the ground.* Those of us who don't stand still much have a more difficult problem.

Actually, the australopithecines had no need to invent a new balancing function for the upper arm. They just appropriated and then promoted what had come into existence for a comparable and biologically important reason: a female monkey carrying her baby had to be able to run along narrow branches and jump from branch to branch without losing her balance. And this system must have come to be as heritable (or "hardwired") as any set of neurologic and biomechanical traits could be, because survival of the young would have depended upon it. For those of us living on the ground now and taking advantage of this gift, the *conscious* experience of upright lifting and carrying—or the deliberately destabilizing movements of dance and floor gymnastics—consists mainly in the registration of either a subjective sense of being balanced, or the alarming realization that we are "losing our balance" when postural reflexes cannot keep up with weight shifts. The emotional feeling of helplessness that almost invariably precedes a fall is justified, since there is rarely anything left for us to do except decide *where* we might prefer to land, or where on the body we would least mind a crop of bruises and contusions.

Our shoulders, in these balancing acts, are not simply carrying the groceries. They adjust the position of the arms just as a highwire-walker adjusts his pole, to abort the fall threatened by each and every step taken in the upright posture. If the crane operator is in the middle of a very advanced applied trigonometry and physics problem, so are *you* when you are carrying something, walking on uneven ground or going down slippery steps. Thanks to the honing of these skills by many generations of your ancestors, you are never obliged to look at an instrument panel.

Having established that the shoulder joint is involved in the body's balancing system (by helping to equilibrate destabilizing weight shifts in the upper body) we should next try to understand, at least in general terms, how this is done. Because of the nearly unlimited freedom of the human body to shift its center of gravity, we would expect the shoulder to be designed to allow as much freedom of movement as possible without losing its ability

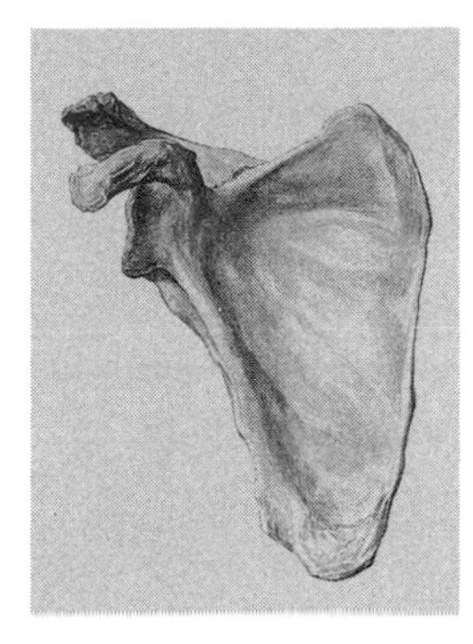
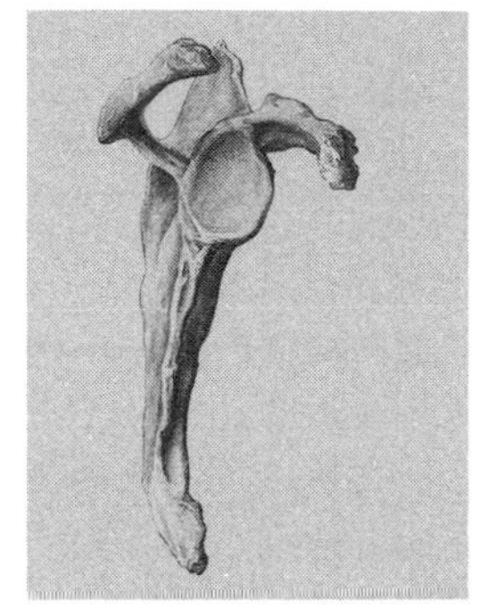
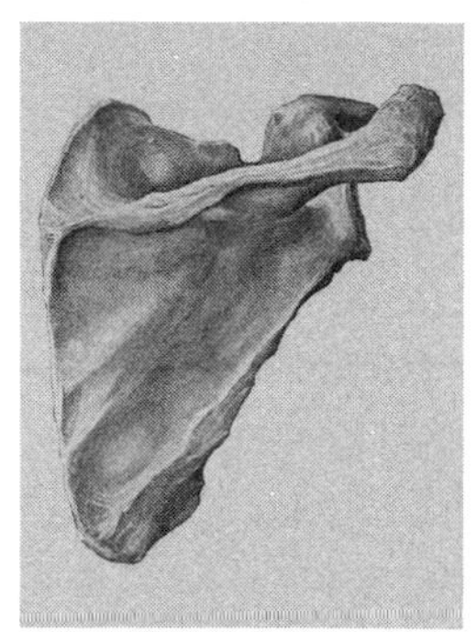

FIG. 3.2 The human scapula, right side, in three views. *Above, left,* the side of the scapula that fits against the rib cage; *center,* a view from the side, showing the wide, shallow socket for the head of the humerus and the two arching projections above it. Together, these (the *acromion* and the *coracoid process*) provide a roof to protect the head of the humerus and restrain it, giving it maximum freedom of movement on its base. *Right,* the scapula seen from behind, showing the prominent spine separating two important groups of muscles controlling movement of the humerus within its socket. (From Werner Spalteholz, *Hand Atlas of Human Anatomy,* 1923.)

to limit movement range, which is to say, to maintain the integrity of the mechanical end points.* This is what we find, of course, but it is accomplished in a way like no other joint structure in the body.[4]

Unlike the leg, whose bony core is firmly planted in the pelvis, the arm is actually held *away*—suspended—from the body by its skeletal and muscular attachments. Anchored at the shoulder to a hollowed-out perch at the end of the collarbone, the upper arm clings to the rib cage as tightly as an abalone clings to a rock. Several layers of interwoven muscles position, hold, or move it—sometimes doing all of these simultaneously. The shoulder girdle muscles, as they are called, are all either directly or indirectly attached to the shoulder blade (the *scapula*), which is the structural heart of the shoulder.

Best seen from behind, or in a mirror, the most prominent feature of this

---

* It is not easy to dislocate the shoulder and it is no fun when it happens. Even this, though, would be nothing compared to complete failure of the restraining mechanism. Remember what happened to the neck of the bewitched little girl in *The Exorcist*?

bone is a gently curved plate about the size of the hand. Its prominent flat portion—the "blade"—is molded to fit tightly to the upper back. On a slender person the shoulder blades suggest a silhouette drawing of two small buffalo heads facing each other—a pair of softly rounded right-angle triangles, snouts pointed down. Near the top of the scapula, on its outer surface, a horizontal *spine* forms a narrow ledge that widens as it approaches the shoulder tip, where it curls upward like a cresting wave. The top of this crest is called the acromion;* the rounded floor of the spine holds a small but very important muscle (the supraspinatus) that initiates movement of the arm out to the side and helps other muscles move the arm farther up, out, and over the head. The only two bones with which the scapula makes direct contact (with which it articulates) are the collarbone (the clavicle) and the upper arm bone (the humerus). This places the scapula in an extremely strategic location, at the center of the only bone-to-bone contact between the arm and the rest of the body. The inner, or body-side, joint is the acromio-clavicular (or *AC*) joint, which is located just above the large, shallow socket (the glenoid fossa) that faces directly out to the side. On the outer, or arm-side, is the *glenohumeral* joint, a ball-and-socket joint not unlike the "joystick" used to control computer games, or the stick shift in a sports car.† The bone-to-bone contact between the arm and the body is completed at the front of the chest where the clavicle meets the breastbone (sternum).[5]

Imagine for a moment that the scapula simply did not exist. You would have a multijointed stick consisting of three sticks of approximately equal length laid end-to-end originating in the middle of the chest: clavicle to humerus, humerus to elbow (which connects the humerus to the radius and

---

* You can easily find this anatomic landmark by grabbing your own shoulder as you might if somebody had just punched you there. Once in place, your hand will find the acromion by sliding a short distance upward toward the top of the shoulder; you will feel it as a flat, bony edge.

† You can't feel the socket itself because it is completely covered by the upper end of the humerus and the big muscles at the top of the shoulder. But you can feel it *work*. Put your hand back on your shoulder (in the "Why'd you punch me?" position) and then move your shoulder the way a baseball pitcher does when he's just starting to warm up. The muscle you feel moving under your hand when you rotate this joint is the deltoid.

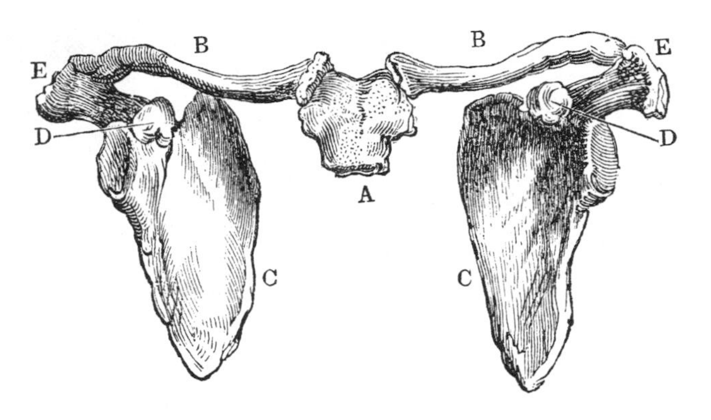

FIG. 3.3 Charles Bell's drawing of the bones of the shoulder girdle. In the center (A) is the breastbone, or *sternum,* which is the only bony contact between the upper limb and the rest of the body. The collarbone, or *clavicle* (B), runs as a strut out to the tip of the shoulder, attaching to the scapula at the top, or *acromion* (E), of its spine. The inside border of the wing of the scapula (C) and the tip of the coracoid process (D) are both important sites of muscular attachment of the scapula to the vertebral column and to the chest wall. (From Charles Bell, *The Hand,* 1840.)

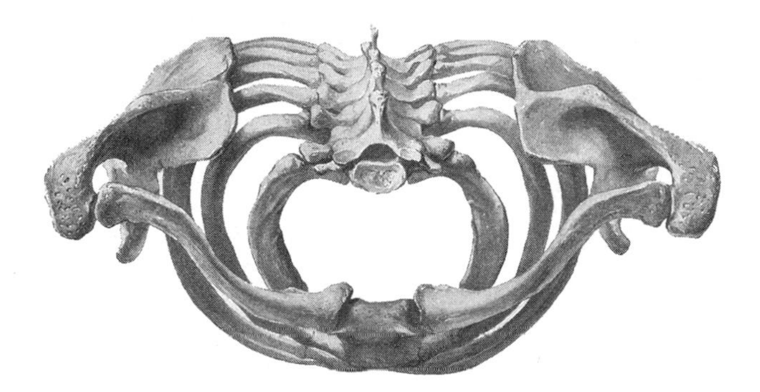

FIG. 3.4 View of the shoulder girdle from above, showing the top of the shoulder joint and the contacts between the clavicle and the sternum (midline) and the acromion of the scapula (at the shoulder). With the rib cage and vertebral column now added, it is easy to imagine that the shoulder girdle could represent a gradual transformation of what was formerly the top rib, now bent outward instead of curving back to attach to the spine. If that were so, the scapula itself would have to represent an expansion and splitting into mirror halves of the topmost thoracic vertebra. This view also makes it obvious why the muscles attaching the scapula to the chest and spine are so critical to the stability of the entire shoulder apparatus. (From Werner Spalteholz, *Hand Atlas of Human Anatomy,* 1923.)

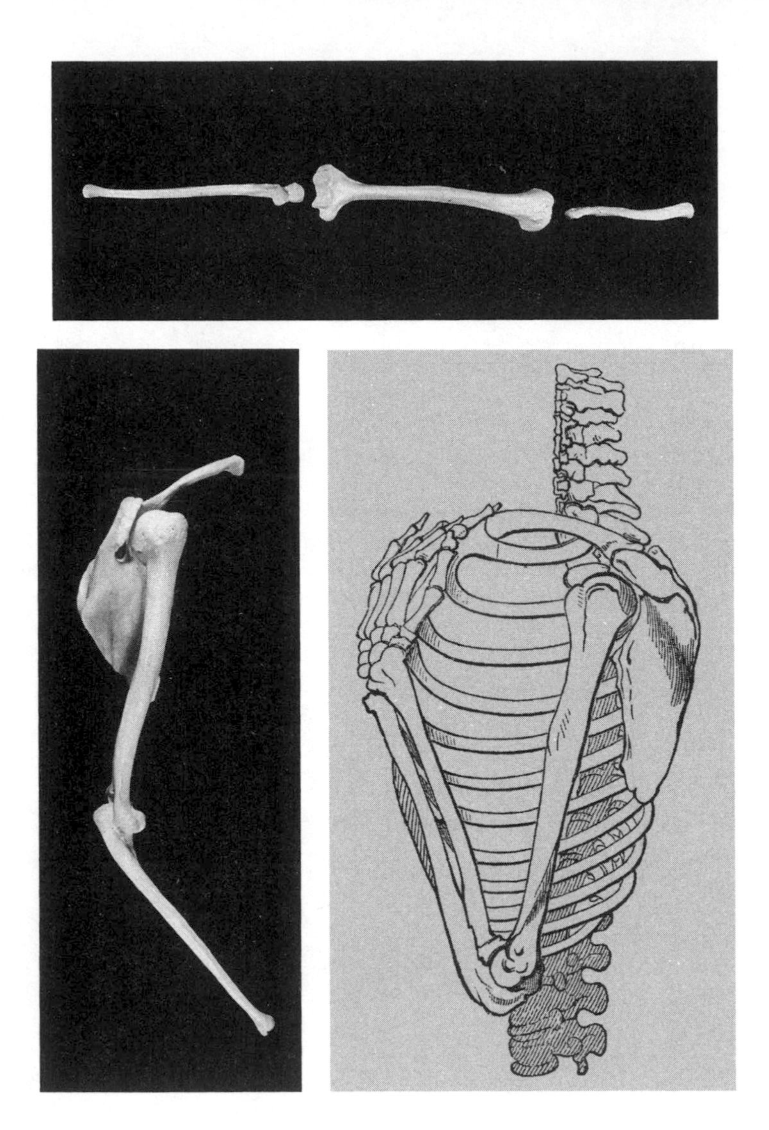

FIG. 3.5 The unique 3-stick linkage of the human upper limb. *Top:* the linear elements stripped down and laid out end to end (*from left to right:* ulna, humerus, and clavicle). *Bottom left:* the same three elements arranged with the scapula placed between the clavicle and the humerus, and the joints connected. *Bottom right:* Charles Bell's charming rendition, adding the chest, the radius, and the hand. Note that Bell has faked the anatomy of the clavicle here, creating an Escher-like illusion that the left clavicle arises from the right side. (Photographs by Diane Hawkey, Arizona State University.)

ulna), and elbow to wrist. Some mechanism must exist to transform three straight, rigid segments into a folding crane that can deploy the hand for whatever task may arise. There are many ancient tasks that rely on this cranelike function of the arm: picking fruit and flowers, climbing trees, burping a baby. And there are a host of tasks newer and more specific to humans: holding a racquet, a bat, or a club to hit a ball; using a hammer to pound nails; maneuvering a small needle in order to sew quilts or reattach fingers; or—no tools required—throwing a baseball overhand. In all of these tasks, old and new, there is a need to maintain body balance as the arm lifts and moves weighted objects. The scapula and its muscular system are silent and indispensable participants in the organization and execution of *all* of these movements.

Each of these tasks imposes an entirely different but equally strict demand for control of each segment of the arm while moving the hand toward the target; in practical terms, because the clavicle is comparatively immobile, this means controlling the movements of the middle (humerus) and distal (radius and ulna) segments while keeping the proximal segment steady. Imagine yourself balancing a broomstick in the palm of your hand—just an ordinary broomstick, upside down. This is a fairly easy trick for most people to manage, but suppose I asked you to balance two sticks (each about the same length as an ordinary broomstick), the second standing on top of the first. You may even have seen someone do this, and if you are a balancing wizard, possibly you can do it yourself. What if the problem were to balance a *third* stick on top of the second one—this would be a *very, very* difficult trick—even if you were to hold the bottom stick (representing the clavicle) tightly clenched in your fist, and I seriously doubt that it has ever been done. What's so tough, of course, is controlling the movement of the lower end of the top stick.*

Mastering the three-broomstick balancing act would be a snap if each contact point between adjacent segments were equipped with a "smart

* An analogous problem faces every truck driver backing a trailer into a narrow driveway. Experienced drivers have very little trouble backing into a driveway when the rig has *two* trailers, but how many drivers do you think can back a tractor and *three* trailers into a narrow driveway?

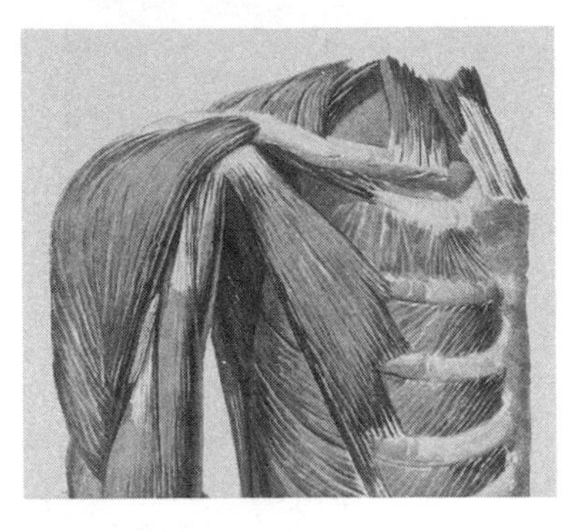

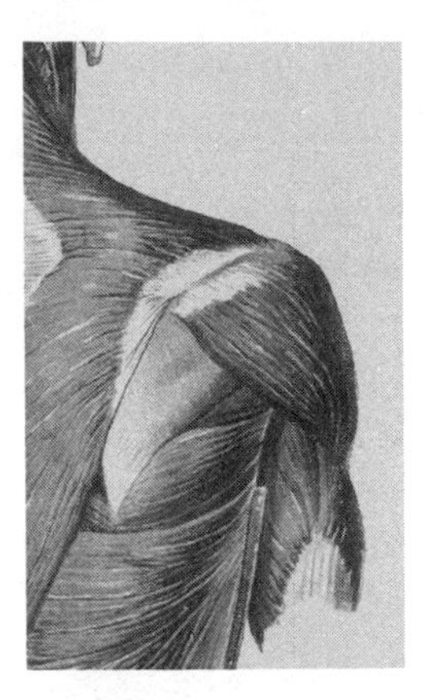

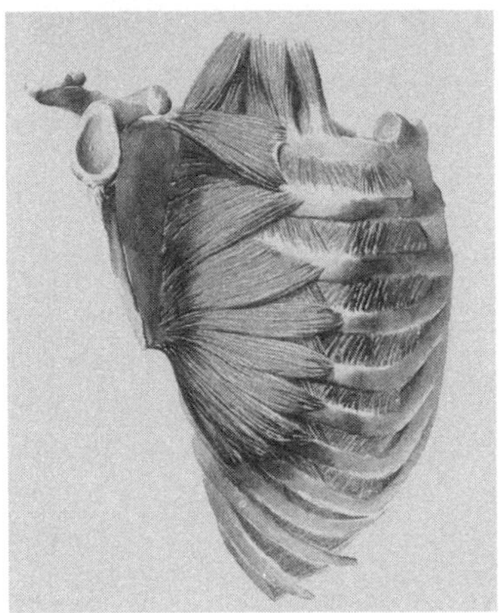

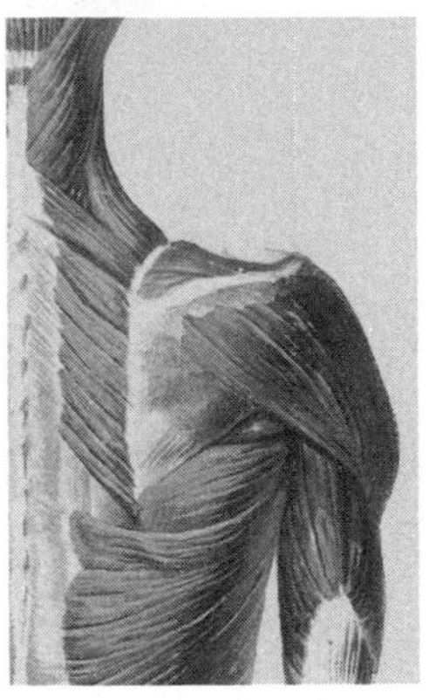

FIG. 3.6 "The upper limb clings to the rib cage as tightly as an abalone clings to a rock." Four views of the shoulder, showing the strategic mechanical arrangements of the muscles on and around the scapula that enable it to act as the "smart coupler" in the shoulder joint. *The two views on the left* show the shoulder girdle from the front; *the two on the right* show it viewed from the back. Note, *in the upper left image*, the muscle running downward from the shoulder across the chest; this is the pectoralis minor, the principal muscle anchoring the coracoid process of the scapula to the front of the chest wall. *Below, left, in a half-right view,* the collarbone and humerus have been removed and the scapula lifted off the chest wall to show the *serratus anterior,* which holds the scapula flat against the chest wall.

The two illustrations on the right show the muscles that both hold the scapula to the spine and lift, depress, and rotate the scapula to aid in positioning and stabilizing the head of the humerus. *Below, right,* the illustration shows the muscles of the scapula after the trapezius has been removed. (From Werner Spalteholz, *Hand Atlas of Human Anatomy,* 1923.)

coupler"—a computerized joint-controller with advance information about the intended move, with current information about how each segment was behaving *during* the move, and with the ability to control adjacent segments to correct for drift during the move. The shoulder apparatus, consisting of the scapula and its muscular attachments, is just such a smart coupler.* The shoulder muscles orient the shoulder joint prior to active movement, and by next controlling the movement of the humerus (in cooperation with the elbow and forearm) deliver the hand, prepared to act, to its target. The muscles that move the scapula, and those that run from the scapula to the humerus and the elbow, orient the shoulder and upper arm in advance of active movement, maintain whatever orientation is necessary to support the acting hand, and contribute to subsequent active movements of the hand by rotating or deflecting the humerus. Shoulder movement virtually always anticipates and then supports movements of the hand, which cannot themselves begin until the hand has been moved ("transported," in the language of motor-system physiologists) to the intended target in preparation for actions taken after contact has been made.[6] Remarkably, there is no conscious mechanism to control the behavior of any of these muscles. It is all done completely automatically.

The shoulder contributes to movements that not only transport but also *orient* the hand. And in shifting our attention to its role in supporting the function of the hand per se, we must understand that there is no such division or segregation of these functions in the body itself; the musculoskeletal system functions in a fully integrated way, so that overall movement is unitary and fluent. In reality, shoulder, arm, and hand functions are exquisitely tuned and responsive to one another in both neuromuscular and biomechanical terms. The kinetic and informational processes take place simultaneously from the body outward to the hand, and from the hand inward to the body.[7]

A particularly important feature of shoulder mechanics impacting on hand movement lies just within the glenohumeral joint. The humerus is capable of two different kinds of movement: from the straight-arm position

---

* So is the elbow and so, too, is the wrist.

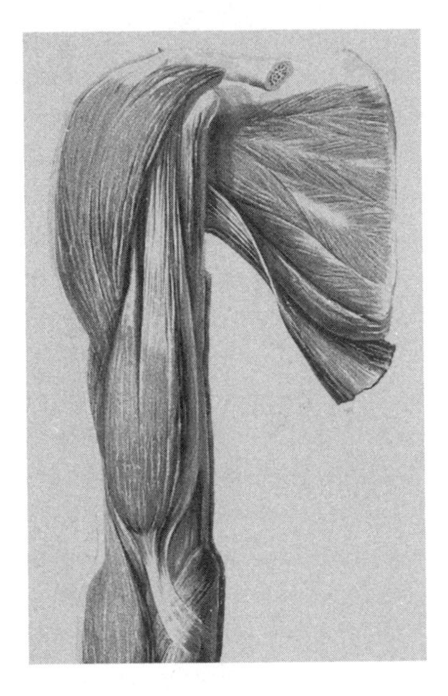

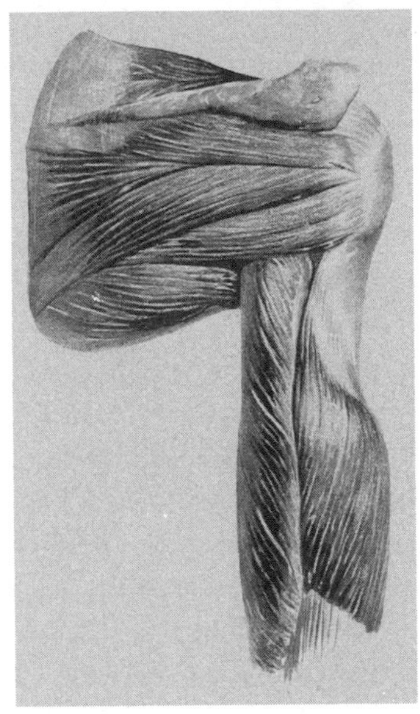

Fig. 3.7 Under the outer cape of muscles holding and moving the scapula, another layer of muscles anchored on the scapula itself attach to the upper arm to lift it forward, backward, to the side, and to rotate it around its long axis. *Left,* view from the front; *right,* view from behind. (From Werner Spalteholz, *Hand Atlas of Human Anatomy,* 1923.)

it can swing in a forward, backward, or sideways arc;* it can also *rotate* (or spin) on its long axis without changing the direction in which the arm is pointed.†

If the arms are kept straight (that is, if the elbows are extended) and held directly in front of the body, they can be placed in a position with the palms flat and facing upward. This is the position of supination, in which the right

* Another way of saying this is that it can move within a plane oriented from front to back (the sagittal plane), in a plane oriented from side to side (the coronal plane), and in planes intermediate between those two extremes. Since one end of the arm is always fixed, the rest of the arm (and the hand) will either lift or drop during arm movement in the sagittal or coronal planes.

† The amount of rotation varies from about one-quarter to about one-third of a circle, depending on where the humerus is pointed. In engineering parlance, the humerus is said to have "three degrees" of freedom at the glenohumeral joint: movement in the sagittal plane, movement in the coronal plane, and rotation on its long axis.

thumb points to the right side and the left thumb, mirroring it, points to the left. Beginning from this supinated posture, the arms can be rotated (the left clockwise, the right counterclockwise) into a palm-down, or pronated, position. In doing this full pronation movement, each hand goes through a 180-degree rotation; and an additional 90-degree rotation, to the thumbs-down ("Farewell, gladiator") position is possible for most people. That comes to 270 degrees of pronation-supination range. One-third of this total rotation comes from the shoulder, two-thirds from the elbow. As you perform this movement you cannot feel the separated contributions of shoulder and elbow; it is experienced as a smooth, unified rotation, just as the rotation of the neck from side to side is felt.

Now, bend your elbow a full ninety degrees so that your hand is slightly above your head. What you will observe is that your forearm, like the minute hand of a clock, easily rotates down and crosses the midline: if you are using your left arm, it goes from the twelve- to the three-o'clock position; if you are using the right arm, it goes (counterclockwise) from the twelve- to the nine-o'clock position. You have achieved all of the possible 90 degrees I said the shoulder could produce in a pronation-supination move of 270 degrees. Return your arms to the starting position, again with elbow bent in the right-angle position. Try rotation in the *opposite* direction. What happened? The rotation stops after about 15 degrees, because the humerus has reached the end point of its external rotation range. In fact, for all practical purposes it was there when you started! The minimal rotation you were able to produce was supplied by the sliding of the scapula across the upper back, pulled by muscles that elevate it and pull it medially toward the spine. You can observe this happen by watching someone else, or looking in a mirror as you do it.

Supination is the arm position for releasing a bowling ball, pronation for dribbling a basketball. If you play Ping-Pong standing close to your end of the table, you will tend to have your elbow bent for both backhand and forehand shots, so almost all of the force you apply to the ball comes from either internal or external rotation of the humerus. The further back you stand, the more likely you are adding both wrist and elbow flexion/extension to your swing, with pronation and supination supplying the underhand or overhand slice.

We began this chapter on the shoulder by noting that an immediate consequence of bipedality was the need to improve upright balance, and we have now seen in some detail how much the shoulder added to the campaign to make bipedality work. A second consequence of bipedality was that the shoulder, already the heart of an overarm propulsion system for moving the body through the trees, was admirably suited to transfer its capabilities to the as yet uninvented world of ballistics—to assist in the propulsion of hand-held missiles toward distant targets. The mostly loveable, social, clever, and opportunistic clowns we imagine chimps to be were destined to be pushed aside by their new cousins—calculating, aggressive, and dangerous hunters. The ownership of the old arm, mounted on a new pelvis, aided by a highly advanced visual system and bipedality, and stimulated by the reality of having no real alternative to living on the savannah, meant that ballistics would become the inaugural centerpiece of a new hominid survival strategy. Having the ability to hold a good-sized rock in the hand, take aim, wind up, and let fly with speed and accuracy meant that there was no need to run down small animals or live on the scraps of food left by other predators. The shoulder, of course, was only part of this complex behavioral evolution, but it was one of the components delivered with almost no need of further modification.* Lucy did not have the hand for extended-arm-length clubbing in close encounters, but the australopithecines *could* have made and thrown spears.

The entire concept of a multisystem adaptation unfolding over time (illustrated by the shoulder-as-rocket-launcher) raises an intriguing question: Is it possible that the (probably inevitable) dissociation of forelimb from hindlimb control in the bipedal brain has consequences and implications we have not appreciated? The long process of adapting the forelimb of a tree-dwelling, fruit- and insect-eating primate to the dangerous situation facing a plain-dwelling hunter ape had profound musculoskeletal conse-

---

* It may be that the visual system lagged most in this transformation; that is a question we will take up in chapter 5 ("Hand, Eye, and Sky"), the chapter on juggling. It is worth noting that ballistic throwing, as best exemplified in modern baseball pitching, is *not* a one-arm performance. The nonthrowing arm is critical to the entire process of balance and rapid weight transfer demanded by the fast overhand pitch.

quences. But more than muscles, bones, and joints were involved in this transformation; not only the physique but the *limb-brain complex* had to adapt to conditions on the ground.

As we have learned, adding weight to the upper body, and then transporting that weight, demands a highly tuned, fast and versatile system for locating and controlling the body's center of gravity. And that system does not behave the same in everyone. Early in my career I spent several years seeing patients in an emergency room, and in that situation came to believe that roofers injure themselves less often than carpenters do, despite the reverse prediction one would make from watching them on the job. Why would that be? I think the reason is that roofing is not less dangerous than carpentry, but so much *more* dangerous that no one attempts it who does not feel completely at home hanging by his thumbs twenty feet off the ground. Presumably, such individuals carry, or have unlocked, whatever genes we still carry that gave arboreal primates the wherewithal to live actively and securely in the trees.

Could this be an example of variable evolutionary lag—an upper-versus-lower limb dissociation in the locomotor systems of the brain of *H. sapiens*? If it is, the effect certainly is not confined to roofers. Dancers, acrobats, and skateboarders seem to be fully evolved biomechanically *and* neurologically as bipeds, while others have retained an aerial bias in the hominid repertoire of automatic balancing reflexes. The latter model of the biped brain lacks a full repertoire of automatic responses *in the legs* to the sudden shifting of a load above the center of gravity.* Have you ever noticed that people who trip look like figures in Chagall paintings? One of my patients described her own fall in just those words once, and I realized she was right: down she went, arms out, clutching for the branch that was left behind *eons* ago. The result of this outmoded reaction is almost invariably a broken arm.

What I am suggesting is that the human brain, especially with respect to the deeply ingrained patterns of motor control upon which survival in the

---

* I feel quite secure on two skis or two skates but have always been completely at a loss on a single water ski, a skateboard, or a surfboard. Bicycles are no problem even for those of us who are obligate wide-stance ground-dwellers because the flywheel effect adds stability even at low speeds.

trees depended for millions of years, continues to be genetically transmitted with a copy of a locomotor strategy more suited to movement on a limb than on a wide, flat surface. Our distant ancestors made no effort to carry great weights with the body; during a fall, all four limbs could be deployed to snag a branch; and there was (usually) plenty of time for—and a reasonable probability of—encountering a branch during the fall. In fact, falling was a *normal mode of locomotion!* That was a *very* sophisticated and highly effective strategy, so quite possibly it is still floating around in the gene pool, surfacing just often enough to keep the roofing business and circuses alive.

The same shoulder that grew up in the trees and helped us not only to defend ourselves but to become aggressive at ground level is now attached to quite a different hand. Lucy's hand represented a colossally fertile initiative in evolution, but we have no way of knowing what its utility may have been to the australopithecines themselves. Certainly its functional utility was unlikely to have eclipsed the shoulder and arm they inherited from the hominid ancestors. This speculation leads me to a radical suggestion: the accelerated development of the hand (and of the brain supporting its new repertoire of movements) in *Homo* seems not so much a *de novo* invention as the completion of what had already been worked out in the rest of the arm (and the shoulder) prior to our arrival on the ground.

The hominid shoulder, then as now, had to adapt its muscular performance to a huge range of motor tasks. *Homo* added watch repair and discus-throwing to the list, but the arm still uses muscles and tendons and ligaments integrated with rigid segments (bones) to orient the arm and hand, and to make adjusting and tension-setting movements to keep the body upright as it does its job. The shoulder-brain still dynamically controls muscle activity based primarily on vision, joint position, and muscle kinesthetic sense. None of that has changed.

If the australopithecine hand ventured out onto the savannah less than fully prepared for the challenges it would meet there, the shoulder and arm and their neurologic support systems were very likely already well *ahead* of the game. The pivoting and sliding movements of the scapula, together with the full supination of the forearm, permitted accelerated overhead throwing of objects held in the hand. There were other changes that had to be made to capitalize fully on this ballistic potential, most prominently an advanced visual-motor control system. And catch up is exactly what the hand (and brain) did. Very quickly—on an evolutionary time scale, that is—the hand and brain not only met but began to redefine the demands and possibilities of a life in which forelimbs had been freed of the obligation to support body weight.

4
Puppet Lessons from Alexandria and Düsseldorf

Our daughters and sons have burst
from the marionette show
leaving a tangle of strings
and gone into the unlit audience.

—Maxine Kumin[1]

Puppets are magical storytellers, and in fact were persuasive envoys of religious thought at least six thousand years ago in India, and for nearly as long in Egypt.[2] Their ability to entertain and to teach begins with face and costume, to which movement—the sign of life—is added to captivate the watcher and awaken the imagination. But puppets are not simply passive or surrogate human narrators. The ventriloquist's puppet is also a lively companion who opens our eyes to the nature of the complex attachments—which are not merely physical—between puppet and puppeteer.

Another puppet, the marionette, simulates the mechanics of muscles and tendons in our arms and legs. Indeed, my original interest in puppets began with the marionette, which I guessed might be susceptible to tangled strings in the way my patients with writer's cramp were susceptible to tangled muscles. As I was to learn, the marionette had more than one lesson to offer about the animation of human limbs. By an odd coincidence, its ancient Greek name (*neurospastos*) memorializes the discovery by Alexandrian anatomists over two thousand years ago that muscles are controlled by the nervous system.

The name *neurospastos* seemed an uncanny coincidence in the narrative of my own research, because writer's cramp is a condition marked by muscles which contract tightly into uncontrolled spasms. But this ancient Greek word for puppet implies no pathology at all. What I had not known before encountering this word was that in ancient Greek a "neuron" was a cord or a fiber. The original marionette was simply a puppet "drawn by strings."

This clarification explained how Greek puppets got their name and in the process corrected a minor deficiency in my own professional enculturation. Although the term *neuron* is part of the working vocabulary of every neurologist, I was unaware that in its original meaning this "cord" or "fiber" was *any* white, sinewy tissue in the body. Until the time of Hippocrates, around 400 B.C., no distinction had been made between the nerves, tendons, and ligaments attached to bones and muscles, because there was no way to tell them apart, nor any reason to suspect they had different functions. But then Herophilus, a pioneering anatomist who was active in Alexandria in the early part of the third century B.C., discovered that some of the neurons connected muscles to the spinal cord. By dissecting living prisoners, he was able to show that when these particular neurons are cut, a loss either of movement or of sensation results.* Contemporary neuroscientists might doubt that so much can have been objectively known about motor and sensory nerves so long ago. But even though Herophilus did not know *how* this functional division operated, there is no dispute that he knew that some "neurons" were connected to the brain and spinal cord, and that he was able to demonstrate in these either a motor or sensory function.[3]

This discovery was more than slightly precocious: *two thousand years* passed before scientists were able to add much beyond conjecture to what the Alexandrian anatomists had established, or to their theory that *pneuma* (air or some other vital "spirit") was carried by way of the nerve to the mus-

---

* Herophilus and his successor, Erasistratus, were not only the first but also the *last* anatomists of ancient Greece to undertake human dissection (and vivisection) for the study of structure and function in the human body. Religious taboos against violating human flesh were powerful in ancient Greece, and subsequent arguments in service of those taboos were universally accepted. For a lucid and enlightening discussion of the "aberration" of the Alexandrian school of anatomy, see Heinrich von Staden, "The Discovery of the Body: Human Dissection and Its Cultural Contexts in Ancient Greece," *Yale Journal of Biology and Medicine* 65 (1992): 223–41.

cle. It was not until the latter half of the second century A.D., when Galen of Pergamum began his comprehensive study of mammalian anatomy, that there was any foundation for new theories to explain the mechanism by which muscles function.*

Galen is credited with an original insight concerning muscular action that has dominated our understanding of voluntary movement since he published the idea in A.D. 157.[4] That principle, based on his claim that a muscle can exert its influence only by pulling (or relaxing) in a straight line, is that all complex movement is achieved by the combined action of *pairs* of muscles aligned in the body to pull against each other, like contestants in a tug-of-war. The "contestant" image proved so apt that it has never been replaced, and is perpetuated in the names, *agonist* and *antagonist,* given to the two muscles of any such pair.[5]

The growth of physiologic ideas, supported by scientific investigation of the mechanisms permitting nerves to activate muscles, did not really begin until the seventeenth century. In 1626 René Descartes proposed that the muscles responsible for eye movement were moved not by a mysterious gas or animal spirit (pneuma) but by a fluid. He proposed that this animating fluid activated muscle by filling it, making it round and thereby shorter.[6] In his model, the main source of the fluid was not the brain but the antagonist muscle. Therefore, the role of the brain and nerves was not to pump their own spirits into the muscles (like the heart pumping blood into the arteries) but to operate a system of hydraulic valves, shunting fluids back and forth between opposing muscles. Descartes's idea of a reciprocal controller was the perfect complement to the mechanical arrangement of agonists and antagonists, and although he was wrong about the hydraulics, the essential accuracy of the underlying principle led Bell to the real truth two centuries later.[7]

It took nearly two more centuries to discover, and prove, where muscle power originates and how it is controlled. In 1640 the English physiologist Francis Gibson showed that a muscle could be shortened by irritation of its

---

* Galen of Pergamum, the Greek (and sometimes Roman) physician, based his entire work on the dissection of animals. It was not until the middle of the sixteenth century that human dissection resumed (this time using cadavers); *De Humani Corporis Fabrica,* the work of the Belgian physician Andreas Vesalius, was published in 1543.

nerve even when the attachment of the nerve to the spinal cord had been cut. This meant that the nerve could be acting only through the release of energy stored within the muscle itself. Still, no one knew how the action of a nerve might release that energy.

Finally, nearly 150 years later, Galvani showed that electrical forces were associated with the contraction of the muscle of a frog's leg, and in so doing he located the last missing piece of the puzzle: regardless of the specific operational details, the control mechanism for voluntary movement had to involve the electrical excitation of a muscle by its nerve. From this point forward, research on muscle control turned to the question of where and how nerves used electrical activity to exert their influence. In very short order, Charles Bell, François Magendie, Edgar Adrian, and Charles Sherrington were able to work out the basic details, and to describe the interactions of sensory and motor nerves at the level of the spinal cord. In plain language, by about 1900 it was no longer a mystery why your leg jumps when the doctor taps your knee with his rubber hammer.[8]

We should now return to the mechanics of joint movement. Suppose you wanted to construct a working human skeleton to duplicate the movements of a real-life upper arm. Suppose further that you knew nothing about the actual arrangement of muscles in the arm; all you wanted was to make your skeleton move realistically. You would probably attack the problem one joint at a time.

The simplest of joints is a hinge that moves only in one plane (or around a single fixed axis, just as a door does). One possible way to control a *skeletal* hinge joint would be to fix two moderately elastic strings (the original neurons) on opposite sides of the bone nearest to the spine, run them outward (distally) across the joint and attach each to opposite sides of the second bone.[9] This arrangement would work well for the elbow, with one string representing the biceps and the other representing the triceps. To bend (flex or extend) this joint, you would just pull on one of the two strings. What happens to the second string in this maneuver? It will have to stretch, and while elasticity remains it will increase its length by precisely the same amount as the pulling string shortens.

The celebrated French physician-neurologist Guillaume Duchenne actually used an articulated skeletal model to discover how the attachments

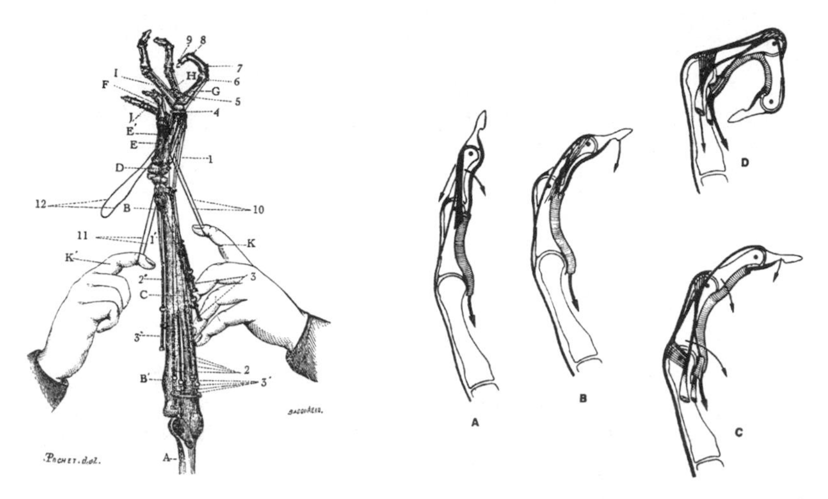

FIG. 4.1 *Left:* The French physiologist Duchenne devised a string-and-pulley model of the human arm "which can be made to reproduce natural motions of the phalanges . . ." (G. B. Duchenne, Physiologie des Mouvements, 1867). *Right:* The French hand surgeon, Raoul Tubiana, describes this mechanism and the sequence of movements it produces in his contemporary encyclopedic surgical text, *The Hand.* (A): Flexion is initiated by simultaneous contraction of the finger extensor and the deep flexor. The resulting tension on a distal oblique ligament initiates flexion of the first (proximal) finger joint. The subsequent release of tension on that ligament (B) allows increased flexion of the distal joint. Next (C), tightening of the tendons of small intrinsic muscles of the hand initiates flexion at the knuckle joint. Finally (D), distal movement of the ligamentous hood of the interosseous muscle allows that muscle to act as a flexor of the proximal phalanx. (From Raoul Tubiana, *The Hand,* vol. I, 1981, with permission.)

of muscles in the hand influence the movements of the fingers. Figure 4.1, from his monograph on the physiology of movement published in 1867, illustrates a test performed with the model.[10] The model makes it abundantly clear why attaching strings to control joints in the arms would soon become very complicated. Of course there is another, much simpler, way to use strings to move the mechanical arm we have assembled: you could attach strings to the wrist, arm, shoulder, head, and knees and pull on them from *above* the skeleton, just as is done with a marionette.

To stand and move, marionettes—like us—must overcome gravity. *We* do it by loading our weight onto an internal, semirigid stack of bones and balancing on a firm surface under our feet. The marionette overcomes gravity by suspension from supports located above the head. In the early Egyptian puppets, a fixed horizontal rod supported the weight of the puppet; the modern, lighter, puppet is suspended from some sort of hand-held device operated by the puppeteer. Puppet-makers, particularly those working in Europe during the past century, have designed many different kinds of these devices, called controllers, but all must translate the puppeteer's movements into realistic head, body, and limb movements in the puppet. The most common of these devices is called a paddle.*

Since movement is critical to the illusion of life, the designers of controllers have increasingly sought ways to allow the puppeteer to duplicate familiar postures, gestures of head, body, and limbs, and locomotion. In the mid-nineteenth century, the details of paddle design and stringing (which the English called "slanging") were closely guarded trade secrets, with the result that very little is known about their evolution.[11] But the principles are straightforward enough. All marionettes have limb segments that swing either in a front-to-back or in a side-to-side plane, or rotate around a long axis. These movements are determined by the action of strings attached to the paddle held by the puppeteer.

A paddle has symmetrical left- and right-side extensions suspended from the central axis—analogues of our own upper and lower extremities—with strings dropped in pairs from various positions on the body and wings (Fig. 4.2). Each matched pair of strings acts on the puppet's joints (say, a knee or elbow joint) in one of three ways: (1) both strings are pulling equally; (2) the left or right string lifts while the opposite string drops; (3) front strings rise or drop while rear strings do the opposite. When only one of the pair is pulling upward, the other must drop by an equal amount, unless the entire paddle itself is being lifted or dropped.

However crude the paddle may seem, it is mechanically ingenious. When the paddle's central axis rotates—think of a real airplane rolling—one wing drops and the other rises by an equal amount. Therefore, on the puppet controlled with this paddle, the lifting of one knee is automatically accompanied by the fall of the other, by equal amounts. By preserving the

* Paddles are usually called "airplanes" when they are oriented horizontally, as is most common in the United States. European paddles are more often vertically oriented.

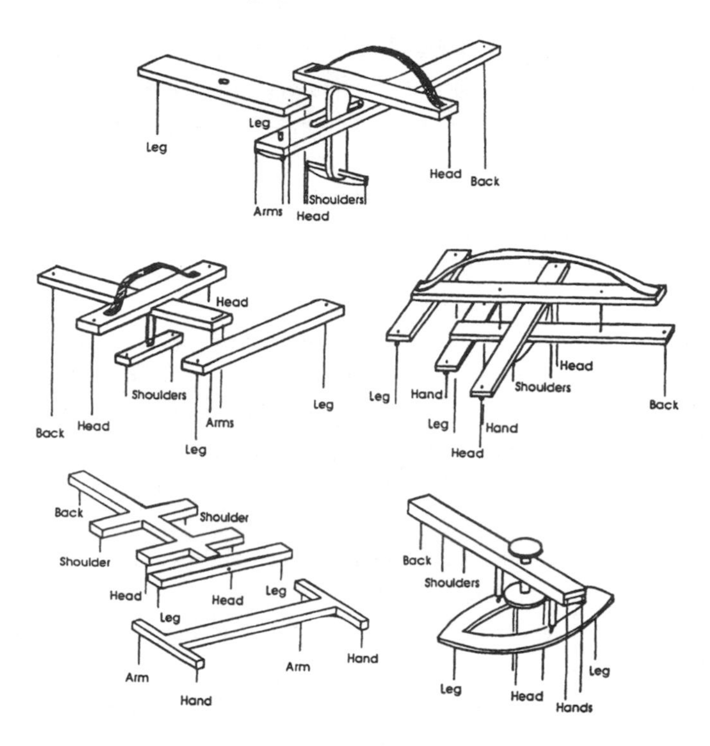

FIG. 4.2 Puppeteers have devised a number of hand-held controllers to permit control of the joints of a marionette. Controllers permit physical separation of the strings (no tangles!), automatic reversal of contralateral limb movements by rolling of the long axis, lifting and dropping movements of head, trunk, and limbs by tilting of the long axis, and discrete movements of any part of the body via separate control of individual strings. This illustration depicts a variety of horizontal, or "airplane," controllers, which are favored in North America. Increasingly, however, puppeteers are favoring "angled" controllers because of their ability to reduce arm, wrist, and hand strain. (From Luman Coad, *Marionette Sourcebook,* 1993, with permission.)

mechanics of reciprocal control, the paddle produces limb movements that are simultaneous, equal, and opposite.

The automatic reciprocal control of the puppet's limb movements through a paddle device replicates two fundamental mechanisms governing the movement of our own bodies. The first, which we have already discussed, is the obligate antagonist pairing of muscles around joints, which Galen recognized and described nearly two thousand years ago. The second principle, discovered in 1829 by Charles Bell, is that the "antagonistic" action of muscles is actually complementary and cooperative. The reciprocal controller is not simply telling either one muscle or the other to turn on; it sends both "on" *and* "off" messages simultaneously, and in a graded rather than an either/or fashion. In this way the interplay of contraction and relaxation of agonist and antagonist muscles results in refined control of limb and body movement.

Sherrington, seventy years after Bell, demonstrated that the spinal cord is the site of this interplay, and that the resulting continuous regulation of acceleration and force of muscle contraction is an essential mechanism underlying rapid and fluent movement. For example, the quadriceps (the big muscle in the front of the thigh) provides critical braking action at the knee when you are going downstairs. You would never arrive at the bottom still standing upright if the quadriceps did not work very hard to limit flexion of the knee with each step.[12] For extremely rapid movements—like the throwing of a baseball—antagonists hold the movement in check while power is being built up in the agonists. (In this way, the movement of the limb can accelerate much more rapidly than it could if muscle activation always initiated joint movement immediately. Arrows fly from bows, and planes are launched by catapult from aircraft carriers, when the restraining force is removed.)

In the simple and idealized case of just two muscles producing flexion and extension of a single joint, each member of the pair of muscles alternates relaxing and contracting with the other: as one muscle begins its contracting phase, its partner must begin its relaxation phase. In your arm, for example, when the biceps begins to contract, the triceps immediately relaxes so that flexion of the elbow can take place with minimal interference.

The obligatory lengthening and shortening of the biceps and triceps as they tug at each other across the elbow is easy to visualize, but the reciprocal interactions and control of agonist-antagonist muscle activity in real life is mechanically far more complex, and physiologically far more fluid, than the description of an either/or contraction and relaxation of the biceps and triceps. As it turns out, however, it is simplicity in the biomechanical principles (as in the physiologic principles) that provides the sturdy but flexible

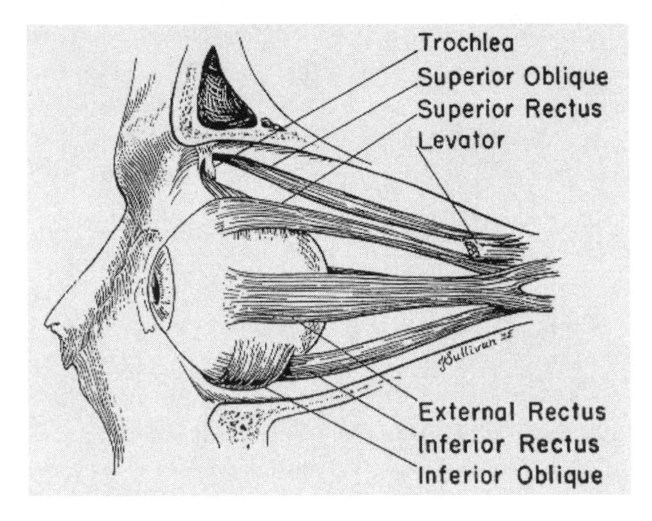

FIG. 4.3 The control of movement by the reciprocal action of agonists and antagonists is a common strategy in the body. The production of rotary movement of the eye is achieved by muscular mechanisms analogous to those at the shoulder, and suggests a complementarity of neurologic computations used for accurate aiming and movement of both structures. See endnote 13. (From David Cogan, *Neurology of the Ocular Muscles,* 2nd ed. Springfield, Ill.: Charles C. Thomas, 1956, with permission.)

platform for the ultimate functional complexity of the system as a whole. And we can see how wonderfully simple the biomechanics are in principle by returning to Descartes and the eye (Fig. 4.3).

Descartes's choice of the eye muscles to illustrate the principle of reciprocal control of muscles acting in mechanically opposed pairs was unwittingly felicitous, because the geometry of eye movements within the orbits corresponds remarkably to that of the head of the humerus within the shoulder joint: the brain points the arm and finger as accurately as it points the eye. In the orbit and at the shoulder, the eye and the humerus are each free to rotate (or swing) in front-to-back and side-to-side planes, and also around their own long axes. And in both cases there is a precise arrangement of muscles aligned and attached to power each of those movements. Despite the fact that muscles can apply their force only in a straight line, the linear force is mechanically redirected (or translated into an angular force, or moment) through levers or pulleys.[13]

So it is also for the rest of the musculoskeletal system, where the basic requirement exists that agonist-antagonist pairs of muscles act at single joints, no matter how complex the movement. As a practical matter, your neuromuscular system is free to compose movements from the whole range of possible joint and muscle interactions in any way it chooses.[14] On a conscious level, you will be focused on locating a target, then moving toward it or manipulating it as appropriately as you can (depending on your intentions). But you do not have to (nor, indeed, can you) decide which specific muscles will do the job for you.

Because of the body's virtually limitless freedom to employ convenient combinations of muscle activity to achieve a desired movement, individual muscles of the body are subject to recruitment by the brain for assistance in an endless variety and range of movements. What begins as a specific behavioral demand—"Catch that ball!"—can usually be satisfied by any of a large set of alternative biomechanical solutions. A specific solution to the problem comes about only because the central nervous system has the capacity to call upon skeletal muscles the way a commanding general calls on his troops, combining the available forces to meet certain tactical objectives as they arise, and continually recombining them into ad hoc working groups based on his perceptions of what each can do and what each mission will require.* A conscious goal ("Grab your coat and put it on") is realized only after it is turned into a biomechanical and neurophysiological battle plan. Ideas are translated into action mechanically: muscles change their length and alter the joint angles between adjacent bones, and off you go.

Cranes and backhoes work on exactly the same principle, substituting a human operator for the brain, motors for muscles, and cables (or hydraulics

---

* This flexibility may seem to be an ideal arrangement, but (just as with a real army) it comes at a very high price. This is an issue of considerable theoretical importance in motor control and we will look at it a little more closely later, but for the moment the point to remember is simply this: the functional flexibility and interchangeability of individual muscles poses a nightmarish challenge to the central nervous system. If the brain prepares for movement like a commanding general who is a brilliant and efficient tactician, this is a general who can never *execute* that movement without first checking to see where all his troops are and what they are already up to. In other words, no movement can begin without updated information indicating which muscles are equipped and available for the task, and what these muscles are doing the instant the command is being issued.

and pistons) for tendons. With experience, the operator eventually "incorporates" all of this machinery and begins to treat the machine's bucket as if it were a spoon in his hand and he were doing nothing more complicated than eating his breakfast cereal. With a puppeteer, of course, there are no motors to assist with the lifting. From a control perspective, crane operators and marionettists confront many of the same challenges, despite the striking differences in appearance (and scale) of the theaters in which they work. The crane operator, as we have seen, "becomes one" with his machine; it is the same for the puppeteer.

The director of the Düsseldorf Marionette Theater, Anton Bachleitner, was born in 1956 in Bad Tölz, a small town near Munich. As a young boy, Bachleitner would watch the shows put on at the amateur puppet theater in Bad Tölz, then go home to imitate what he had seen the puppets do. He taught himself to build puppets at home, then worked as a helper for the man who made puppets in the theater workshop. It was there that he learned to draw and carve puppets, and by the time he was eleven he was spending all his free time in the shop. He performed in his first show when he was fourteen. The next youngest performer was twenty, and the oldest among the ten performing at that time was "maybe sixty or seventy." His parents initially supported his interest, but as he grew older and his interest increased, their enthusiasm waned. His father was a carpenter and wanted his son to learn a "real" trade. In fact, Bachleitner did work as an apprentice carpenter for several years in his early twenties (not with his father), but in his mind he had already settled on wood carving: "Carpentry and wood sculpture are not the same, you know. Many people think that if you can work with wood you can make carvings, but that is not so."

This skill has proven of considerable value to Bachleitner's work because the design and carving of the puppet's head and hands is of central significance for the dramatic impact of the figure. Both the expressiveness of the face and the nuances of head and hand articulation convey important emotional meaning. The balance of the head begins with the design of the face—including the size and shape of the chin, ears, and nose—and is aided by the placement of weights in specific locations to encourage movement in a particular direction.

The placement of strings is crucial to the illusion of realistic movement,

FIG. 4.4 Puppets and cranes both present problems in control of movement at a distance. Regardless of the scale of the movement, refined control demands that the operator be able to see and feel what is happening at the end of the strings. *Above, left:* Anton Bachleitner and his colleagues operate their marionettes from a bridge above the stage. (Photo courtesy the Dusseldorf Marionette Theater.) *Above, right:* a steelworker during the construction of the Empire State Building, New York City, in 1930. (Photograph courtesy the Avery Architectural and Fine Arts Library, Columbia University in the City of New York.)

as is the design of the paddle. Bachleitner has developed his own paddles (called in German *Spielkreuzer,* or "play crosses") and stringing arrangements to permit better control of the puppet's forearm and hand movements.[15] Bachleitner explains:

> A human puppet has between nine and twelve strings. There are three in the head—behind the nose and behind the ears. There are also strings to the shoulders and to the knees, one to the middle of the low back, and

> each hand can have two. There are none on the feet. The two strings to the hand go through a small ring before attaching near the wrist. This allows you to turn the hand with the individual strings.[16] Sometimes there are strings for special effects; sometimes you want to move the mouth, for example. In order to have any movements of the hand itself—grasping of the fingers—you have to have a puppeteer on the stage who is dressed in black.*

The puppeteer works from a movable bridge above the stage, leaning forward, with his abdomen firmly planted against a rail to keep himself steady. That, by the way, is why the puppeteer can't do the puppet's voice. I asked Bachleitner to describe how puppeteers learn how to operate marionettes:

> It takes at least three years of work to say you are a puppeteer. The most difficult job technically is to be able to feel the foot contact the floor as it actually happens. The only way to make the puppet look as though it is actually walking is by *feeling* what is happening through your hands. The other thing which I think you cannot really train for, but only can discover with very long practice and experience, is a change in your own vision.
>
> The best puppeteer after some years will actually see what is happening on the stage as if he himself was located in the head of the puppet, looking out through the puppet's eyes—he must learn to be *in* the puppet. This is true not only in the traditional actor's sense, but in an unusual perceptual sense. The puppeteer stands two meters above the puppet and must be able to see what is on the stage and to move from the puppet's perspective. Moving is a special problem because of this distance, because the puppet does not move at the same time your hand does.
>
> Also, there can be several puppets on the stage at the same time, and to appear realistic they must react to each other as they would in real life. So again the puppeteer must himself be mentally *on the stage* and able to react as a stage actor would react. This is something I cannot explain, but it is

---

* On his recent Asian tour, Bachleitner saw his first Japanese Bunraku play, a traditional form of puppetry in which three men usually control one puppet. The master controls the head, shoulders and right hand, one assistant moves the legs, and a second assistant moves the left hand.

very important for a puppeteer to be able to do this. The problem is greater with certain plays, where the puppet may fly, as often happens in operas, or may drop through a hole, as in *Der Golem,* or do something else that is unusual. These are situations where the danger of tangles can be very great.

As revealing as Bachleitner's description of marionettes and his life with them turned out to be, our discussions had not prepared me for what I experienced as I watched him work. Bachleitner seems to float as he animates Golem (the central character of a classic Yiddish play adapted for the Düsseldorf Marionette Theatre). The weight of Golem's life is deeply set in the lines of the puppet's sad face; his exaggerated nose, looking almost like the bow of a ship, inches forward in the dark until it is caught in a single beam of light. The slightest movement of this head creates a new sculpture. The head tilts slightly—Golem yields wearily to the thought of some new trouble. The head turns just enough to suggest a glance over his shoulder. The shoulders rise slightly—a shrug. Slowly, one arm seems to lift, the palm turns upward, signaling a decision. "I must go." Hesitantly, one leg answers. Again and again Bachleitner and Golem confirm the unity of movement, thought, and emotion. How does the illusion of life and gravity become so powerful in this tiny form the moment it appears on stage?

FIG. 4.5 *Der Golem,* the central character in the Düsseldorf Marionette Theater's rendition of a classic play, exemplifies the unusual accomplishments of the theater's director, Anton Bachleitner. The face was designed and carved by Bachleitner to take advantage of lighting effects, and in his creator's hands on stage the puppet comes poignantly and convincingly to life. (Photo courtesy the Düsseldorf Marionette Theater.)

Suspended beneath the trained hands, eyes, and mind of a real master, the movements of a puppet so closely mimic those of a living person that it is a shock to see the puppet alone, off-stage, collapsed, nothing but an inanimate doll. As in real humans, life flows into the puppet's body along barely visible cords that have been called neurons since they were first found in humans. The puppet moves across a tiny stage, alive but ignorant of the true source of its animation, oblivious to its own powerful effects on the thoughts and emotions of those who are watching. A mesmerized neurologist concludes that Bachleitner and Golem are actually communicating with each other and that their bond is intimate. At this moment the technical questions simply evaporate: they *are* each other, so it is no use asking who is really pulling the strings.

Thanks to the genius of scientists like Herophilus and Bell and Sherrington, we now know that the neuron is not just a string but a lifeline to and from the spinal cord and brain. It neither pulls on muscles nor inflates them but, rather, sends the instructions and receives the reports required to balance the contractions and relaxations on which coordinated movement depends. Thus, thought becomes action, and action becomes thought. By an enchanting coincidence of semantics, it is a "neuron" that controls the arm, whether it is that of a puppet or of a human. And although the influence of the neuron on the marionette is mechanical and on the human electrochemical, the grace and eloquence of the resulting movement is quite the same.

5
Hand, Eye, and Sky

Each time he spins it,
it lands, precisely,
at the center of the world.

—Octavio Paz[1]

The push-button type of one-to-one correspondence does not exist and the cerebral motor area organizes responses by deftly adjusting and balancing between resultant external forces and the manifestations of inertia, constantly reacting to proprioceptive signals and simultaneously integrating impulses from separate central subsystems, so that ten successive repetitions of the same movement demand ten successive impulses all different from each other.

—Nicolai Bernstein[2]

Juggling, an amalgam of gymnastics, mime, comedy, and magic, has been a popular form of public entertainment since the beginnings of recorded history. Jugglers performed at the 40,000-seat Hippodrome in Constantinople during the reign of Justinian I, some fifteen hundred years ago, and contests of juggling skill were a popular Native American sport long before the arrival of Europeans.[3]

Since juggling does not *accomplish* anything—at the end of a typical routine, the objects being juggled are back where they started—we might ignore its popularity and dismiss the practitioners of this art as mere curiosities. But juggling invites us to consider the transformation of an originally combative skill into a form of play. It also raises a question about throwing's complementary skill: When did human ancestors start *catching*, and why?

All of us have spent time trying to coax (and coach) our bodies into doing something we did not already know how to do, or could not do well. Perhaps you recall from your own childhood the first time you ever *practiced* a skill you had decided you wanted to learn. Do you remember your first attempts at skating, riding a bicycle, jumping rope, swimming, or tying your shoelaces? Whatever it happened to be, the chances are good that you watched someone else do it, tried to imitate what you saw, failed, and then—with or without help, "practiced" until you were ready for someone to watch *you.* What is practicing, and why does it help?[4] Beneath any question about the pragmatics of learning—what is the nature of a successful rehearsal strategy?—lies a far more difficult question: What constitutes "learning" in the nervous system itself? Can neuroscientists distinguish the pre- from the post-juggling-instruction brain?[5]

I do not remember exactly when I first tried to juggle, or why, but I do recall understanding from the start that juggling is not only a physical challenge but a procedural riddle. The riddle at its core is this: How can you continuously repeat the exchange of three balls between the left and right hands without ever holding more than one ball in either hand? In its only irrefutably convincing form, the solution to the riddle is found in the domain of physical experience—talk runs a distant second to just doing it. In that respect, juggling can be meaningfully compared to sex.

Genuine interest in juggling is uncommon in children before they reach their early teens. Very small children may enjoy watching Dad clowning around as a juggler, but that is because he will drop what he throws and as a special bonus may even break something in the process. Likewise, to a child's eye, watching a real clown tossing large, colorful objects into the air is a visual joke—many common objects (cats, for example) belong right-side-up and on the ground and are a wondrous sight airborne and upside-down. When children are drawn to the suspension and circulation of multiple objects in midair, the attraction comes from a danger fantasy: eggs are on the loose and mayhem is imminent.

The idea that juggling is a mystery or a trick doesn't come until we are older, after nearly a decade of training in hand-eye coordination has given us our own repertoire of smooth and practiced moves, or until a certain amount of instruction in sports, music, and games has given us our own slant on the difficulties and pleasures of physical skill. No child really "gets" juggling before moving from the hesitation of the first year's see-reach-touch-grasp experiences to the confident playing of catch-it-on-the-fly ball sports.

Nothing could be intuitively more clear than the critical dependence of the skill of juggling on hand-eye coordination; what is not so obvious is the degree to which almost all physical skill flows from the maturation of motor skills under the guidance of both visual and kinesthetic monitoring. It is a delightful surprise to see how clearly this point was made by Charles Bell in his treatise on the hand:

> This faculty of searching for the object is slowly acquired in the child: and, in truth, the motions of the eye are made perfect, like those of the hand, in slow degrees. In both organs there is a compound operation: the impression on the nerve of sense is accompanied with an effort of the will, to accommodate the muscular action to it.[6]

Bell packed quite a lot into this brief statement: he asserted that both the hand and the eye develop as sense organs through practice, which means that the brain *teaches itself* to synthesize visual and tactile perceptions by *making* the hand and eye learn to work together. According to Bell, the learning process must involve the correlation of sensory information from retinal (light) and cutaneous (tactile) receptors with what he called a "muscle sense."[7] In other words, the brain actively orients the receptors in the eye or the hand toward a target of interest, and then moves them precisely during a process of exploration. The resulting image constructed by the brain must of necessity be based both on the messages from retinal and/or skin receptors and on the record of guided eye or limb movements occurring during the collection of the sensory data. This is an *extremely* sophisticated concept, and contemporary research into the functional organization of perception has strengthened it.

Charles Sherrington later suggested that the most sensitive portions of the skin at the tip of the thumb and index finger were treated by the brain in much the same way as the most sensitive part of the retina (called the macula) is. Sherrington postulated that the first stage of exploration is not under voluntary control: in vision, the eyes move automatically toward the target and lock on to it before searching out its identifying optical features; in touch, an unknown object is grasped and "instinctively" oriented within the hand, or the hand itself is oriented over the target surface, so that its most salient topographic features come into contact with the sensitive fingertips for closer inspection.[8]

One extremely important difference between the eyes and the upper limb as movable explorers has to do with the biomechanical (hence "computational") complexity of moving the sensor (retina or fingertip) to the target of interest. The eye, assuming the head to be in a favorable position, need only rotate toward the target to bring light from it to the retina. The hand, however, is located at the end of a complex biomechanical linkage and must actually *get to* the object to be touched. The body may first have to be moved toward the target. Once there, and taking into account its configuration as it approaches the target, the reaching arm has wide latitude in the combinations of joint angles and contraction-relaxation patterns of trunk and upper extremity muscles that can be assumed in order to bring the hand itself into contact with the physical target.

The infant, as Bell recognized, is not given this skill but must learn it. The long march toward physical and mental agility begins while the infant is still on its back, where it must remain until the muscles of the neck can hold the head still while the eyes search and study. Newborn babies will reach toward objects placed in front of them by the time they are two weeks old. When a baby's neck muscles are strong enough to support the head, reaching rapidly becomes accurate in an upright posture. This may seem a small technical point, but it is not: the hand works under the guidance of the eye but cannot do so until the head has become a dependable platform for the visual system. An interesting experiment has shown that when the head is supported in an upright position, babies between five and eight weeks old will reach with the accuracy of babies *twenty* weeks old. As French physiologist Marc Jeannerod says: "This observation emphasizes the role of head position in the building of an efficient reference system for directing movements toward extrapersonal space."[9] That system must also, of course, lead to perceptual stabilization of the visual world.

The coupling of hand and eye movement is an enormously complex learning task in which the child must be intensively engaged before it can ever hope to pry its bottom off the floor. Before the baby can (or should) stand and walk, its brain must develop and integrate a multisensory reference system to track limb movements on an ongoing basis—the nervous system must know at all times where each hand is with respect to the midline, with respect to its root at the shoulder, with respect to the other hand, to the foot, to the mouth, and to the eyes. Inevitably, this same process helps to establish a coordinate system for external objects located in three-dimensional space. Where is the hand with reference to mother's breast, a rattle, a handle, or the puppy's ear? Increasing the range of exploration permits the calibration of bodily movement against space and objects in the real world. And while all this is going on, the hand is learning to operate at the end of the arm, beginning to capture objects of interest and to bring them closer to the body.

Before the fingers begin to work independently, two critical and apparently separate events in neuromuscular development are necessary: the arm must have learned to move to a target under the guidance of the eye, and the hand must have learned to orient and shape itself in preparation for grasping the target. The first of these stages is normally complete before the age of five months, and the second before the age of ten months; after that, the hand is ready for a lifetime of physical exploration.[10]

The truth is, there is no point lifting the baby's little bottom off the ground until the brain is prepared to confront the explosion of visual-spatial information that will result. The newly mobile brain will then discover through direct experience the consequences of its own incessant relocation in space, of the destabilization of horizontal and vertical landmarks, and of the independent movement of targets that have their *own* trajectories and that can and do change course whenever they please. Once the brain takes off on its own two feet, there are immediate and continuous changes in the physical state of the body, as well as in relations between the body and the world, and these will have to be recomputed many times every second.*

---

* As the body matures and begins to move faster and in more complex and improvisational ways, it becomes increasingly critical for the brain to "defend" the perceptual constancy of the visual world. The next time you happen to see a film of a cheetah running, notice that despite the violent moves of the body, the head itself remains as steady as the Rock of Gibraltar.

Learning how to catch a ball dropped into the hand demands close to a year of preparation and represents a huge milestone of neurologic development. Getting good at catching, learning to move toward and capture smaller balls that move faster and faster—learning to anticipate movement, to plot and execute interception strategies—will take years of trial-and-error experience. Small wonder it takes a while just for the *idea* of juggling to make sense to a child. The *act* of juggling takes considerably longer to make sense of and a major portion of a performer's lifetime to perfect.

Several years ago, I met a remarkably gifted juggler who turned out as well to be a perceptive and articulate guide to the pragmatics of juggling. Serge Percelly, a Belgian whose unlikely specialty is juggling tennis rackets, was performing during the 1993–94 season of the Big Apple Circus in New York City.[11]

Percelly followed his parents into circus life. His father was a clown and his mother was a high-wire acrobat whose career brought her own family into its fifth generation of professional circus life. Serge grew up in Geneva, Switzerland, where he lived with his grandparents until he was fourteen, seeing his parents only during the few weeks each year when the circus came there to perform. He remembers helping out in the circus as a young boy and being captivated by the atmosphere. Even if he was just moving chairs or helping build a ring it was a magical place to be. When his parents were traveling, he thought of them constantly. Dreaming about the circus was in fact just another way for a young boy to dream about the parents he wanted to be with:

> The saddest moments always in my life, too, were when I was in Geneva and the circus came to town. The circus grounds were on the grass, and when the circus leaves you can still see *everything*. You can see where the caravans were, because the grass gets white. You can see the circus ring with the sawdust there—all these things. The next day when the circus was gone, I used to come back from school and sit exactly where our car-avan had been. I used to sit there, and in my head the whole circus was still

> there. I could still see everything! I used to spend hours there, just sitting on the grass, watching everything and saying to myself what they were doing.

When he was fourteen, his parents let him go on tour with them for the first time. That was when the juggling started, and it was triggered (as was Bachleitner's immediate desire to become a puppeteer) simply by seeing a performance.

> I saw one of these jugglers and I was just fascinated—*amazed*—by it, and knew it was something I wanted to do. I was not only watching, I was really hypnotized—it wouldn't go out of my head. I didn't know if I had the talent, but after seeing him, before even trying, I already knew that was what I wanted to do. I knew it without knowing if I *could* do it or not. I didn't even wait to buy balls—I started with stones. It was so fast. I've been caught in it from that day until now.

Percelly recalls that he began juggling with the benefit of an athletic prowess gained from two years of playing tennis, and with what he believes is an unusual way of watching a moving ball. He says he was told early on that he had both the movements and "an eye for tennis."

> Most tennis players look at the ball until it hits the racket, and I never did that. I could see exactly where the ball was going without fixing the ball. You have to do the same thing in juggling. You have to fix a point somewhere, where you actually see a bit of everything, and that comes only with practice. You always know right away if it's a good throw or a bad throw when you do it, and you know if you're going to catch it or not.*

* Serge's strategy seems to defy Sherrington's postulate that visual fixation is not a matter of choice. However, it strongly supports a surprising finding of an MIT study (see Austin, pp. 144–148, and comments later in this chapter) namely, that it is actually possible to juggle when as little as the top *one inch* of the ball trajectories are visible—equal to less than 1/20 second of viewing time!

It was his father who suggested that Serge move his rackets from the tennis court to the circus ring, mainly because of the audience's love of novelty and the demand for the unusual in every circus act. He remembers very clearly the struggle to learn at the beginning.

> I remember exactly, even now. You rehearse, you do a trick, you do another one, they're fine. You do it again, it's good, you do it again, it's better, and so on. The next day you can't do it at all! You've practiced the whole day, you did it ten, twenty times the day before. The same thing keeps happening until it slowly starts to get better. The work gets less intense and the average gets slowly, slowly better.

This opinion of a working performer is shared by Howard Austin, who was awarded a doctoral degree in electrical engineering and computer science from MIT for his analysis of juggling:

> The most striking observation is the discrete nature of the overall process. It is not surprising to find that the learning occurs in (or between) a series of distinct trials. That's to be expected, since the basic plan is to go until you miss and then start again. However, it is unexpected, shocking even, that trial-to-trial variations of over 1000% are frequently observed. . . . Even if we adopt the more reasonable criterion of comparing the current trial to the best previous attempt, variations of 400% and more are common. For example, on trial 28 John threw 11 tosses, but his previous best was 3 tosses. . . . Almost every subject had a breakthrough which doubled or tripled their previous best effort. . . .
>
> These breakthroughs, exhilarating when they occur, are an extremely interesting puzzle. They are counter to a large body of psychological experiments which support the theory that learning is a gradual, continuous process. The present data considerably alter that view in that, while the overall direction seems to be one of gradual improvement, the improvement takes the form of distinct episodes between which there may be dramatic differences in performance.[12]

This mode of learning may be common to many forms of acquisition: there *will* be good days and bad days, with *wild* and largely unforeseen fluctuations in performance any time a new skill is being learned or an old one is being modified. This seems to be the natural mode of progression in a motor skill.*

We should stop here for a moment and take stock. We know from observations of young children that a great deal happens during the first year of life to prepare a young child for walking and maneuvering among objects that move independently in the environment. The first essential milestone in this development is the creation of a stable platform for the visual system (the head supported on the neck) so that the eyes can reliably guide movements of the upper limb. No young child can become a successful hunter (or third baseman, or skater, or tennis player) until the brain becomes adept at manipulating an elementary physics equation (distance = speed × time) needed to solve "relative motion" problems.† It is in the child's earliest experiences in practical physics—watching, locating with both hand and eye, and then intercepting moving objects—that the nervous system builds its own unique library of solutions to the computational problems presented by coordinated movement.

The critical importance of such a library means that a baby's instinctive attraction to movement, followed by the impulse to reach and then to grasp (catch), is one of the earliest maturational imperatives in the human ner-

---

* It may also account for a strong aversion among teachers and students to the analysis of both success *and* failure, not to mention a strong propensity toward superstitious attitudes about study. An actress once said this to me: "You do something very well, try to figure it out, and it goes away. That makes you believe that you shouldn't *think* about what you're doing. Analyzing success just leads to mistakes."

† A relative motion problem is exemplified by a common situation in ship navigation. I am steaming at night and spot a light 20 degrees off my starboard bow. My radar tells me there is a reflective target (another ship, probably) where the light is, and that it is 10 miles away from me. Fifteen minutes later the light is still 20 degrees off my starboard bow, but the radar now says it is only 8 miles away. No matter what speed or direction either of our two ships is actually going, our *relative motion* is straight toward each other at a speed of 8 knots, and if one of us doesn't change course or speed within the next hour there is going to be serious trouble. Exactly the same problem, with different numbers, faces an outfielder chasing a fly ball, only in this case if there is *not* a collision (between his mitt and the ball) there will be serious trouble.

vous system. Without ever being told to do so, or having rehearsal strategies explained to it, the baby will play this game and rehearse these movements endlessly, as it gives itself things to do with its body that are more and more difficult. These games are created by the nervous system in order to teach itself a concrete, experiential sense of physics.

One of the most important physiologic consequences of these accumulated body lessons in physics is the establishment of an internal temporal reference system—a biologic clock—and juggling makes it clear just how important an accurate internal clock is. The juggler throws a weighted object that will rise and fall a certain distance, and he must time the preparations for a catch accordingly. Whatever is going up is going to come down, and nothing will change its rate of fall. This seemingly trivial fact leads, however, to a somewhat surprising finding in the MIT study: the accuracy of the toss depends on when you let go. As Austin puts it, "Timing seems to be so critical that even what looked like errors in space (the direction of the toss is wrong) usually have to do with an error in the timing of the release." What does this mean? It means that you will never be able to throw accurately enough to hit a target unless your own internal clock, which controls the timing of muscle activity, is perfectly calibrated. Based on what we have learned about the normal behavior of babies less than a year old, that clock is not set or calibrated until the head is under control. And when that time comes, the setting and calibration is organized through active movement, including the catching games the child plays.

Percelly and Austin also agree that simple repetition does improve some aspects of performance, but much more is required for expert performance. The development of any high-level skill requires intelligent rehearsal: repetition according to a well-designed plan. As Austin observed,

> Simply *telling* someone the ideas *won't do.* No matter how sincere the inquiry, a great deal of practice, and a special kind of practice, is necessary for real understanding. This is no drawback, since many people currently invest large amounts of motor skill practice time with no noticeable results.[13]

Percelly learned the hard way:

> The hardest part for me was that I had nobody to teach me, which sometimes made it a problem to get better technically. Some tricks are quite complicated, and you can watch a juggler as many times as you want but it goes too fast and you just don't get it. So I was blocked sometimes, really hooked in a state where I couldn't go any further.

Austin addresses the issue of formal teaching in his study, using an artificial-intelligence model for comparison. He came to this conclusion:

> The most powerful technique for making global programs changes is to call on a specialist who can give expert advice on the problem at hand. The expert can be the coach, teacher, or some other external (to the learner) source or another internal program. . . . The ability to find, or give yourself good advice then becomes a critical part of the motor programming process and hence of athletic ability. It is important to note that in the present theory, advice techniques, skill models and various other HIGH-LEVEL EDITOR features play the dominant role in the construction of new motor programs. *These activities are inherently intellectual in nature and hence lead to the claim that so-called physical skill is largely mental activity.*[14] (Italics added)

Teachers, in other words, have essential information about the best way to sequence the elements that must be mastered in order for the skill to develop without interruption; they will steer you around the potholes and save you from endless repetition of the predictable mistakes.[15]

What about the practicing to retain a skill that has already been learned? No one has any idea why this is true, but extreme demands in performance accuracy seem to come with a very high maintenance cost. The impression gained from Percelly's description of his own experience is that the body, like a good violin, has to be kept in tune, and the body's clock has to be recalibrated and the muscles (and eyes) synchronized perfectly with it just before a performance:

It's amazing how, after all these years, it still can go wrong. Sometimes, maybe even an hour or an hour and a half before a show, I can feel incapable of doing more than warming up. It can seem both physically and mentally *impossible* to do more. Sometimes you really think you're not going to make it. But you keep doing your warm-up, you do it slowly, do a little bit more, a little bit more, and you finally reach your own level; finally you get to it.

I do every trick behind the curtain to get warm. Some days I just need to do them two or three times. I see them clearly and they're fine. Sometimes I have to do them ten or twenty times and they *still* don't feel right. A year ago, for a full week my foot tricks just weren't there. Not that I was dropping, but it was *unclear,* and I always needed that little bit of luck to get it. It could have gone entirely wrong, or could have gone well. I wasn't controlling it at all.

Sometimes even *after* my show, when I did everything fine, I can pick up five rackets and I can't do it. Maybe it's the muscles—and maybe also the tension—that just leave me. The rackets go everywhere and I can't fix anything. The eyes have to stay in practice, too. When I stop for one or two days, there's no problem with my body, my arms are okay, but my eyes have to get used to seeing all of it again. The eye thing is very, very important.

What could be the explanation for this seeming need to "rehearse" the eyes in order to maintain juggling skill? In talking about early-life establishment of hand-eye coordination, I did not raise specifically the issue of maintaining visual fixation when the head itself is moving. Austin refers to the work of Emilio Bizzi (also at MIT), who established that eye movements are corrected for head movements through a system of feedback with the part of the middle ear that acts like a gyroscope.[16] Conceivably, this system is so finely tuned that it needs frequent recalibration.

There are other possible explanations as well. An extremely active area of research at the present time concerns the specialization of function of brain cells and connections associated with visual control of movement. For example, some brain cells respond only to light having a particular orientation or moving within a specific velocity range. The brain and eye, in other words, can respond to, or track, rapidly moving objects without actually "fixing" them, just as Percelly says he does. Other cells, called bimodal cells, are a critical link in the coordination of eye and arm movements. As Marc Jeannerod points out, "these movements require mechanisms for coding target position on the retina, eye/head position, head/body position, arm/body position, and arm position with respect to the target."[17]

In his new book, *The Cognitive Neuroscience of Action*, Jeannerod explores an extremely important concept now taking hold among researchers in this discipline. The new conceptualization, which began to take shape in the late 1960s, pictures human vision as being dominated by processes that took their origin in the separation of the orientation and identification functions of vision. Most animals use vision to detect movement (which often signifies the approach of either a threat or a meal) and to guide the motor system so that movements related to the target will be directionally appropriate. This orienting system, which is very old, constitutes what is called the *dorsal* visual system. Its main task is to process visual data in order to create what might be called movement shapes (what Jeannerod calls "dynamic form in relation to motion"). In higher primates and humans this system is essential both for simple target identification and tracking—so that the arm and hand can be moved to intercept the target—and to guide the preformation of the hand so that it can perform an anticipated task as soon as it makes contact with the object. This new model of the dorsal system was "no longer based on the modalities of visual coding of the movement but rather on the modes of representation of the *goal* of the movement."

Although functional distinctions between the human dorsal and ventral visual systems are not well defined, the connections of the ventral system are such that it could uniquely support behavioral exploitation of the novel grips that became available after ulnar opposition was introduced into the hominid hand. In the ventral system, information about object color and surface features gains access to linguistic areas of the temporal lobes, and Jeannerod has dubbed this the "semantic processing system"—an information channel concerned with "manipulating, identifying, or transforming objects."[18]

The ventral system, in other words, could assist in high-level integration of vision with motor actions dependent upon recognition (or "labeling") of distinctive object properties, or in the advance planning of precision hand and finger grips and movements, related to the performance of cognitively complex tasks. The appeal of this concept is that it suggests the emergence of specializations in the major subdivisions of vision related to an expanding repertoire of hand skills. The dorsal system is known to assist in the approach phase of object-oriented movements, and it may also act as an information channel for manual performance related to familiar tasks. The ventral system, then, could be the preferred channel for visual and sensorimotor information related to the use of unfamiliar objects, or objects about which there is an interest in acquiring greater detail or more precise categorization. Information carried in this channel, which connects occipital and temporal cortex, has access to and could thereafter influence the entire cognitive and behaviorally generative machinery of the brain.*

Percelly's first real break in show business came at the age of eighteen, when he performed at the Circus Festival, an important show and talent-search held in Monte Carlo every year. He didn't win any prizes, but people remembered his unusual act, and a number of job offers came as a result. But because he was not a prize-winner at Monte Carlo, he was forced to work for a time in out-of-the-way places. He used these jobs to test his abilities and to try out new routines, and although at the time it sometimes felt like a "down and out" period for him, his view now is that this was when he learned to be not simply a juggler but a performer. Part of that lesson involved discovering what did and did not excite his audiences.

---

* It should be understood that the techniques for studying the neurologic control of hand movements are very different in humans and in nonhuman primates, making direct comparisons impossible. See Jeannerod, 1997—especially chapters 3 and 5—for a more complete discussion. One innovative method for studying brain activity in relation to hand movement, positron emission tomography, suggests just how much of a computational demand is imposed on the brain by the control of individual finger movements. Scott Grafton, John Mazziotta, Roger Woods and Michael Phelps at the UCLA School of Medicine have shown that the "simple" task of tracking a moving dot on a computer screen with one finger activates major portions of the sensorimotor cortex and anterior cerebellum. The activation pattern in supplementary motor cortex was even further enhanced when subjects were given time for movement preparation. See "Human Functional Anatomy of Visually Guided Finger Movements," *Brain* 115 (1992): 565–87.

> I remember trying six rackets. I worked on that for a long, long time. I had more blisters on my hands than I have now, and I started to get good at them after two years. I could do them, it was good, but I still hadn't put them into the act. I was working in Germany at the time, I was in Düsseldorf, and one evening when there was a very nice crowd, I said, "Why shouldn't I do six now?" I felt good, I did six, and I thought, "They're going to go wild," but they didn't. It was a flop! I tried it a few more times and it was *always* a flop. It didn't matter how well I did it, they just didn't react. So I said, "Just forget it."
>
> That's the kind of thing nobody can tell you. You have to learn to be a performer by working with audiences. You can be very good when you're young, but you're not necessarily a performer then. I could have done all this ten years ago, but I would never have enjoyed performing the way I do now. I know people want to see something, but first of all they want to see *somebody* who is having fun, who makes it interesting as well.

Another point of remarkable and somewhat surprising accord between Austin the theorist and Percelly the practitioner has to do with the matter of making mistakes. Percelly, talking about his own development, said this:

> I also made a lot of mistakes, but when you finally learn something that way, you learn it *twice* as much, better than if you would go through it without making mistakes. The mistakes will still come later on. So it turned out to be a very good thing for me; in the beginning it was very hard, but now it's a good thing for me.

Intuitively, one might suppose that practice pays off by making movement more and more precise: you learn to toss the balls to exactly the same height all the time. That, however, turns out to be a terrible mistake, because this kind of practicing inevitably leads to serious limitations in a juggler's development. An inflexible routine built on the expectation of a long sequence of perfect tosses would be extremely vulnerable to deviations in the behavior of the object being juggled. Austin refers to this flawed strategy, common among beginners, as the "one-height bug." The problem, he says, is that "if a small deviation occurs in the height of a given toss, the mismatch between expected and actual position frequently disrupts the pattern enough to cause a miss." As "preventive medicine," he suggests the student practice a wide range of juggling heights and intermix the heights during a particular run. He also asserts: "It should be distinctly surprising, even to a computational theorist, that getting better at juggling for the most part involves building better and more sophisticated error compensations."[19]

In other words, getting better means increasing the repertoire of things that you do when something goes wrong. Percelly agreed, wholeheartedly and philosophically:

> Of course! So you learn to make the mistakes just another part of what you're doing. You never make mistakes in the same spot. Even if it's in the same trick that you miss, it's always going to be a different moment of the trick. Always. For me, with all these years of practice, I never drop at exactly the same moment. Never. So it's always a different situation, and you have to have a different way of dealing with it. You have to deal with it because it's a *big* part of juggling, because you drop more than you can imagine.

Percelly makes clear that the basics of the motor skill—mastering the trick of the ball toss and exchange—are the easy part. Learning what he needed to know at that level took him all of thirty minutes, more or less. But then he invested another decade (or, with time out for meals, sleep, and recreation, an additional 40,000 half-hour increments) of hard work, thought, additions, subtractions, triumphs, and disappointments before he reached a point at which he knew he was secure and that what he was doing was not only technically proficient but *intriguing*. Only then was he willing to say, "It's all right. It's okay. I like it."

Percelly had now explained to me how he became interested in juggling, how he taught himself, how he used his experience to prepare himself for the real world, and how he continues to maintain his skills. My final questions to him had to do with his own inventiveness as a performer: Where do the ideas for a performance come from, and how do they get into his act?

> Nobody really *invented* juggling and most of it just comes around by practicing. But as you do it more, you get the little ideas that just come with the practicing. When you practice seven hours a day, like I used to, you get to a certain state of mind where you just do whatever goes through your head. You just try anything. You don't really care what goes where—you just *try* things—you experiment.
>
> But what I really like is not actually a particular trick; it's putting it all together. You can't really see the act change, but each year you can tell how much more successful it gets. And that may happen not because you put something in the act that's really difficult, but because you put something in the act in exactly the right way—in a way that makes it more interesting, not only for me but for the audience as well. I'm just trying somehow to *do* the act that I would have loved to *see.*

Juggling, with the considerable help of Serge Percelly, Howard Austin, and Marc Jeannerod, has taken us directly to the frontiers of research in neuroscience. It also confronts us with a premise about human learning that is so deeply embedded in educational theory and practice that it is almost never critically examined. It asks us to question the premise that intelligence is a purely mental phenomenon, that the mind can be educated without the participation of the body. Learning how to juggle, certainly as Percelly describes it, challenges the mind-body dichotomy in an extremely interesting way.

6
The Grip of the Past

THE FATHER BY INSTINCT SUPPLIES THE ROUGHNESS, THE STERNNESS WHICH STIFFENS IN THE CHILD THE CENTRES OF RESISTANCE AND INDEPENDENCE, RIGHT FROM THE EARLIEST DAYS. OFTEN, FOR A MERE INFANT, IT IS THE FATHER'S FIERCE OR STERN PRESENCE, THE VIBRATION OF HIS VOICE, WHICH STARTS THE FRICTIONAL AND INDEPENDENT ACTIVITY OF THE GREAT VOLUNTARY GANGLION AND GIVES THE FIRST IMPULSE TO INDEPENDENCE WHICH LATER ON IS LIFE ITSELF.

—D. H. Lawrence[1]

DAVID HALL IS A MINISTER NOW LIVING near Mount Shasta in northern California. He is also a carpenter, a hunter, and an unscathed, honorably retired rock climber. David was the first real, live rock climber I'd ever met, and it was that experience that triggered our first interview. What I really was after was a hand story, because—with an appreciation of biomechanics born of a decade of taking care of musicians—I wanted to know how the hand could be engineered for two such utterly different tasks as rock climbing and piano playing.

Consider pianists: the optimal posture of the hand at the keyboard is a very gentle curve, just as it appears when the arm is relaxed and hanging to the side, or as it would appear resting on top of a basketball. Keystrokes are ideally made from a fulcrum at the knuckle joint, and flexion at the two joints in the finger itself should be kept to a minimum. For the pianist, the development of what is called *touch* is a matter of the utmost importance: a pianist's career can rise or fall entirely on his or her reputation for having hands able to coax musical nuance from keys, hammers, and strings. Touch—the genius of the fingers—is life and death to a pianist.

For a rock climber, life and death also ride on the fingers, but for a different reason and in an entirely different way. Here the requisite skill is *grip:* specifically, the ability to flex the finger joints and hold them in that position against the pull of the full weight of the body. The force needed to depress a piano key far enough to produce sound is less than three ounces (not quite a fifth of one pound), and the force transmitted through the finger pads in a climber's three-finger hold might rise to something in the vicinity of 80 pounds per square inch. In a rapid musical passage the fingers may strike single keys in succession at rates close to 20 keystrokes per second, where, by contrast, we can guess that the 80 lb/in^2 might have to last as long as 40 seconds before alarming sensations in the hand and arm would commence to draw the climber's attention away from the beauty of the canyon below. So—and please understand that I am simply dramatizing a point—to make a first rough approximation, timing and force in these contrasting tasks are being regulated at opposite ends of performance scales whose magnitudes differ by a factor of 800. How could such ranges of performance be built into the same upper extremity?

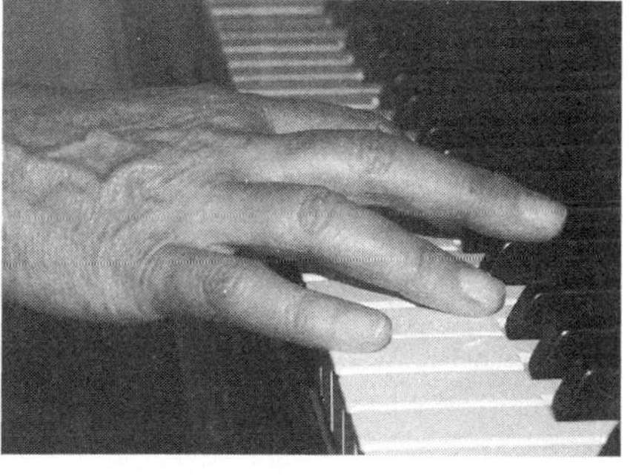

FIG. 6.1 Two radically different tasks ably undertaken by the human hand: slow movement, high forces; rapid movement, low forces. In both cases, there is a very low tolerance for mistakes. *Above, left:* a rock climber enjoys a breathtaking view of the landscape below. (Photo courtesy Tyler Stableford.) *Above, right:* the author's wife indulges a long-standing affection for Chopin's music.

Rock climbing and piano playing are carried out by what is, on paper at least, exactly the same physical apparatus. In reality, of course, the combined effects of innate anatomic variation compounded by different developmental and training regimens mean that serious pianists and serious rock climbers have very different hands. Among serious classical pianists, even playing the guitar is seen as exposing the hand to unacceptable risk of injury, or worse, the risk of acquiring gestures that will disturb the fluency of keyboard techniques.

David grew up in Colfax, a small town in eastern Washington State. His parents raised six children in their home on a small farm about twenty miles outside of town. His father had grown up in a family that migrated from Tennessee in a wagon train and had homesteaded near Colfax. "They were pioneer people, very tough." And David fit the mold; even as a very young boy, he displayed unusual physical propensities.

> When I was not even able to walk yet, my older brother was always pushing me to do things he couldn't do. My mother tells horror stories of finding me in high places before I could even walk. When I was eight or nine months old I was able to pull myself up onto the top of the dresser. I still couldn't walk but could somehow pull myself up on the chest of drawers. One thing I was always able to do that no one else did was to climb the side of our barn, which was several stories high. When I was seven years old I would climb hand over hand to the peak, maybe forty or fifty feet above the hayloft. I'd go to find birds' nests and that kind of thing. I scared the living daylights out of my parents, but I liked doing it and I didn't think it was unusual.
>
> I guess my hands were unusually strong. I milked cows, and I think that was part of it too. In grade school I was just a scrawny little kid but I had big arms. I was in the library once with my arm up on the table and the librarian looked at me and said, "You look just like Popeye!" My forearms were always overdeveloped, I think, even though I didn't do anything special to exercise them.
>
> I suppose it was genetic—a predisposition toward having tremendous arm and hand strength compared to body size. Thin upper body, small

> legs, and this arm and hand strength—I'm sure this was passed on. I'm a little more like my dad than the rest of my brothers are. I remember when we would go to the park my dad would do chin-ups. We would be doing two-handed chin-ups while he was doing one-handed chin-ups. Our goal was to be able to do one-handed chin-ups just like him.

David's father was not merely strong. Like the other pioneers in his family, he was *tough*.

> Dad's never taken pain medication—that would be a sign of weakness and there's no need for it. If he gets his teeth pulled, or cavities filled—no pain medication. I watched my dad drill a hole through his hand one time—it was a half-inch bit. I watched him pull that thing out, wrap a towel around his hand and finish the project with a half-inch hole punched through his hand. And never even flinch. I grew up seeing this kind of thing; from the beginning of my conscious life, I was always exposed to extremes.[2]

David's mother came from a very different family.

> They were more academic. My grandmother had a master's degree and had traveled to India to teach English. She didn't marry until her mid-thirties. My mom, she was sixteen when she and my dad married. She lost her first child and she gave birth to my younger brother, who's number four, when she was twenty. So that was a rough beginning.

Although the difference in family attitudes about education had not made much difference early in his parents' marriage, well before his teen years his mother felt increasingly isolated and unfulfilled. The marriage began an accelerated decline when his father bought a small farm in Idaho.

> It was supposed to be dad's big break, but it didn't work out. We went through a couple of bad winters, a couple of big freezes, and he ended up losing that farm. It was a major trauma and just too much, I think, for my parents' relationship. We were very isolated there, living in a three-room

> house with maybe 800 square feet for the whole family. One winter—I think it was 1968—we were snowed in for three weeks straight. It was so extreme that it finally just overwhelmed my mother. My mom was very verbal, my dad wasn't. She would torment him and bring him to the point where he would lose control. He never touched a drink in his life, but his temper and his anger were bad in those situations.

The family moved back to Colfax, and David's parents were divorced.

> I was twelve or thirteen then. All six children were told to go with my mother, but my older and younger brother and I just told the judge we wouldn't do that. So the three older ones moved back to the farm, living alone and trying to lick our wounds and recover. It was a real hard time, a kind of overload for all of us. There were reasons for what happened, but as a kid of course you can't process that. It just seemed like a lot of chaos and I was bitter and angry for a long time.

During high school David kept his feelings to himself, and one can imagine that very few people around him had any inkling about his turmoil. In fact, his teenage years were a time of continuous growth for him, and there were many successes. Having been an avid boyhood builder of doghouses and an attentive student of machinery maintenance on the farm, he was building and installing wood cabinetry in people's homes and rebuilding engines in the machine shop by his junior year in high school. David was also a high school and college wrestler. He did not consider himself as powerful as other wrestlers in his weight class, but he managed to compensate for the disadvantage.

> I developed a theory in high school that when I touched somebody with my hands, when I was wrestling, the first touch had to send a powerful message to the opponent to make him realize that he could not defeat me. I remember thinking that if my hands communicated to my opponent that I could not be dominated, then I would mentally beat him. I learned to do a quick little move to the head gear and give a little *pop!* to the head. It

didn't *seem* like much but it was enough to throw the person totally out of whack. And usually it would make them angry.

The other thing I would do was reach for the guy's head, and grasp it in the back—we call that a tie-up. I would grab the neck, and my elbow would go into the sternum at the same time. *Bump!* It was the impact. Nothing that you would see, but a very powerful statement being made. Also, when I grabbed the neck I would pull the head forward hard enough that usually we would butt heads. I wasn't trying to butt him hard, but there would be an impact, something saying, "You better be afraid." So he got my message on his ear, on the middle of the chest, and in the middle of the head, kind of like "snap, crackle, and pop." And instead of being in control, he'd *lost* control before he even started.

It would be different if I'd had other physical attributes, if I was stronger or faster—but I wasn't. I didn't have that. I had hands. I had balance. And I had a mental and psychological edge.

Since David grew up on a farm, it should be no surprise that his life involved considerable contact with animals.

I raised horses, and much of my relationship with my horses happened through my hands. I had a little Appaloosa, raised several foals out of her, and I remember touching her and being close to her and guiding her with my hands. When you work with a horse, your hand is what controls the whole horse. Horsemen talk about people who have "light hands." You can guide a horse and communicate with a horse through a very delicate hand. You can build trust with your hand and your touch more than you can with any amount of whipping or yelling or anything else. There is tremendous power there.[3]

It is now easier for David to talk about the dark side of this same period, which he has spent a great deal of time reconstructing and analyzing.

I realize now that there was a death wish in my life from the time I was twelve years old until I was seventeen or eighteen which had to do with not being able to process all of that pain. I felt, and I think my father sometimes felt, too, that it would be easier to end the pain than to continue in it. The only way I could do that was to put myself at risk, put myself out *so far* that I would inevitably kill myself. So I was reckless—*really* reckless. But instead of killing myself, in the process of doing all the stuff I tried, I found out there was very little that I couldn't do, and before long I *thought* I could do anything.

As I grew older in high school, on trips, camping, I would climb the face of a rock without thinking about it at all. I would see a route and would climb it when no one else would even *consider* climbing it. To me it was just what I wanted to do, a normal, easy thing. I wasn't thinking about taking risks, but what I was doing would be utterly terrifying to my friends.

In his college years, David learned more formally about rock climbing and did a great deal of hunting, hiking, and backpacking. Working in summer camps was something he found he particularly enjoyed.

I did a lot of backpacking trips with delinquent kids. We went up into the Sierra with a group of kids who had been incarcerated. I liked exposing young people to new experiences in a new environment. As I reflect on this—the use of the hands, the introduction of new skills, the rock climbing and the spelunking, or the camp skills involved in setting up the camp and cooking—I'm sure what we were doing awakened something inside these kids that had been dormant. They'd been fed and coddled by their parents or by society. They'd never washed their own clothes or cooked their own meals and they'd never done anything that would give them a sense of satisfaction. So we worked through their hands just to give them another idea of what they could do, or what they might be.

In order to understand better the development of David's manual skills, we need to know something about the human grip—what it really takes to hold on to the side of a cliff. John Napier introduced the terms "power grip" and "precision grip" in 1956, in a paper whose purpose was to simplify the method for clinical evaluation of hand function and the effects of hand

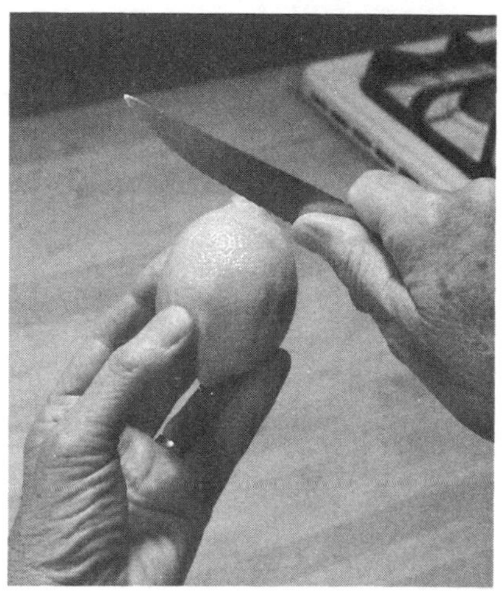

FIG. 6.2 Napier's power and precision grips at home. *Above, left:* the hammer is held in an *oblique squeeze* grip, with its handle buttressed against the palm by fingers and thumb. The shadow emphasizes a special advantage of this grip: the object held in this grip extends the reach of the arm. Also, as Napier noted, when accuracy of movement is required in a power grip, the thumb tends to be drawn close to the side of the hand, or *adducted. Above, right:* the lemon is held in a *precision* grip between thumb and fingers. The separate and powerful long flexor of the thumb is essential to the combined stability and maneuverability of this grip.

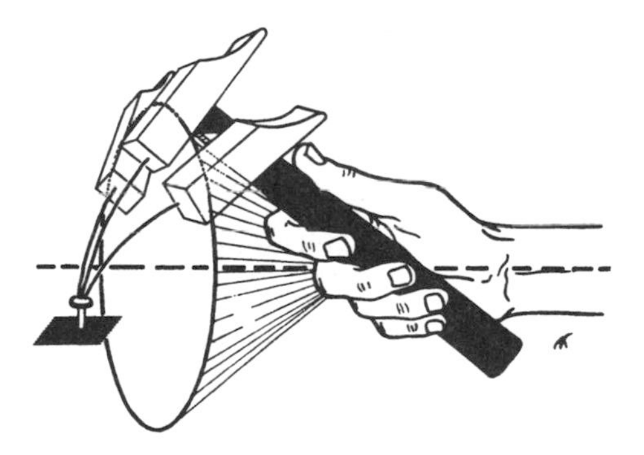

FIG. 6.3 Additional advantages (and complications) of the oblique squeeze grip. Aiming the hammer at a target is aided by the positioning of the thumb (see Fig. 6.2), and by the capacity of the forearm to rotate around its long axis (i.e., to *pronate* and *supinate*). The totality of this movement capability began in the trees with brachiation and was completed on the ground with the addition of an enlarged thumb and increased ulnar deviation. (Figs. 6.3 and 6.4 from Raoul Tubiana, *The Hand,* 1981, with permission.)

injuries. First, he separated "prehensile" from "non-prehensile" movements.* Prehensile movements are those in which an object is held partly or wholly within the hand; non-prehensile movements are those in which the object is manipulated by the hand or fingers but not grasped. Combing one's hair is an example of the former; typing and playing the piano are examples of the latter.

Focusing on prehensile movements, Napier defined the power grip as any holding posture using the palm as a buttress. Any other grip using any combination of thumb in opposition to fingers was called a precision grip.[4] Napier realized that for humans it was not the tool but the *task* that dictates the grip. Regardless of the shape of the object being held, its intended use automatically defines an expected range for control of both the force and amplitude (that is, power and precision) of movement. The gripping posture is anything but haphazard: it is in fact a highly precise registration of neurologic preparations for the biomechanical requirements of the task.[5] For example, a fast and forceful hammer swing requires maximum pressure to be applied to the handle by the fingers and the thumb, which must squeeze against it to lock it into the palm. Once the swinging movement begins, sensory monitoring of the hammer will be needed more to reduce slippage between the skin and the handle than for fine-tuning its movement within the hand.[6]

Controlling a hammer within a power grip turns out to be no small matter. As Jeannerod says:

> Lifting an object implies a sequence of coordinated events where the grip force (to grasp the object) and the load force (to lift the object) vary in parallel. The grip force/load force ratio must exceed the slip ratio, itself determined by the coefficient of friction between the skin and the object surface. . . . The respective contributions of anticipatory mechanisms and of reflex adjustments to the accuracy of grip force have been extensively studied. It appears that the adaptive changes in grip force are strongly dependent on tactile afferent signals.[7]

---

* The term "prehension" comes from the Latin word meaning "to seize."

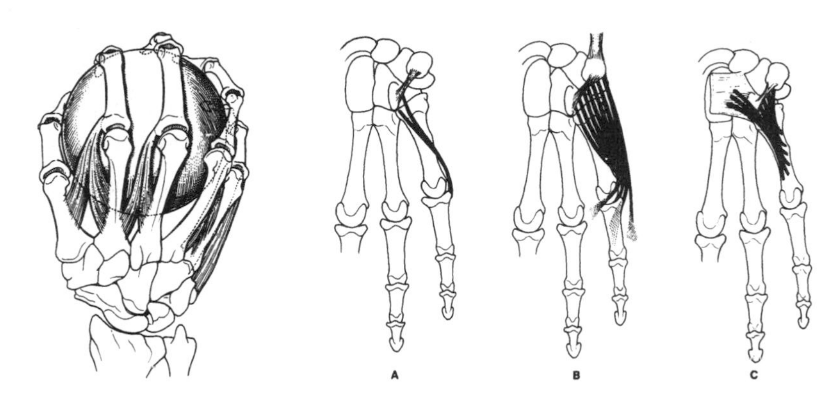

FIG. 6.4 The ability of the hand to conform to large, spherical objects is due in part to the action of small but powerful intrinsic muscles of the hand that help to maintain its arch. *Above, left:* the right hand seen from its dorsal surface, showing the dorsal interossei. *Above, right:* the right hand from its palmar surface, showing the ulnar intrinsics. These muscles are essential to ulnar opposition and are active in the oblique squeeze grip.

In 1970, Charles Long and his colleagues at Case Western Reserve University, in Cleveland, carried out an exhaustive electromyographic mapping of hand and forearm muscle activation in a variety of handling tasks, including the upswing of a hammer stroke. In virtually all instances of power grip, all of the forearm muscles were active, in combinations that varied only slightly with the particulars of the task. The activation patterns of muscles located within the hand itself—the "intrinsics"—varied considerably with the specific task.[8]

There is yet another interesting and important set of findings for us to consider, which hints at unheard-of precocity in the boy who remembers scaling furniture before he was walking. The development of the precision grip in children was reported in a series of papers published between 1991 and 1995 by H. Forssberg and his colleagues at the Karolinska Institute, in Stockholm. The most recent of those papers reported a study of the maturation of the ability to control grip in relation to frictional forces (the problem being as described by Jeannerod, above). They found that frictional adaptation begins to develop at the age of two years but progresses slowly and is compensated for by the use of excessive grip force, thus giving the lift a "high safety margin." (This harks back to the earlier discussion about "representations" in chapter 2, "The Hand-Thought-Language Nexus.")

> The lack of anticipatory control in the youngest children in the unpredictable series may indicate an immature sensory representation of the friction and an immature capacity to store and/or retrieve sensorimotor memory information related to objects' frictional properties. . . . The youngest children apparently need several lifts to establish a useful memory representation related to the frictional condition, in contrast to adults, who need one lift.[9]

We can't be sure what David was really doing with his hands before he was two, but we can guess that if he *was* hoisting himself to the top of the dresser before he could walk, he must have had a prodigious capacity to increase the safety margin in his grip force/load ratio.*

There is one form of the power grip that requires neither the palm nor the thumb; this is the so-called hook grip, familiar to all of us who carry briefcases and suitcases, and since it is the common grip used for brachiation, it probably represents an undifferentiated version of the prehensile power grip, to which the refinements of thumb and ulnar opposition were later added. David reassures us that even though humans no longer employ this mode of ambulation, rock climbers have shown that the skill can be reconstituted:

> There are two kinds of finger holds you use in rock climbing: you have nubs protruding out, and then you have cracks, or finger jams, and the nubs can also be toeholds. Hanging on, either inside a crack or on to a nub, a lot of times you have to "chin" your weight on it to climb up. So we'd practice by using people's doors. I remember going to people's homes and

* It may be that David also inherited the gift of storytelling from his pioneer ancestors, but I don't think we should doubt his veracity on this point. The practical lesson here for parents seems useful: when your little boy drops his glass of milk, you should be pleased. He's not being clumsy, he's *calibrating!*

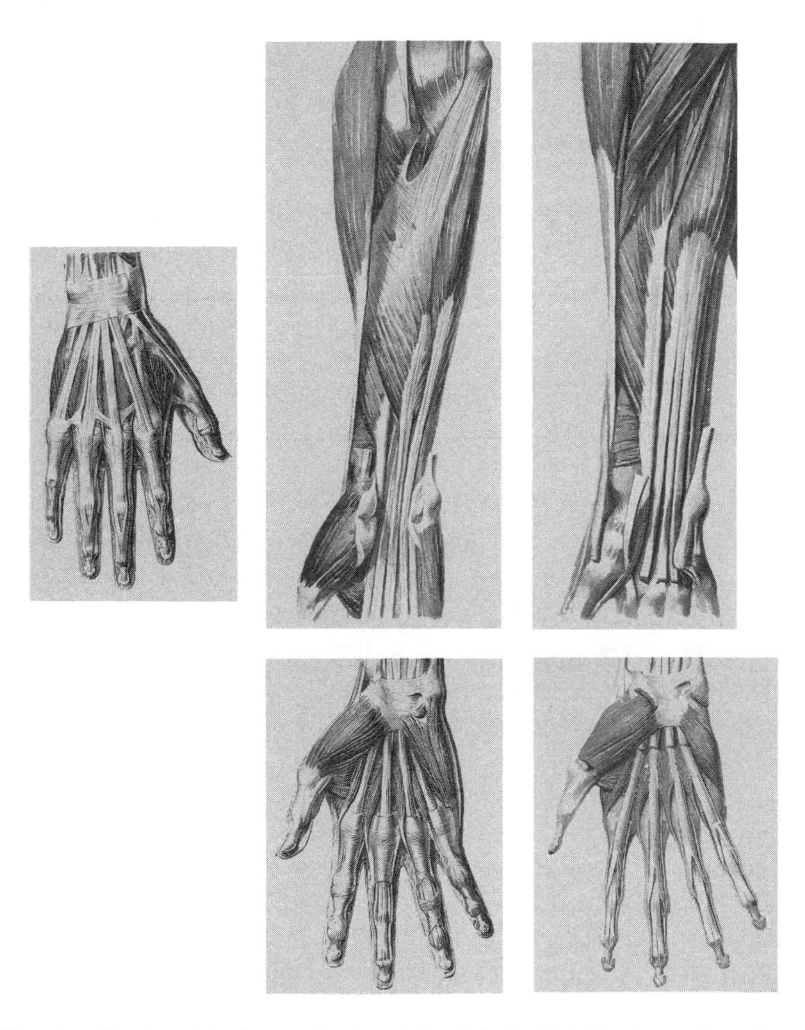

Fig. 6.5 Muscles of the forearm, hand, and thumb (palm side). Two layers of muscles on the flexor side of the forearm enter the hand through a small passage in the wrist called the carpal tunnel. *The top, middle drawing* shows the superficial finger flexors, their tendons extending into the hand shown just below (*bottom, center drawing*). *The upper right drawing* shows the deep layer of finger flexors with their tendons extending into the hand just below (*bottom, right*). *Top, left:* the back side of the hand, showing the extensor tendons and the small intrinsic muscles of the hand (diagrammed in Fig. 6.4, *above*). The functions and interactions of these muscles, all of which are involved in both power and precision grips, are extremely complex and their performance is highly subject to training effects. (From Werner Spalteholz, *Hand Atlas of Human Anatomy,* 1923.)

testing out their door casings—to see if it was strong enough to hold my weight—and then doing the chin-ups just to keep my fingers strong.[10]

It's worth noting that Serge Percelly, who we can calculate has juggled for thousands of hours over the years he has worked on his show, found it easier to juggle behind his back with his left arm because it was less "muscley." He practices long hours and has done so for years, but without any tendency to increase muscle bulk or power as a result. What practice does for him is to increase the smoothness and the variety of movement, and the finesse with which he absorbs bad tosses and "drops" into the flow of his performance.

As a rock climber, David Hall confronted a different task. He was not launching tennis rackets and then catching them but suspending his body weight with his fingers, sometimes even lifting himself with just the fingers of one hand. To do that, he exercised his forearm and hand muscles, but he did so for *strength.* Although exercise for both men focused on arms and hands, muscles and movements, development in each took its own separate path. Serge was after quick, repetitive, precisely timed movements, and David was after sustained power.

As a child, David was taught to ignore both physical and psychological pain. Indeed, his father's stoicism almost certainly contributed to the end of the marriage, plunging the entire family into an abyss of pain that took David years to recognize, react to, and resolve. When he was a teenager, David's rock climbing carried a hidden meaning, which he later came to recognize as a death wish. Eventually, as a youth counselor and a minister, he was climbing to help others gain strength and overcome fear in their own lives. This transformation of an innate physical talent that was first expressed as a childhood game became in adolescence a powerful context for the expression of both rage and helplessness, and finally was adopted as the catalyst for teaching young people about life. Here a "merely" physical interest matured and was integrated into profoundly important intellectual and psychological processes.

And what of David's employment of his hand with friends, opponents, and even horses to signal strength, confidence, and gentleness? No one ever *told* him about this very direct means of communication, or offered lessons in its application or refinement. Once he *discovered* it, being not only a thoughtful but an inventive observer, he created his own personal dictionary of manual semantics.

David's story anticipates an issue we will take up later, concerning societal strategies for raising and educating children. Parents today tend to be concerned, if not obsessed, with getting a child moving in a certain desired direction as soon as possible. If getting ahead is the new purpose of American life, getting your child ahead of the rest of the kids is its sacramental corollary: the right toys, the right preschool activities, so many hours of this, so many hours of that, somehow beating the timetable of the public school system. In David we see an earlier educational model, one rooted in life's immediate circumstances, very rich in rewards for self-reliance and invention. David grew up where "farm work" was an open-ended, loosely structured plan providing real-life demands (and real hardships) that produced many branchings, many unexpected experiences, many opportunities for a young child to explore and pursue interests on the basis of native curiosity. Is the model outmoded?

And we return to a question raised above, in the Prologue. Does inheritance determine not only bodily expressions of gender, size, and coloration but the preferential development of specific groups of muscles important in particular physical skills? Is it just a developmental aberration that David could power-lift himself to the top of a dresser before he could walk? Or does neuromuscular "prewiring" express its blueprint for motor skill into preferred performance profiles based on enhanced potential in specific grip and movement synergies? Did David lean toward climbing and carpentry because a powerful grip was part of his personal genetic package, or were his "Popeye" arms simply created at the milk pail?

David's story, like those of Anton Bachleitner and Serge Percelly, reminds us of the evolution of self in our own lives: the mysterious process through which body and spirit unify, inexorably asserting unique themes of personal growth and learning. There will always be trouble—and opportunity—in our path, so that each of us must perfect our own juggling act on the road. The act will be a mix of the old and the new—standards from the *Homo sapiens sapiens* repertoire infused with family and tribal routines and spiced with a few improvisational numbers. For humans, this "act" of living has increasingly come to mean the certain occurrence of unexpected and transforming new behaviors. Given a creature that loves to play, sharpens what it knows how to do through endless practice, and *will try anything that comes into its head,* one must expect that new skills, or new combinations of skills, will be regularly devised, and will be made part of the repertoire if they prove useful and can be taught (or passed on in some other way) to others.

7

The Twenty-Four-Karat Thumb

THE HAND GIVES THE UPPER LIMB ITS IMPORTANCE AND ORIGINALITY.

—Raoul Tubiana[1]

THE MOVEMENT OF THE THUMB UNDERLIES ALL THE SKILLED PROCEDURES OF WHICH THE HAND IS CAPABLE.

—John Napier[2]

WHAT IS IT SPECIFICALLY about the human hand that differs from its primate antecedents? The existence and persistent influence of an ancient blueprint for a modern-day rock climber is not so hard to accept, but where do puppeteers and jugglers, surgeons and engine mechanics come from? How can primate and simian evolution have yielded such a kaleidoscopic roster of skilled doers without knowing we were coming or what we would need or want to do when we got here?

The evolutionists' answer to this question is clear, and there are no exceptions: an unbroken procession of failures in the integrity of our common genetics program, acting over a period of countless eons, made room for all of us in this life. Selection by fitness is the engine for transforming random genetic mutation into a highly successful long-term survival strategy when environmental conditions fluctuate unpredictably over time.[3]

Given the stunning shift in the fortunes of *Homo* associated with the receipt of a new hand and new brain to go with it, the real shock in the anthropologists' answer (discussed in chapter 3, "The Arm We Brought Down from the Trees") is its footnote-like character. The features that distinguish the human hand from the hand of apes are nearly imperceptible and in fact were recognized by anatomists only very recently; while there has been no dispute that human *use* of the hand is unmatched in the animal world, no one thought there was anything special about its design.*

Sir Charles Bell set the stage for modern studies of the structure and function of the hand when he said, "It is in the human hand that we have the consummation of all perfection as an instrument."[4] One hundred years later, Frederick Wood Jones declared that the true explanation of the power and versatility of the hand was to be found in the *brain:* "It is not the hand that is perfect, but the whole nervous mechanism by which movements of the hand are evoked, co-ordinated, and controlled."[5]

Jones was right about the brain but wrong about the hand. As anatomist O. J. Lewis explained in 1989:

> It is commonly believed by anatomists that the human hand is essentially primitive; yet the hand has its full quota of features . . . which are finely attuned to its specialized role as a delicate manipulative organ. . . . In the emergent hominoids, there must have been progress towards enhancing the overall grasping repertoire of the hand.[6]

Indeed, it was not until anatomists and anthropologists had come to appreciate the critical role played by small anatomic variations in "the grasping repertoire of the hand" that the importance of distinctive features of the human hand were noticed. Prior to that time, understandably, the story began and ended with the thumb. As John Napier had put it:

> The hand without a thumb is at worst nothing but an animated fish-slice, and at best a pair of forceps whose points don't meet properly. Without the thumb, the hand is put back 60 million years in evolutionary terms to a stage when the thumb had no independent movement and was just another digit. One cannot emphasize enough the importance of finger-

---

* In a serious discussion about evolution, the term "design" is a battle cry. Bell, writing before Darwin, meant quite specifically that the condition of the human hand was evidence of an intelligence outside of man who was, in his words, the "anonymous author" of the design. Richard Dawkins argues the case against design in *The Blind Watchmaker* (1987), and Daniel Dennett does the same in *Darwin's Dangerous Idea* (1995).

> thumb opposition for human emergence from a relatively undistinguished primate background.[7]

But it was actually Napier who first saw *past* the thumb; his landmark work on the relationship between structure and function in the primate hand alerted other anthropologists to the critical importance of grip in the behavior of evolving primates. His influence was reflected in the early attention given to Lucy's arm by Donald Johanson and Owen Lovejoy. As Lovejoy noted when he first saw this specimen:

> Although it has the fully opposable thumb of humans, the muscles at the base of the thumb appear to have been small. That means that precision gripping between thumb and finger was probably excellent, but that power gripping which involved the thumb and the entire hand was poor.[8]

In his landmark work on precision and power grips, Napier had established that manipulative hand movements were not arbitrary but choreographed to achieve a continuous and integrated solution to the biomechanical and neurophysiologic constraints of any movement. His model went far beyond revealing the physical rationality of these movements: he showed that despite their enormous variability in posture, force, speed, duration, and trajectory, the underlying control *principles* are simple and elegant.

It is not clear that Napier—or anyone else, for that matter—realized how profoundly important his discovery of this principle really was. But his is genuinely one of the shining examples of beauty and power in modern biologic thinking. Articulated hand control is a recent accomplishment of the entire evolutionary experience of the primate upper limb and as clear an example of functional revision through the distilling process of selection as one could hope to find.

Following Napier, Mary Marzke has been highly influential in focusing the anthropologic search for clarification of the mechanics behind grip and its role in manipulation, throwing, and striking. Beginning with her analysis of the wrist and hand bones from Lucy, she has produced a meticulously documented evolution of hominid hand function associated with discrete structural modifications.[9] Marzke has concluded that Lucy had the capacity to grip small stones in novel ways that would have permitted their manufacture and use as small, simple cutting tools. The three crucial grips were:

- *Pad-to-side grip,* with the tip of the thumb pressing against the side of the index finger—not a new grip shape, but probably functionally altered because of muscular changes that made it more powerful. The benefit: a tight grip on a small, edged stone used for cutting.
- *Three-jawed chuck,* or "baseball" grip, with the thumb, index, and middle fingers held as though cupping a baseball—a new grip made possible by greater freedom in thumb and index-finger movement, with improved shock-absorbing capacity at the base of the middle finger. The benefit: ability to hold a hammer stone securely while pounding another hard object, or to secure a stone in the hand in advance of throwing.
- *Five-jawed cradle grip,* with the hand open, and palm and fingers up—not a new grip. The benefit: ability to support an object in the palm while moving it with the thumb and four fingers.*

The earliest functional changes in the hominid hand (that is, Lucy's hand) initiated a radical transition from its ancient role in suspensory locomotion and knuckle-walking to a new role in object handling and use. Marzke attributes the accumulation of progressive changes leading to the modern human hand to prolonged and expanded experience with, and dependence upon, stone tool manufacture and use. She has identified a total of eight features that distinguish the human hand from nonhuman primate species, noting that these morphological features "form a pattern which is favorable to achieving the precision pinch grips and precision handling that in our experiments appear to be essential to effective tool making."[10] The new features include a longer thumb, widened pads on the fingertips, modifications of the muscles at the base of the thumb, a stronger and more independent

---

* The human five-jawed cradle—now the five-jawed *chuck*—is a major advance because although it permits the same cradling of objects in the hand, the augmented manipulative range stemming from having both thumb and ulnar opposition increases enormously the options for precision handling of small and irregularly shaped objects.

FIG. 7.1 Three grips shown by Mary Marzke to have been within the capabilities of the hand of *A. afarensis* (Lucy). The right hand, *top,* is holding a sharp stone in a pad-to-side pinch grip; the right hand, *below,* is holding a hammerstone in a three-jawed chuck (the same grip used for throwing a stone of that size); the left hand, *below,* is holding a large stone in a five-jawed cradle grip. Note in the top photo that the stick is being held in an oblique squeeze grip, which Lucy's hand was not capable of. Notice also the complementarity of function of the left and right hands. (From Mary Marzke, "Evolutionary Development of the Human Thumb," 1992, with permission.)

long flexor of the thumb, and small but important changes in metacarpal and wrist bones.

After analyzing experimental stone tool manufacture and use in humans, Marzke notes that "forceful and accurate blows by a hand-held stone require control of the hammerstone by firm precision grips which assure both retention of the stone in the hand and fine adjustments in its orientation by the thumb and fingers. . . . A secure grasp and controlled maneuvering of stones by the thumb, fingers, and palm are facilitated by a unique pattern of hand proportions and joint-and-muscle configurations that permit cupping of the hand and the formation of a wide variety of grips." Among the most important new grips available to humans are those in which the palm is incorporated as a passive buttress, grips in which the fingers actively squeeze an object against the palm, and finger grips which maneuver the object as it is being used (precision handling with translation and rotation). The oblique squeeze grip has the additional advantage of extending the long axis of the arm, making clubbing or hammering more forceful.[11]

Marzke summarizes the effects of these changes:

> Effective manipulation of prehistoric stone tools required the ability to conform the hand to the shape of the natural objects and to tolerate the large stresses that accompany repeated percussion involving objects held, without hafting, in the hand. It is these activities, carried on for several million years, that must have molded the human hand. Tools today are being designed for a hand that evolved in adaptation to prehistoric tool use and tool manufacture.
>
> Most of the unique features of the modern human hand, including the thumb, can be related to the specific requirements of these grips for placement of the hand segments, exertion of force and tolerance of the stresses that would have been incurred with use of these grips in the manipulation of stone tools.[12]

Using the hands of apes and monkeys as a base of comparison, Marzke has identified specific modifications in hand structure that lengthened the list of grips used by man's earliest bipedal ancestors. Relative to the fingers, the human thumb is longer than that of chimpanzees; it is based on a widened saddle joint; there is a slightly altered contact surface between the second and third metacarpals (the hand bones supporting the index and middle fingers) and the trapezium, trapezoid, and capitate (the three bones that form the base of the index finger's metacarpal at the wrist). The com-

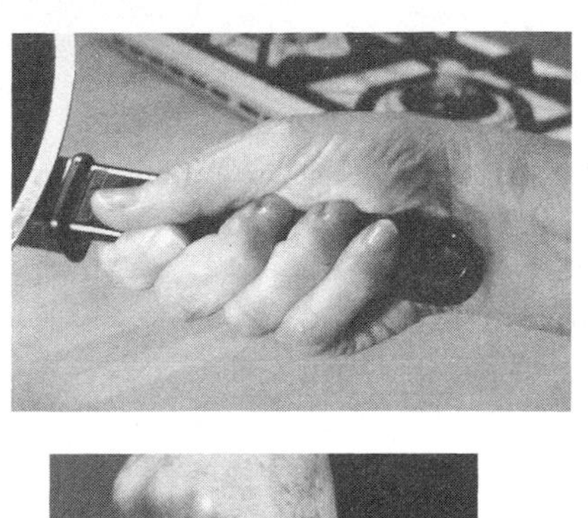

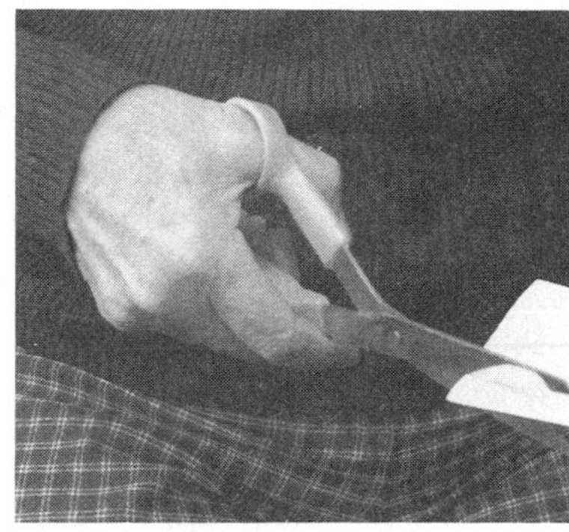

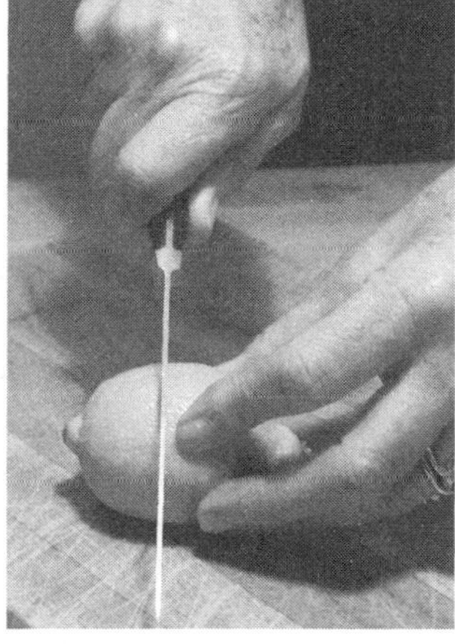

FIG. 7.2 Marzke had noted that in early stone tool manufacture, "A secure grasp and controlled maneuvering by the thumb, fingers, and palm are facilitated by a unique pattern of hand proportions and joint-and-muscle configurations that permit cupping of the hand and the formation of a wide variety of grips" (see text). The continued elaboration of tools and tasks has steadily increased human exploitation of this potential. *Top, left:* the oblique squeeze grip (identical to that used in hammering) is adapted to lifting, swirling, and flipping of the skillet and its contents. This grip also explains why, in a pinch, the skillet can be used effectively as an assault weapon. *Bottom, left:* The long flexor of the thumb presses the handle of the knife against the fingers, whose flexion locks the handle against the palm. Note that the extreme tilting of the proximal phalanx of the index finger helps align the knife with the long axis of the forearm; note also, the left hand has now substituted the table top for the thumb as a stabilizer in a power-precision grip, making rotation of the lemon by the fingertips much easier. *Top, right:* The thumb is moving in opposition to the fingers, but in this case contact between thumb and fingers is projected by the scissors to an object (paper) *outside* the hand. *Bottom, right:* Threading a needle is possibly the ultimate expression of the human precision grip, in which small objects are controlled between the tips of the thumb and one or more fingers. Note in all of these examples (except the holding of the skillet) that *both* hands make essential and complementary contributions to the control of objects in the hand.

bined effect of these changes is to make the radial side (the thumb side) of Lucy's hand very close to the human hand in its capacity to conform to a variety of object sizes and shapes, to grasp them securely, and to manipulate and hammer with them.[13]

This increasingly coherent evolutionary story is both satisfying and risky. The risk stems from its potential to charm us into supposing that events were moving inevitably in our direction—that some unseen pressure was gathering steam, and the steady accumulation of prehominid changes was preparing the way for us. While it is not only acceptable but obligatory to search for relationships between these events, any inference of purpose behind these changes—"preadaptations" leading to the morphology of the human hand, or to us—is unwarranted because we are telling our story from the vantage point of several million years of intervening history. It is nearly impossible to escape the corrupting bias created by what we already know. The truth is that the arrival of the opposable human thumb was a complete *nonevent* when it happened. It was a turning point from *our* perspective because—and only because—we happen to be here now to talk about it. As Marzke says:

> If the skills of humans were compared with those of other animals in performing human activities of daily living, one might discern evidence of a progression from primitive to advanced, with the thumb representing the summit of an evolutionary process. . . . However, the human thumb emerges as a compromise at one point in evolutionary time, a locomotor organ that has been transformed into a manipulative organ by stone tool use. It is a mosaic of primitive and unique structures, the functions of which are facilitated and constrained by its links with the rest of the evolving hand and brain.[14]

It would be difficult to know how to even begin to calculate the extreme improbability that such minor modifications in an old hand could somehow have triggered the sequence of events we now associate with our own history. In other words, if the creation of the hand was improbable, everything that had to happen since that time to capitalize on the hand was equally or even more improbable.* Sometime between the appearance of the australopithecines and the first *Homo* species, stone tool manufacture and use apparently came into regular practice: using Lucy and *Homo habilis* to bracket the emergence of this transitional (and transforming) behavior, that happened roughly during the span of time between 4 and 2 million years ago.

By almost 2 million years ago, *Homo erectus* had appeared; thus began the worldwide migration of our thumb, its owners and their stone tools. *Homo erectus* used stone tools, built enclosures, and used fire. This certainly looks like progress, or evidence of direction. But the more we know, the more the wondrous "preadaptations" of the hand appear to be precisely what Stephen Jay Gould called "an odd arrangement."† As James Shreeve, author of *The Neandertal Enigma*, writes:

> [The stone tools made by predecessors of *erectus*] were little more than sharp flakes or lumpish rocks with an edge whacked away. In contrast, *erectus* made a variety of more laboriously produced implements, typified by the symmetrical Acheulean hand ax.‡ These beautiful teardrops of quartz or flint enter the archeological record as early as a million and a half years ago, along with the first signs of *erectus* skeletons. The trouble—from an "all-important" standpoint—is that the species then goes right on making the same hand axes and other tools for the next million years. From this point of view, the whole Acheulean period associated with *erectus* represents not the arrival of a rich and resilient intelligence, but, as one archeologist has called it, a period of "unimaginable monotony."[15]

By the time of *erectus,* the hominid hand had been *stretched,* but, as has already been pointed out, we still do not know when its present architecture

---

* Richard Dawkins's latest book, *Climbing Mt. Improbable* (1996), provides the necessary check on unrestrained anthropocentric sentiment. For example, he explains how probabilities exclude certain adaptations we might deeply wish for—the ability to fly (without props, that is).

† You may want to take another look at the epigraph (by Gould) that stands at the head of chapter 1 ("Dawn").

‡ Archaeologists refer to the entire collection of advanced stone tools attributed to *H. erectus* as the "Acheulean tool industry," named for the finds made at Saint-Acheul, in France.

was completed or exactly why those modifications (especially the isolation of the long flexor of the thumb and ulnar opposition) were worth keeping. Although evolutionary theorists rightly caution us to curb our passion for finding purpose in evolution, there is no reason to be slack in our admiration of what actually exists. The thumb, like the shoulder, is an astonishing example of versatility realized through structure. Its most important movement is well worth looking at.

Two problems had to be solved in evolution to give the thumb its ability to move in opposition. First, it had to be made long enough to reach the other fingertips.[16] Second, its attachment to the wrist, and the muscles and tendons moving it, had to be modified to permit repositioning of the thumb so that the tip of the thumb actually *could* make pad-to-pad contact with the tip of each finger. This movement can be broken down into separate components (see Fig. 7.3):

- the first metacarpal abducts on the basal joint of the thumb;
- the thumb column (metacarpal and two phalanges) pronates during abduction, producing what Tubiana calls *anteposition;*
- the wrist extends, *projecting* the thumb column distally;
- the two distal joints of the thumb (the MP and IP joints) flex, bringing the tip of the thumb into contact with the tips of any of the four fingers.[17]

Since any developing grip might be used for power or precision (or some combination of the two) and since the size of the object to which the grip must conform is variable, the thumb may align itself in any of the possible permutations of pronation, flexion, extension, adduction, or abduction. Its ability to do this, of course, is as much dependent upon the muscular arrangements as on the placement of bones and ligaments and the orientation of the joints. The arrangement of muscles pulling the thumb can be pictured as being something like the arrangement of the long ribbons around a maypole. The thumb is not a rigid pole, however, and it is the balanced interplay of opposing pulls and the wavelike flow of forces circling the thumb that produce its complex movements. There are eight (sometimes nine) muscles attached to the thumb (Fig. 7.4).

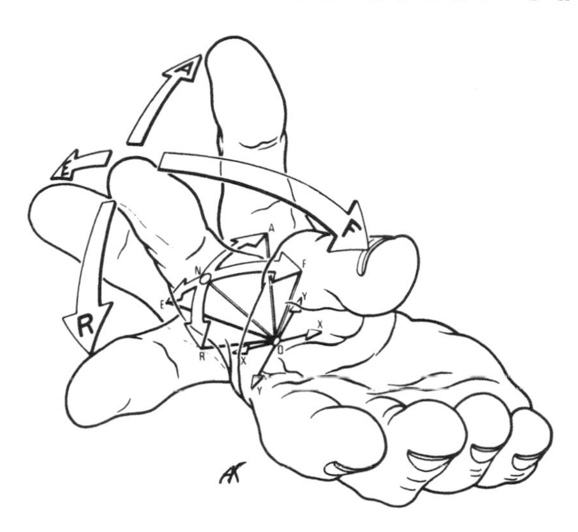

FIG. 7.3 The simplest movement of the thumb column is flexion and extension (shown by the F/E arrows). Movement of the column in a plane roughly perpendicular to the plane of the hand is more complex because the thumb also rotates (pronates and supinates) on its basal joint during this movement. Tubiana calls this movement (which would normally be called adduction and abduction) *anteposition* and *retroposition* (shown by the A/R arrows) to make explicit the movement of the thumb in relation to the plane of the palm. The most complex movement of *opposition* occurs when flexion and anteposition are combined with extension of the wrist, causing the thumb to *project* distally to a position so that its tip can meet the tips of the fingers. (From Tubiana, *The Hand,* 1981, with permission.)

Four of these muscles originate in the forearm: the tendons of three of these enter the thumb on its extensor (thumbnail) side and function primarily to move the thumb *away* from the position of opposition. (These are the long and short extensors and the long abductor of the thumb.) They help to shape the curve of the thumb and to widen its separation from the index finger so that the hand can accommodate larger objects; they also pull the thumb outward to stretch and flatten the hand (as when a pianist reaches for an octave); they participate in the "brushing" or "twiddling" movements in which the thumb tip sweeps from the ulnar to the radial side of the hand (as in the emblematic gesture suggesting that an exchange of money is needed to move things along); and, of course (in baseball), these muscles help the umpire inform a batter or a runner that "You're OUT!"

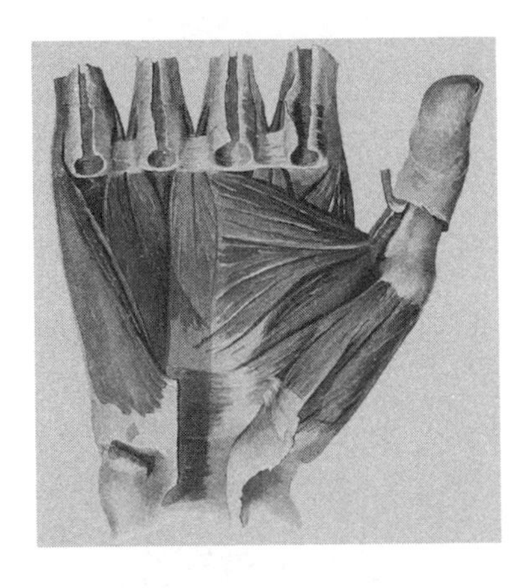

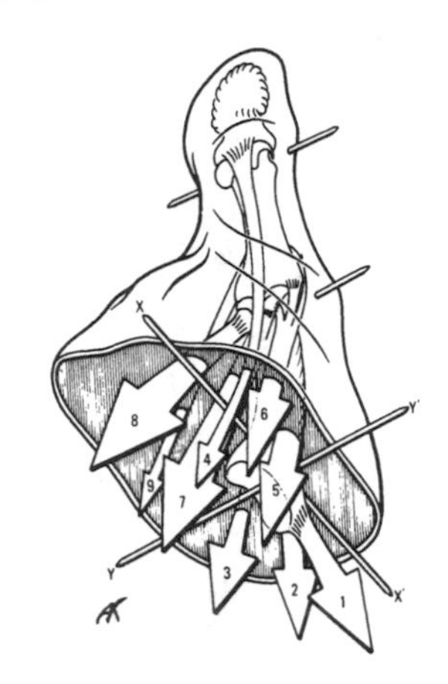

FIG. 7.4 The complex movements of the thumb result from the arrangement of nine muscles that enter the thumb from the forearm (extrinsics) and from the hand (intrinsics). In the diagram *above, right,* Tubiana shows these as separate force vectors arranged around the three joints of the thumb, all of which move independently in flexion and extension. The joint at the base of the thumb, a saddle joint, also moves in abduction and adduction, thereby permitting the thumb to rotate. (From Tubiana, *The Hand,* 1981, with permission.) *Above, left:* the actual anatomic arrangement of the intrinsic muscles. (From Werner Spalteholz, *Hand Atlas of Human Anatomy,* 1923.)

The fourth muscle arising in the forearm (the long flexor) enters on the flexor side (the pad side) of the thumb; it is the single most powerful muscle of the thumb. Its two main jobs are to help lock the thumb over the backs of the other fingers in an oblique squeeze grip (as in holding a hammer), and to aim the tip of the thumb straight at the tips of the other fingers and *hold* it there in a tip-to-tip opposition pinch grip. The remaining muscles of the thumb are located entirely in the hand itself (for which reason they are called *intrinsics*). Three of them are the secret weapon of opposition: one muscle (the *opponens*) initiates the move, and the other two (the *abductor* and *short flexor*) complete it, positioning the thumb tip so that the long flexor can apply force to the other fingertips. The fourth intrinsic is called the *adductor,* whose main role is to pull the thumb tight against the index finger, as it does in the side-to-side pinch grip.

The thumb is the only digit in the hand that has this freedom to rotate or swivel; it is also unique in that all of its movements can take place independently of those of any of the other fingers: as everyone says, the combination of strength, independence, and versatility sets it apart. Because of its unique capabilities, as we are about to learn, the thumb, if need be, can carry on a solo act.

George McLean was born in Gilroy, California, and lived on a small ranch there until he was about nine years old. His father, commander of the National Guard unit there, died when George was two years old. He remembers his childhood as being rather solitary.

> We lived outside of town, so I took the bus to school. I wasn't very social because I got up early, walked to the bus, went to school, took the bus and walked back home. There was no one within at least a mile to play with. I think because of the times and because my father was in the military I had a lot of lead soldiers. I used to build forts when I was eight or nine, and I remember that I didn't like pieces if they weren't in scale. I used to like making model airplanes, and I might even put an airplane in a fort I was building, but only if it was the right size. I would design whatever I was building ahead of time, then put it together; if a toy was not the right size then it couldn't be a part of the whole thing.

After his father died, an aunt from South Dakota came to help out, as George's mother was a nurse and not often home. His aunt stood in easily for the father he had lost.

> I followed her around from the time I got up until I went to bed. She was always showing me how to do things—*build* things. She'd been married a short time and had worked in a bank in North Dakota. She had had a homestead there which had started out as a tar-paper shack. She lived out in the country and rode a horse to town. She was used to raising animals, and when she came to live with us she took care of the ranch until it was sold, about four years after my father died.

The family moved from Gilroy to San Jose, California, where George remained through high school. He remembers enjoying high school shop classes because the work seemed "logical" to him. After abandoning an

Fig. 7.5 George McLean had been trained as an artist to draw using movements of the whole arm. This pen-and-ink drawing of a garlic, made after the loss of all four fingers of his right hand, shows that delicate and precise motor control is a potential capability of the entire upper limb.

attempt in junior college to major in both language and chemistry, he took an aptitude test and was told that he should consider taking courses in art. He followed this advice, enrolled in Stanford, earned a bachelor's degree in art, and then went to Los Angeles to study automotive and industrial design. After a brief stint in the navy he returned to Stanford, where he began work on master's degrees in both art and art education. During this period, in 1959, disaster struck:

> I was working on a home-shop do-it-yourself project and one day I reached down to turn off the machine—a saw—in the normal way, just as I'd done before on another piece of equipment. Then I realized that the power switch was on the other side, so I proceeded to reach down, underneath, and put my hand *through* the blade. Of course I realized immediately that I'd done something *very* stupid. As soon as it happened I closed my eyes, grabbed my hand to stem the blood flow, and got my brother-in-law, who was there with me, to put a tourniquet on and call the ambulance. While he was doing all of that, I asked him, "There's just one thing I want to know—do I still have my thumb?" He said, "Yes," and I said, "I *thought* I saw it. Okay."

He was thinking at that moment that if he still had his thumb he would be able to hold something; he would have the ability to grip. He *did* have his thumb, but the remaining four fingers on his right hand were gone. He was taken to a hospital, operated on, and the next day a surgeon visited him to explain what would happen next: he would have to learn to use his left hand. To begin training his left hand to take over for the right, the surgeon gave him "the old drill" of maneuvering a half-dollar across the fingertips.

> It took me half an hour to learn to do it in both directions without dropping it. In the second half hour I was able to do it with a dime. Breakfast that morning was a boiled egg in a cup, still in the shell. I was damned if I was going to call in anyone to help me, and I figured out how to crack that shell and get the egg out. I was pretty determined.
>
> Within a day or two I started to write with my left hand and I found that I could write upside down and backwards. The slant was slightly different but otherwise it wasn't that much different from writing with my right hand.

George wanted to be able to do everything with his left hand that he had been able to do with his right hand before. And that included drawing upside down.

> In school, to sharpen our ability to draw accurately and in perspective, we would design a car for someone so that it was right side up when the other person looked at it. It definitely helped us develop accuracy in spatial relationships. I should add that one of my favorite subjects in junior high school was typesetting. Typesetting is done upside down and backwards. I was very good at that.

After recovering from the accident, George worked as a freelance artist, lived in England for a year, and in 1962 began teaching art in a junior high school in Palo Alto. His students persuaded him to start a jewelry-making class, and very quickly something inside seemed to just kick in and take over. Three years later George opened his first jewelry business and he has been making jewelry since then—a period of over thirty years. In 1979 he joined the faculty of a prominent training school for jewelers, the Revere Academy in San Francisco, and later became its program coordinator. The first time I met George McLean was at the school, where I not only heard much about his work but was also able to see him operate with his own tools.[18]

> I became a jeweler about five years after the injury to my right hand. As soon as I started using my right hand again to grip, I realized how important my art training had been: we had been taught to draw using major arm muscles. We didn't even do tiny details with the fingertips. That meant that I was already more or less trained to hold on to something with my thumb.
>
> What I found is that the right hand is the tool holder and the left hand is the manipulator. You have to hold the tool exactly so that it does the same thing every time. You rotate the work to the place where it should be.
>
> If I'm doing very accurate work, I seat the hammer against the chest. I very seldom hammer with my left hand. It's hard because you need to grip the object you're working on. Learning how to control the tools used in jewelry-making can take as little as two or three months. And it takes in quite a few senses. It takes in the sense of hearing, as well as the physical sense of how a tool is touching something else. For example, with a torch you are concentrating on heating something without melting it, so you're visually focused tightly on what's going on there. But you are also sensing the sound that the torch is making. You can hear a change in the mixture of oxygen and gas, depending on where the torch is. Of course vision plays a very big part, but there is also the physical feeling when the tool you're holding is in the proper position and is making the proper contact.
>
> When you're filing or hammering, there's also the sound. Hammering on a piece of metal is like ringing a bell. A person needs to have feeling for all of those things. You know, tools are very sensual things, and *using* them can be. The filing, the polishing; drawing is very physical, sensual. Filing is almost like petting a cat.
>
> I've always found that making the simplest ring—just a tube, even—when I'm doing it and doing it well, I can feel high afterwards. I can just concentrate on that and shut out the rest of the world. I like to make things, and that's the way I've been since I was probably four years old. I

was always encouraged to draw, and there were always materials available and tools to experiment with and work with.

Most of the people I have trained already have a significant experience in jewelry. Whatever it takes to be competent, what I call tool sense, is probably already there. And it takes someone with a lot of patience, who has a good ability for self-criticism, is somewhat introverted, and is happy working alone and really focusing on what they're doing.

Bachleitner, Percelly, and Hall, each at a young age, demonstrated notable precocity in a specific physical skill, and their work as adults seemed strongly anticipated in what they were "naturally good at" as young boys. George, by contrast, did not discover jewelry until he was in his thirties. But one does not have to look very far beneath the surface to see that the strong interests and affinities he displayed as a young child converged as he became older, and that by the time he discovered goldsmithing he was, in a very real sense, already a jeweler.

From the time he was a young boy George loved to build, particularly elaborate models, always with constructions insistently made to scale. In doing so, as we shall consider in the book's final chapter ("Head for the Hands"), he was emulating the childhood experience of a group of highly successful professional sculptors.[19] He used drawing and sculpting tools to realize mental shapes and images, and he found working with tools to be an overtly sensual experience. He has a particular skill with small tools and almost always relies on his right hand to control them despite its lack of fingers. He enjoys a highly developed visuospatial ability in both his reading and his execution of drawings, and a very particular set of geometric images consistently serve as the raw material through which he exercises his strong desire to create. He is comfortable working in solitary and meditative settings, and he is a patient and imaginative listener and interpreter. George became a highly successful jeweler after an accident that on its face *should* have immediately and completely disqualified him from what became his life's work. His description of the path he followed to his career makes it clear what he still had in his possession, including both of his thumbs, after his accident.

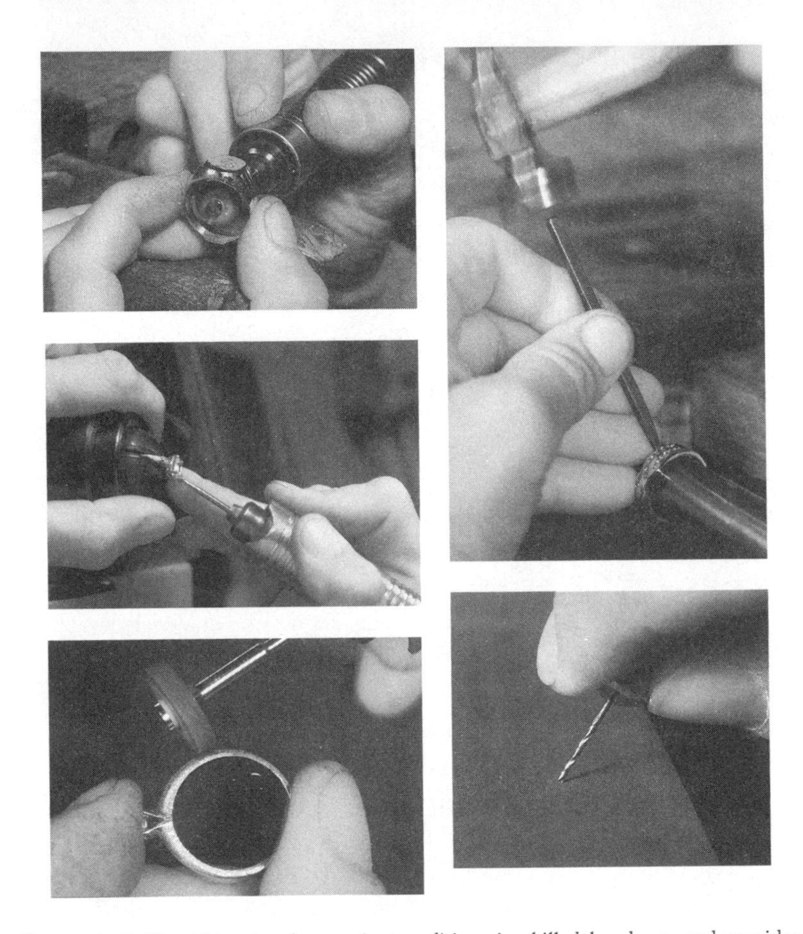

FIG. 7.6 Goldsmithing involves ancient traditions in skilled hand use, and provides exceptional examples of the exploitation, elaboration, and blending of the hand's capacity for precision and power gripping and handling. As these illustrations make clear, the addition of power tools further extends the reach of the hand's ability to work with small objects. *Top and middle, left* and *top, right:* Note the use of refined "bracing" by the palm and the tips of extended fingers. Dentists use comparable techniques in the repair of teeth, and when they fashion new teeth out of gold their work is indistinguishable from that of the goldsmith. Microsurgeons extend these same techniques to the manipulation of bodily tissues through the use of operating microscopes and the special tools of microsurgery, but continue to manipulate their instruments with unassisted precision grips (see Chapter 14, "Hidden in the Hand"). (Photographs courtesy Barry J. Blau, photographer, and Alan Revere. The lower, right-side image, showing the hand-sharpening of a small jeweler's drill, is from Alan Revere, *Professional Goldsmithing,* New York: Van Nostrand Reinhold, 1991, with permission.)

There are many possible reasons why George's constellation of skills led him to goldsmithing; the gradual maturation of his work into a lifetime career, which took on its own directions and meanings, is another question. After a number of conversations with him, I believe he is most drawn to work that permits him to transform important, and sometimes extremely private, feelings and memories of his clients into tangible symbols.

> Jewelry is often connected with ceremonies in life. That's why the wedding ring is such a mainstay for jewelers. I've found designing for an individual takes a lot of listening, trying to find out what they feel—often it's something very personal that's happened to them or to the partner or between them that they want incorporated. That can be magical, and also very secret.

In discussing his work, George referred to a "visual vocabulary," as if jewelry might be the accomplishment of an occult, nonverbal language. Based on his description, there can be little doubt that the pieces he produces flow from internalized rules for combining materials and structural elements in a way that acquires meaning and has come to bear his personal stamp. How does this process differ from a writer's, whose stories are borne by words joined according to a grammar that governs their relationships? Are the skilled hand movements of the artisan gestures that write their meanings into solid shapes, as the writer writes words onto paper? How far might such an analogy go? Indeed, might such an analogy find a counterpart in the actions of the brain itself?

George's story gives us three major issues to consider next:

- the transformation of two hands into a complementary partnership, with lateralized specialization of function (handedness, or hand "dominance") becoming nearly ubiquitous as a stable and lifelong behavioral trait;
- the transformation of the hand into an articulate organ of expression;
- the appearance of complex sensori- and perceptual-motor linkages of hand and brain, normally associated with task-specific learning, sug-

gesting the interaction of both hereditary and experiential factors in skill aptitude and achievement.

These issues bring into relief a fundamental premise of Darwinian thought, that structure and function are interdependent and co-evolutionary. The brain keeps giving the hand new things to do and new ways of doing what it already knows how to do. In turn, the hand affords the brain new ways of approaching old tasks and the possibility of undertaking and mastering new tasks. That means the brain, for its part, can acquire new ways of representing and defining the world.

Bearing that thought in mind, we turn next to the most intractably baffling of the myriad consequences of hand-brain coevolution, which cognitive scientists almost uniformly consider to be the most consequential of human prelinguistic adaptations. George set the stage for this part of our story when he said, "What I found is that the right hand is the tool holder and the left hand is the manipulator. You have to hold the tool exactly so that it does the same thing every time. You rotate the work to the place where it should be."

We turn now to the mystery of human handedness.

8

The Right Hand Knows What the Left Hand Just Did

THE DEVELOPMENT OF LATERALIZED FUNCTION HAS BEEN SUGGESTED AS BEING IMPORTANT TO NORMAL LANGUAGE, COGNITIVE, SOCIAL, EMOTIONAL, PERCEPTUAL, AND MOTOR DEVELOPMENT, AS WELL AS TO THE EARLY DETECTION OF CERTAIN KINDS OF ABNORMAL DEVELOPMENT AND DISABILITIES. THE TRUE RELEVANCE OF LATERALIZATION OF FUNCTION WILL BE ASCERTAINED ONLY BY CAREFUL RESEARCH THAT ACKNOWLEDGES THAT ASPECTS OF THIS FUNCTION DEVELOP THROUGHOUT THE LIFE SPAN, BEGINNING AT BIRTH, IF NOT BEFORE.

—Gerald Young[1]

THE HUMAN HAND HAS AN EXTREMELY LONG HISTORY both as an implement of social interaction and as the object of social attention. Nuances of meaning not conveyed by speech are communicated by gesture in every culture and language. Ritual codes pertaining to membership, rank, and duty are acknowledged and enforced through signs, salutes, and postures of the hands. Adornments of the fingers and hands (gloves, rings, tattoos, decorated fingernails) convey intimate and sometimes complex messages about personal identity and attachments. Social customs and meanings attach not merely to the hand but to the left and right hands specifically. Societies everywhere, it seems, have found (or invested) in this anatomic pairing so powerful and evocative a symbolism for fundamental contrasts in human nature that almost anyone, anywhere, can easily produce the list on the following page.

LEFT	RIGHT
profane	divine
sinister	benign
unlucky	lucky
impure	clean
gauche	adroit
slow	quick
weak	strong
clumsy	dexterous

While right-handers in our society may take this kind of polarization lightly, its real meaning and importance are clearly understood by left-handers everywhere. I was particularly struck by the story one of my own patients told me several years ago. He was a man in his early seventies, recovering from a stroke that had left him with a clumsy right hand. What caught my attention was not the immediate hand problem (he had in fact already improved considerably) but the far greater difficulties he had experienced with his hands during his early life. He was barely a year old when his father, an English merchant seaman, had detected in him a tendency toward left-handedness. Hoping to prevent a lifelong struggle with this "defect," the father had his son's left hand bound up in a cloth and directed his wife to oversee the corrective effort. My patient said he had no specific recollection of this period in his life but realized it later because of what happened.

Initially at least, his father's strategy succeeded, and by the time he was ready to begin school, he had become fully right-handed. Unfortunately, he was found to be a slow reader in school, and his penmanship was poor. His first teacher berated him for laziness, and he was required to stay after school to improve the appearance of his writing. Later that year his right hand was severely injured; he had tripped while crossing a street and fallen in front of a car, which ran over the hand. While his injuries might have been far worse, the damage to his hand prevented him from using it for several months. The silver lining to this cloud, of course, was that for the first time in his life he was permitted to write with his left hand.

Almost immediately his penmanship improved, as did the rest of his schoolwork. Since he had no recollection of his preschool conversion to right-handedness, the change struck him as miraculous. After he recovered from the injuries to his right hand, he continued to write with his left hand; his father was no longer at home, and his mother did not object.

At the beginning of the next school year, his mother moved her family to another village. His new teacher was a strict disciplinarian, who, like his father, was strongly opposed to left-handedness. Incredibly, for the second time in his life he was prevented from writing or holding any tools or implements with his left hand. By the end of that school year he was once again a right-hander and once again the object of ridicule because of poor performance.

He told me that the next few years were a time of growing frustration and alienation from school. Finally, when he was about ten years old, he ran away from home. When he was brought back, he was transferred to another school and placed with a woman teacher who was warm and sympathetic. Remarkably, she concluded that his school failures were due to the prejudice against his left-handedness, which she felt could and should be restored. Unfortunately, neither her ingenuity or perseverance in that attempt, nor his own eagerness to cooperate, were rewarded. The door had closed, leaving him at the age of ten with a makeshift and quirky ambidexterity, equally awkward with both hands. This situation never improved. He told me he was certain he would have gone much further in school—that he might even have earned a degree in engineering—if he had not had to contend with the early and intractable difficulties in reading and writing and with the deep feeling of alienation he developed toward school.

Even though our ideas about hand dominance may be unclear, personal experience teaches us that our own nondominant hand is somehow disabled compared to our dominant, or "good," hand. Most of us remember a left-handed kid who always seemed to be having trouble in school. Why should that be? Has something in the remarkable capabilities of the hominid hand been lost as a consequence of evolution? And if the contrary is true—that we have *gained* skill compared with nonhuman ancestors—why does only one hand seem to be the beneficiary? And what does schoolwork have to do with all this?

My patient's odd and unhappy story poses a set of related questions: Why do we almost always use the same hand for tasks demanding manual skill? Wouldn't it make more sense if we were ambidextrous? Even if there is good reason for one hand to be our preferred hand, why should that be the right hand for the overwhelming majority of us? And if handedness really does influence reading ability or intelligence, how does it do so?*

Since most of us discover and adapt to our own handedness early in life, we have little reason to puzzle over this propensity any more than we puzzle over the color of our eyes. But handedness *is* uniquely human, ranking with speech and tool use as a distinctive behavioral trait of *H. sapiens.* We and we alone, *as a species,* exhibit a strong preference for the use of the same hand—left *or* right—in a variety of manual tasks.[2] We also divide the labor between the hands in bimanual tasks in a stereotyped fashion, and are strongly inclined toward the exclusive use of a "dominant" hand in those one-handed tasks requiring precise control of a tool.

The ubiquitous *social* pressures to be right-handed reflect, embellish, and reinforce (as does the list above) what is fundamentally an expression of *biologic* influences and processes. And whatever its real impact has been and continues to be, the social stigma for left-handers simply diverts our attention from the real questions, which leap out at us from a list of right-handed versus left-handed attributes purged of insinuations concerning the manners and character of left-handers. This revised list describes quite specific performance measures by which the two hands seem objectively to differ, and calls our attention to something fundamental in human makeup that begs for explanation—our "manual performance asymmetries":

NONDOMINANT HAND	DOMINANT HAND
clumsy	dexterous
slow	quick
weak	strong

* A further, and unexpected, new question, has to do with longevity. Several recent attempts to determine the distribution of right- and left-hand dominance as a function of age unexpectedly found that almost no one over the age of eighty years is left-handed. Various experts are scrambling either to validate or to refute this claim, and (in the event it turns out to be correct) to explain it. For a review of this fascinating topic, see M. Peters, "Handedness and Its Relation to Other Indices of Cerebral Lateralization," in Davidson and Hugdahl, eds., *Brain Asymmetry* (Cambridge, Mass.: MIT Press, 1995).

Even this short list (which no longer asserts that the right hand is the "good" hand) may overstate or misrepresent real differences. But at least it gives us qualities that can be objectively tested. If these, or other, measurable differences could be shown to be significant and consistent, we could then decide whether or not to side with psychologist Michael Corballis, who has dubbed *Homo sapiens* "the lopsided ape."[3] All we know for certain at this point is that for a very long time both social and scientific views of handedness have amounted to little more than a stew of assorted myths and prejudices, into which chewy nuggets of serious scientific speculation or observation are occasionally dropped. We still know essentially nothing about the history of our special trait or its neurologic foundations. But human manual lopsidedness is fast becoming a major theme of inquiry in primatology, anthropology, genetics, neurology, developmental psychology, and linguistics.[4]

There is more and more evidence in biologic research that handedness may be nearly as old and as influential as bipedalism was in shaping human development and orienting our subsequent history. There is at least some reason to believe that right-hand dominance is rooted in primate bipedalism; that is, it may have begun when a particular line of apes preferring a terrestrial life on two, not four, legs began to define a new environmental (hence behavioral) niche for itself. After Ralph Holloway first suggested that throwing might have been an early stimulus to hemispheric specialization, neurophysiologist William Calvin offered a provocative solution to the emergence of handedness—in particular the extreme bias toward right-handedness—by suggesting that females were the principal hunters and that they hunted carrying their babies in their left arms, where the sound of the heart would soothe and quiet the infant.

In advancing this intriguing (but untestable) hypothesis, Calvin—like Holloway—sought an explanation of handedness that could be linked to the evolution of lateralized specialization in the *brain,* which is an important possibility to consider since it is now known that the left hemisphere (which exerts a powerful influence on movement of the right arm) could provide the temporal precision needed for the kind of complex sequenced movements that the dominant hand seems especially suited for.

Calvin was suggesting that habitual right-arm throwing, which would demand, or benefit from, an unusually accurate "clock" in the left hemisphere, is an ideal candidate for the behavior that made the clock—and the advantaged right arm whose precisely timed movements it supported—worth preserving genetically.[5] While he may or may not have been correct that Lucy and her sisters and daughters hunted with babes-in-arms (and in their *left* arms specifically), he is almost certainly correct that the explanation for human handedness lies in novel behavior or behaviors among early human ancestors that improved their chances for survival on the ground.

It is actually quite easy to see that the establishment of a lateralized habitual upper-limb skill could have conferred a critical survival advantage on the australopithecines in the competitive environment they inhabited and that it *must* have occurred if marksmanship was their basic weapon. This change would have required neither a nursing mother nor any new property in the primate brain: everything required was already there—including an environment indifferent to the outcome.

For most, if not all, primates, voluntary sequential movements of the limbs, no matter how complex, are executed with increasing fluency and precision the more they are repeated.[6] If the survival of early hominid groups depended upon hunters who were proficient rock throwers, then any groups whose hunters learned to throw habitually *with the same arm* would have gained a significant competitive advantage; such hunters (all other things being equal) would have most quickly optimized the path to speed, distance, and accuracy in throwing, which is what is needed for bringing down prey, or for outdoing or even eliminating the hunters of competing hominid groups.*

Of course, the establishment of a strong hand preference, left or right,

---

* They need not have deliberately adopted this strategy or conceived of what they were doing as practicing, in order for the behavior to have had this result. Obviously, if a skill practiced with a single limb could be transferred without further effort to the opposite limb, lateralization of skill would not occur. While some degree of transfer *does* occur (though not to the same degree in each direction) the effect is not strong enough to overcome the essentially exclusive benefit of one-sided practice to the limb doing the practicing. Apparently, transfer of at least one skill learned with the *dominant* hand shortens the learning time needed for the same skill by the *nondominant* hand, but not vice-versa! See U. Halsband, "Left Hemisphere Preponderance in Trajectorial Learning," *Neuroreport* 3 (1992): 397–400.

for *any* reason, even if unrelated to throwing, would have become an advantage in a throwing war.[7] In either case, the outcome would have been the same. Given the abundance of rocks, tasty targets, and rivals, and given also the sophistication and programmability ("plasticity") of the primate sensorimotor cortex—which came already wired to handle the special biomechanical array of limbs and pelvis with which Lucy was equipped—whatever ambidexterity our ancestors may have enjoyed prior to that time was likely doomed to extinction. At the very least it would have had to give way to at least one major heritable manual performance asymmetry: the overarm throw.* Note that this particular trick, then as now, would have served left-handers and right-handers equally well.

We cannot know what actually did launch the march toward handedness, but there is at least some evidence that another distinctive and important manual performance asymmetry had appeared among the hominids a mere million years after the time of Lucy (nearly two million years ago) in East Africa. That evidence has been found in collections of stone flakes generated by stone tool manufacture.

During a seven-year period beginning in the late 1970s, anthropologist Nicholas Toth examined stone flakes (fragments created when a stone core is struck with a hammerstone) recovered from research sites in Koobi Fora, Kenya, and Ambrona, Spain. The Kenyan site has been radiometrically dated to approximately 1.4 to 1.9 million years ago, when presumably both *H. habilis* and *H. erectus* were present there; the Spanish site is estimated to be between 300,000 and 400,000 years old, and its artifacts are attributed to *H. erectus*. Toth and his colleagues had devised a technique for distinguishing flakes created by a hammerstone held in the right hand from those struck by a hammerstone held in the left hand. Based on his own experiments with making stone tools (called "knapping"), he had learned that the surface patterns of individual flakes reveal which hand was holding the hammerstone during the strike. Regardless of the purported handedness of Stone Age knappers, when cores are being used randomly, the process will yield on average equal numbers of left-oriented flakes and right-oriented

* Why heritable? See endnote 7.

flakes (representing only about one-quarter of the entire crop of flakes, the orientation of the remainder being indeterminate). However, when a core is being repeatedly struck to produce multiple flakes in sequence, the preponderance of the crop produced by right-handers will be "right" flakes, with the contrary result for left-handers. In his own knapping and that of his colleagues, Toth had found that right-handers produce a 56:44 ratio of identifiable right-to-left flakes; when they analyzed over 1,500 flakes from Koobi Fora (the site dated between 1.9 and 1.4 mya), they found a 57:43 right-left flake ratio, identical to that of experienced modern right-handed knappers. Of course, these figures imply nothing about other possible manual asymmetries; nor do they indicate that handedness as we understand it existed during the early Pleistocene epoch.[8]

Why would handedness be an advantage in stone tool manufacture? Well, if training just one upper limb for aiming stones to be thrown toward a distant target was a good strategy for hunters, lateral consistency in holding and aiming hammerstones was likely to have been an equally good strategy for knappers. Holding a hammer or a projectile in the hand, taking aim, and then propelling the arm rapidly forward toward a target is complicated enough physiologically so that it demands rehearsal: present-day pitchers and diamond cutters alike spend a great deal of time working their way up to the big leagues, and they work long hours to stay in practice once they get there. A bull's-eye is a bull's-eye, whether the target is nearby or at some distance.

There is something else about this ready-aim-fire trick that should alert us to the special requirements for control it might present. During the preparation phase (the windup) the arm is "cocked"—which is to say, moved to a ready position, then forcibly restrained while chemical energy in the muscle is released to increase muscle tension. Just as with a bow and arrow, or a slingshot (propulsive analogues of throwing), movement onset must be delayed until enough *potential* energy has accumulated in the contracting muscle to permit rapid acceleration of the limb at "launch." Once the brakes are released and the move begins, the *proximal* limb muscles instantaneously relax.

But the *distal* muscles, which must keep the handheld object in place through at least part of the swing, cannot relax at movement onset. In fact the hand must *tighten* its grip during limb acceleration or control of the handheld object could be lost. In the case of a rock being thrown, the next critical control event would be a precisely timed release—a relaxation of grip—just a few thousandths of a second after limb release. If the rock was being used as a hammer, grip would have to be maintained and forearm and hand muscle tension set to a "shock-absorber" configuration. Imagine what an extraordinary control problem these tasks would present to the nervous system: in each case, distinct contraction and relaxation profiles would have to be created for the upper arm and for the lower arm and hand; in a certain sense, two segments of the arm would have to be treated as though they were not even the same arm.*

The control of an invariant sequence of interacting muscle contractions and relaxations requires nothing less than perfect regulation of the start and stop times of all these events, which in turn determine the movement of each of the involved limb segments, and which regulate even their contractile state through the course of the movement and at its endpoint. There are no points for near misses, and the penalties for mistakes can be quite high. *Apparently,* regulating the timing of all the necessary "on" and "off" switching is a large part of what is so difficult in perfecting movements that are rapid, brief, and (of necessity) extremely accurate. *Perhaps* the control problem is further complicated by the need to segregate commands sent to proximal limb segments from potentially incongruous or conflicting commands simultaneously directed to distal segments. Small wonder the move demands practice, or that the nervous system seems to "memorize" its own instructions to the muscles when the move has such a complex and—once perfected—inflexible activation scheme.[9]

Should it happen that improved control for a single skill provides a faster learning curve when comparable skills or movements are carried out by the same limb, then the probability increases for an uneven assignment of other jobs to the two sides. The probability also increases for the generation of a

---

* Although this description of a physiologic sequence is my own invention—and it is merely an educated guess—the notion of proximal-distal segregation of upper limb function is an idea that has been looked at in some detail before. For a review of one such proposal—and two rebuttals—see endnote 17 (the Healy, Peters, and Steenhuis papers).

special-purpose neurologic strategy being called up within the brain whenever any task is encountered that would optimally be solved with that strategy. The brain, in other words, might teach itself to look out for (or identify) all tasks with performance profiles *A*, *B*, and *C*, and simply "call up" routine *A*, *B*, or *C* to take care of business as the appropriate occasion arises. It might in effect create a "chip" (like a math coprocessor in a computer) to standardize and speed common operations of a particular type. Of course, any such "chip" would disappear with the death of its owner. But if such a special skill were to retain its survival value over the requisite number of generations of the species, the special status of members having a high level of that skill would favor increased expression and propagation of the gene in the population at large.[10]

If it is true that our handedness simply reflects the cumulative evolutionary effect of environmental demands and opportunities for a new class of upper-limb movements in the hominid line, we must ask ourselves, beginning with the australopithecines, what happened to the nondominant arm and hand in this process. Did it stagnate? Was it "dumbed down" somehow, in order to guarantee the emergence of a manual performance asymmetry? Or did the nondominant upper limb itself become specialized in some way that we have overlooked? It is worth recalling, in that regard, that, from the earliest days of informed neurologic pronouncements on the cerebral hemispheres, the nondominant hemisphere (which is most often the right hemisphere) has been treated as if it were superfluous. The contemporary view, however, is that the two sides of the brain function as a coordinated whole and that whatever differences there may be in their specific functions, they are complementary: it is a true marriage with a division of labor between left and right. Perhaps the same is true of the two hands.

Informal surveys of adults concerning handedness, using self-descriptions, usually produce estimates of approximately 9:1, right-handers to left-handers, with the proportion of left-handed males slightly greater than that of left-handed females. But serious attempts to define handedness precisely or to estimate its true incidence cannot depend upon the labels people apply to themselves. In 1971 R. C. Oldfield published a survey of handedness among 1,100 young adults based on ten questions about habitual hand use (including three tasks requiring the use of both hands).[11] The responses to this survey led Oldfield to propose a "laterality quotient" to describe the *strength* of handedness based on the number of tasks done with one hand or the other. Other studies of this kind showing comparable variability in hand preference or skill led researchers to infer that handedness exists along a spectrum from strongly right-handed, through ambidexterity, to strongly left-handed.[12]

Theoretical support for this idea came in the form of a theory that there is a "right shift" gene, which, while not explaining the behavioral details of handedness, provided at least a theoretical explanation for human variability in handedness. That explanation was compatible with both the statistics and the seemingly paradoxical occurrence of opposite handedness among identical twins.[13] By the early 1980s, however, a number of investigators had begun looking more seriously at the possibility that handedness was a more complex phenomenon than anyone had ever imagined.

The picture that emerges from the most recent work is that almost all of the longstanding generalizations are flawed, and that even the "enlightened" physiologic view of hand skill (dexterity) as a one-dimensional, lopsided trait cannot account for the facts. Students of the problem have agreed that closer scrutiny of handedness among nonhuman primates—particularly the relation of posture and reaching patterns to hand use—and studies of chronology of stabilization of hand use among children are needed.[14] An improved accounting of the relation of neurologic syndromes of altered hand function, or "apraxia," to theories of handedness is needed.[15] Needed most of all, however, is a comprehensive descriptive and analytic science of hand movements and use, and of the biomechanical and neuromuscular events responsible for skilled movement.[16]

The last of these specialized areas of investigation, kinesiology, yields *the* central question about handedness: How much do we know about the specific movements associated with tasks assigned to, or performed best by, the left and the right hand on one-handed and bimanual tasks? The answer is: almost nothing.

Following Oldfield, a number of researchers began to look at hand preferences and performance in *specific* unimanual and bimanual tasks. Initially, no attempt was made to describe the movements or to make inferences about the underlying biomechanics or neuromuscular contributions to those movements. But even at a superficial descriptive level, the fundamental distinctiveness of manual performance asymmetries was becoming evident, as was the extreme difficulty of making meaningful generalizations about handedness and manual skill in humans.

The principal strategy used to discover how dominant and nondominant hand skills differ physically from one another has been to ask how they group themselves among individuals; specific skills that are strongly associated with the same hand are likely, from the control point of view, to be different versions of the same motor skill or to have analogous control requirements. For example, the question might be asked whether writing, opening jars, and hammering nails are *always* done with the same hand in every individual. If they *were* (and they are not), these skills would be extremely likely to have similar requirements for strength, agility, endurance, and so on, so that an analysis of the entire collection would provide no more information about "dominant" neuromuscular performance than would the analysis of any one of them.

The most consistent results of all these studies can be summarized briefly:

- Writing (or drawing) is strongly related to a wide variety of skills demanding skilled use of small tools manipulated using a precision grip; self-definitions of handedness are invariably based on the "writing/drawing cluster" among both right- and left-handers.
- Throwing is strongly correlated with other lateralized "whole body" skills in which the dominant limb swings from the shoulder and upper-body rotation and opposite-side arm movements contribute to acceleration and provide counterbalance. Throwing arm choice is strongly correlated with same-side preference for kicking; grip strength is greater on the same side, and the thumb-tip is usually slightly wider.*

---

* Swinging a golf club, a baseball bat, and a sledgehammer are in a special class of bimanual whole-body movements, where normally the dominant throwing hand is placed closest to the striking end of the implement. But there is greater variability of hand placement in this class (and in less strictly one-directional stick use, like sweeping and raking leaves) than there is in throwing; as neurologist Norman Geschwind said, "Switch-hitters are more common than switch-pitchers."

- Among right-handers, throwing and writing (drawing) are almost invariably done with the same hand;* among people who write or draw with the left hand, approximately half throw with the right arm. When writing and throwing are dissociated in left-handers (i.e., left-handed writers), strength and thumb size are greater on the right side, and the right foot is usually preferred for kicking.[17]

An extremely important dissent from the inferior-superior and "just different" theories of hand performance asymmetries was published in 1987 by French psychologist Yves Guiard. Impressed by the degree to which bimanual skills are required for daily activities, Guiard proposed that a more realistic (and revealing) approach to this problem would be to conceptualize the two hands as a partnership. The question should not be which hand is dominant, but how the two hands interact, or complement each other's action in a given task to achieve an objective.

> If we think of a violinist, for example, it is rather obvious that neither of his or her two hands is dominant in any clear sense: Each of the two manual roles (grossly speaking, manipulating the violin, manipulating the bow) is crucial and difficult. To deal with laterality in the context of bimanual differentiated tasks, one definitely cannot resort to the current notion of hand preference, for the simple reason that no hand can be excluded from the task. At the same time, it would be arbitrary to say that these roles differ in difficulty.[18]

In writing—as consistent a unilateral skill as is known—Guiard showed that the nondominant hand plays a complementary, though largely covert, role by continuously repositioning the paper in anticipation of pen movement. Even when *no* movement seems needed (as in dart throwing), the passive hand and arm are probably crucial in counterbalancing the move of the active arm and hand.

---

* A right-handed writer who throws with the left hand is generally considered to be a biologic left-hander who was converted to right-handed writing, as my father was. His handwriting was completely illegible, but people always assured him that this was the sign of a real doctor.

Looking beneath complementarity as a principle, Guiard went on to show that the physical characteristics of movement, and the requisite sensory control mechanisms supporting each, were different. Specifically, movements of the dominant hand tended to be lower in excursion (amplitude) and faster in repetition rates than those of the nondominant hand in a given move. Said technically, the spatial and temporal scales of movement of the two hands were different, the dominant being "micrometric" and the nondominant being "macrometric."

Another important difference between the two hands in a shared task, after scaling differences, is that the nondominant hand (as when adjusting paper position in writing) "frames" the movement of the dominant hand: it sets and confines the spatial context in which the "skilled" movement will take place. In sewing, for example: "Unlike a clamp, the postural (left) hand ensures *plastic* stabilization, that is, produces steady states that are subject to frequent alterations thanks to which the position and orientation of the fabric always remain appropriate to the action of the right hand."[19]

Finally, said Guiard, the framing, stabilizing activity begins in one hand *before* the action of the other member of the pair. In summary, Guiard proposed, there is a logical division of labor between the two hands that appears to govern the entire range of human bimanual activities. For a right-hander: "Right hand motion is built relative to left hand motion, corresponds to a temporal-spatial scale that is comparatively micrometric, and intervenes later in bimanual action."[20] Put another way (and to paraphrase this chapter's title), *the left hand knows what the right hand is planning, and the right hand knows what the left hand just did.*

What is perhaps most intriguing about this conceptualization is that it maps almost perfectly onto the structural changes that distinguished the earliest hominid hands from their closest simian relatives and that created the opportunity for new *grips* and new capabilities in this hand. When John Napier proposed that prehensile hand grips be classified along a power-precision axis, he hinted (but only hinted) at the neurologic implications of such a particular organization of hand function, saying "it is in the elaboration of the central nervous system and not in the specialisation of the hand that we find the basis of human skill." We now must recognize that a major part of this neurologic elaboration has something to do with handedness.

Recall that Marzke found in early hominid hands a structural basis for three distinctive grips useful in stone tool manufacture: the pad-to-side grip (thumb pad against the side of the index finger); the three-jaw chuck (thumb, index, and middle fingers formed into a baseball pitcher's hold); and the "cradle" precision grip, in which the thumb opposes the tips of the other fingers to support the weight of an irregularly shaped object, and is able to turn it from below. Taken together, in fact, these three grips provide everything needed for *all* of the manual skills necessary for survival in circumstances where stones are the only raw material available for tools.

- The three-jaw chuck is used by the dominant hand for throwing and for hammering.
- The pad-to-side grip is used by the dominant hand for cutting, using a blade tool, and by the nondominant hand to maximally expose the surface of a larger stone core to percussive blows of a hammerstone while securing it against the force of such blows.
- The cradle grip is used by the nondominant hand to hold and manipulate stone cores.*

Thus, from the beginning of the terrestrial life of our earliest ancestors, a complete collection of the minimal requirements for a survival strategy based on the use of stones was present *for the first time* in a simian hand. Already at this stage, in addition, the habitual use of stone tools would have favored not only strong lateral preferences in hand and arm use (confining ready-aim-fire moves to a single arm), but the development of a division of labor between the hands, particularly in tool use, as described by Guiard. By the time of *H. habilis,* 1.9 million years ago, further modernization of the hand had taken place, the most salient change being more powerful precision handling movements. By that time there was a separate long flexor for the thumb, which added greatly to the power available to the thumb as a

---

* The cradle, or "precision handling," grip could also have been used to hold fruit or other pieces of food as they were being positioned for eating (think of yourself eating an apple); the pad-to-side grip could also have been used by the nondominant hand to move and hold anything being cut by a stone blade, just as a modern butcher would use his nondominant hand when separating meat from skin.

buttress.[21] While there are many unanswered questions remaining, one is of surpassing interest: Why is the comparatively new skill of writing so strongly lateralized? Why are we so inept at writing with our nonwriting hand? Among many possible explanations, two seem to me to be the best candidates and equally worth pursuing—and they are not mutually exclusive.

First, consider that from the standpoint of muscle synergy, many examples of dominant hand activity combine a mixture of "tonic" (slow, postural) and "phasic" (fast, brief, and usually repetitive) muscle contractions: maintaining pressure on the tool while at the same time moving the wrist and fingers in rhythmic sequences of alternating movements to produce stereotyped trajectories. This is quite a complex act because engaged along with forearm and hand muscles, whose job is a *sustained* contraction, are muscles that must contract and relax over and over again. The solution the brain adopts for overcoming the complexity of these muscular tasks is to automatize them—to create simple micrometric movements, memorize them, and repeat them without variation. Once the movement is learned, very little sensory monitoring is necessary.

By contrast, the nondominant hand must move in supportive anticipation of the action of the other hand, conforming its movements both to the behavior of an external object and to the actions of the other hand, to ensure that the object and the handheld tool will intercept at the intended time and place. The motor demand is for a wide variety of improvised hold and move sequences, carried out in patterns that are not as strictly rhythmic or stereotyped as those of the dominant hand. The sensory control of the nondominant hand requires accurate feature detection and analysis, and integration of that information with projections of opposite-hand movement.

In writing, as in stone tool manufacture, the dominant hand's performance is micrometric, rehearsed, and for the most part *internally* driven ("pre-programmed).* The performance of the nondominant hand is macrometric, improvisational, and *externally* driven.

---

* One explanation for the extreme tendency of writing to lateralize, and the reason why control transfers so poorly to the nondominant hemisphere, is that the underlying movements have become linked to another rather special operation of the dominant hemisphere: writing movements have been somehow "captured" by the "language chip." This intriguing possibility is

It is interesting to consider the task of handwriting in this context. While the path your pen traces on a paper does not look like a collection of identical hash marks, it turns out that this is essentially what writing comes down to. You can see the deep structure of handwriting, physiologically, simply by practicing row after row of connected circles (which is what children used to do in penmanship class). The conversion of those rows and rows of O's to real words and sentences happens when small variations in the *amount* of verticality or laterality of the movement are introduced into the movement. The extreme orderliness and predictability of individual handwriting is due to this mode of generating and controlling the movement. Take a good look at what your hand is actually doing the next time you write. For a good laugh, try slowing down the rate at which you sign your name—by about half—and see if you recognize your signature.

We still have not explained why these critical functions are most often controlled by the left side of the brain.[22] This, without question the paramount organizational distinction of the human brain, remains a mystery. It has been widely suggested that in earlier primate brains the right brain had already established a special capacity for visuospatial processing, and that therefore a different mode of neural processing (sequential as opposed to pattern-based) would most naturally come to reside in the opposite hemisphere. If that is true, it would seem to represent an extremely serendipitous compromise, because it would juxtapose a right-arm throwing advantage with a left-visual field sighting and tracking advantage—nothing could be more useful to a pitcher! Whatever the truth, when we landed on the ground, there was room for a new, comparatively low-level computer job (subcontracting repetitive, stereotyped movements of the hand and fingers) and that job was assigned to a (then) comparatively idle left hemisphere. Perhaps the low-level computer rather unexpectedly taught itself some new tricks and stole the show.

---

thought by many to explain the otherwise purely happenstance fusion of lateralized writing with lateralized language function in the brain's "dominant" hemisphere. Among the proponents of this idea is Doreen Kimura, who originally suggested that these functions were comparably dependent on sequencing operations. See D. Kimura, "Neuromotor Mechanisms in the Evolution of Human Communications," in D. Steklis and M. Raleigh, eds., *Neurobiology of Social Communication in Primates: An Evolutionary Perspective* (New York: Academic Press, 1979).

9
Bad Boys, Polyliths, and the Heterotechnic Revolution

NOBODY FIXES ANYTHING ANYMORE. NOW YOU JUST BUY PARTS AND CHANGE THEM.

—Jack Schafer,
in his garage, May 24, 1993

FOR THE LATE-ARRIVING VISITOR to the Decade of the Brain—now winding down, or just cooling off—prospects for a pleasant sojourn are not good at the moment. Something akin to the original Tower of Babel seems to have come into existence, with many tongues being spoken and very little communication taking place. We now insist that we will never understand what intelligence is unless we can establish how bipedality, brachiation, social interaction, grooming, ambidexterity, language and tool use, the saddle joint at the base of the fifth metacarpal, "reaching" neurons in the brain's parietal cortex, inhibitory neurotransmitters, clades, codons, amino acid sequences, etc., etc. are interconnected. But it is a delusion. How can we possibly connect such disparate facts and ideas, or indeed how could we possibly *imagine* doing so when each discipline is its own private domain of multiple infinite regressions—knowledge or pieces of knowledge under which are smaller pieces under which are smaller pieces still (and so on). The enterprise as it is now ordered is well nigh hopeless.

The strong feeling that we are trapping ourselves in this kind of noisy blind alley came over me recently as I read through an extended critique of a fascinating paper published in 1991 in the prestigious scientific journal *Behavioral and Brain Sciences.* The paper, authored by Patricia Greenfield, a professor in the department of psychology at UCLA, was titled "Language, Tools and Brain: The Ontogeny and Phylogeny of Hierarchically Organized Sequential Behavior."[1] Not the most irresistible title, but it caught my eye, nonetheless, and it proved a rich and exciting paper to read. Taking the idea of human intelligence as a biologic unfolding (along the lines suggested by Jean Piaget, Lev Vygotsky, and Noam Chomsky), Professor Greenfield reviewed her own long experience observing children as they work on simple hand-oriented problems—eating (or just playing) with spoons, stacking blocks, nesting cups, or copying geometric patterns of sticks. She proposed that the human brain organizes and oversees the child's interactions with objects almost exactly the same way it organizes and oversees the production of speech. These two specific skills (manipulating objects and manipulating words), and the developmental chronology associated with the child's mastery of those skills, proceed in such transparently parallel fashion that the brain must be: (a) applying the same logic or procedural rules to both; and (b) using the same anatomic structures as it does so.

Among several tests Greenfield describes is one she first reported twenty years ago and which she believes not only unmasks a "silent" hierarchical rule generator (a "combinatorial strategy") in the human brain but gives us important information about how it works. This rule generator overcomes every child's naïve behavior so methodically and stereotypically that there just cannot be any argument about its presence or influence: the observed mental growth of a child follows the unfolding of a set of rules that manifest *in the brain* as genetically determined neural configurations and *in the child* as predictable behavioral propensities. How do we know that we know this? If you tell me how old a kid is, I will tell you how she will solve Greenfield's stick puzzle when we put it down in front of her.*

Professor Greenfield has been doing this for over two decades, and her findings are genuinely intriguing. Here is how one of her early tests works: a set of twenty identical small sticks is laid down on a table and made into patterns of connected boxes; the child is given a corresponding set of sticks

* The children given this task, as reported in the Greenfield study, were six, seven, and eleven years old.

and asked to copy the forms. No child does the test exactly the same way, but each age group goes about solving the problem in a way that Greenfield shows to be consistent with her hypothesis. At age six, no matter where the child begins the copying task, she always places a stick somewhere in contact with one end of the *last* stick she put down, if such a place is still available. Every such move extends an existing path and, without disrupting the physical continuities she is creating through her own physical action, brings her pattern closer to that produced by the tester. The sequence of her steps conforms to a simple, rather confining, but *successful* principle and implies that somewhere in her brain resides a rule for constraining the potentially unmanageable complexities of the task. The beauty of this rule is that she does not have to solve the problem all at once. She need only begin, and she can begin anywhere. The unfolding sequence of her moves defines, and in fact itself becomes, the solution to the problem.

Watching her movements but having no way of seeing inside the child's head, the observer nevertheless discerns the existence of rules guiding the child's hand and makes predictions based on that inference. When the prediction is fulfilled, the observer says—in effect speaking for the child—"*This is how I do it.*"

The theoretically naïve child moves along what Professor Jeanne Bamberger at MIT calls a "felt path."[2] In the backs of our minds we are thinking (as will every scientist baptized at the fonts of anthropologic and Darwinian thought): *This is exactly how chimps would do it; she is going through the chimp stage!* It is not a bad analogy, and it amuses us.* What *we* know that the child herself does not yet know is that she is about to turn on her afterburners and will leave the chimps and her own "chimp stage" in the dust. At this tender age she is still *unevolved* in her mental development.

The test becomes more interesting as it is administered to older chil-

* This is, for an extremely important reason, *only* an analogy. Humans may mimic but cannot "recapitulate" the behavior of animals from whom they are not descended. So although at some stage of development a human child may "think just like a chimp," this statement is equivalent to saying a bat "flies just like a swallow." The wings of bats and swallows are comparable in their exploitation of the same laws of physics to achieve airborne propulsion, but the evolutionary routes by which each arose are unrelated.

dren. The seven-year-old child no longer places each stick in contact with the last stick. She may jump around, working on one section of the problem for a while and then switching to another, generating a rough symmetry as she advances, never letting one area get too far ahead of the others. She proceeds in this fashion until she is finished. Even though no two children follow the same path at this age, their solutions overall tend to have this character. Whatever precise sequence they follow discloses the influence of an abstract, *hierarchically* organized, process that has pushed the felt-path controller aside and is doing things its own way. The seven-year-old, manifesting maturational changes in her brain, approaches the problem as an *architectural* one. Aided by an internal representation of the target pattern, she no longer looks to the sticks she has placed (and is still placing, one after the other, on the table) for the answer to the question, "Where next?" With the tester's pattern in her head, or in easy view, being used by the part of her brain that generates the movements of her arms and hands, this child is showing off her budding cognitive architecture. Hierarchical representations are reshaping both her perceptual processes and the control of her movements. Fasten your seatbelts because the afterburners are coming on.

An eleven-year-old child given this task matches the test (or "target") pattern accurately; yet as you watch her, you might worry that something has gone awry—the child's performance is not *orderly*. Greenfield, however, does not fret about this lack of order, which signifies to her neither a perceptual aberration nor a lapse in rule following; on the contrary, this child's behavior signals a critical advance in cognitive function. She has become an *improviser*. Her use of hierarchical pattern thinking is now so secure, so integral to her technique, that she has been set free. The eleven-year-old child is so well equipped for this task that her performance can be compared to that of a tour guide at the Pentagon, or even that of a great jazz player. She is so confident in her internal representation of the pattern of sticks—which will be scrutinized by her brain's Generator of All Possible Next Moves—that she can wander around inside the problem and goof off any way she likes. She knows what the whole puzzle structure looks like and where she is in it; she can even close her eyes and see all of that; and she knows how to get to the doors. She is now unequivocally behaving with *intelligence!*

Greenfield concludes:

> Evidence points to the linked ontogeny of object combination and sound combination programs in early development. . . . After about two years of age . . . language and object combination begin to develop more autonomously, each ultimately generating its own special forms of structural complexity.*

Let me briefly recapitulate our own felt path to this point. We have now considered, from a variety of perspectives, the origins of hominid divergence from simian roots; following Darwin's lead, we have focused our attention on the upper limb that was freed of its obligation to support the body against gravity, and on the new realm of exploration that that change offered to primates moving from the vanishing forest into the open and dangerous savannah. This hand could, and did, evolve under the influence of novel opportunities. No longer did this hand simply sport an opposable thumb: it transformed itself into an organ with a vastly increased manipulative range, with the capability to grasp and control objects of various sizes and shapes, powerfully or delicately—and constructively.

By including the hand in the question of how the human brain evolved as it did, we have allowed ourselves to see the true richness of possibilities induced by the mutual interdependence of structure and function in evolution. We have seen that the hominid brain, like the hand, was free to "exper-

---

* There is considerably more to Greenfield's paper, adding cross-cultural comparisons and a critical discussion of recent work on neurologic mechanisms related to language and object manipulation. We will consider these in greater detail in chapter 10 ("The Articulate Hand"). For more detailed information about object manipulation by children, see Greenfield and Schneider, "Building a Tree Structure: The Development of Hierarchical Complexity and Interrupted Strategies in Children's Construction Activity." *Developmental Psychology* 13 (1977): 299–313. As Henry Plotkin points out, the relation between object manipulation and the emergence of "causal understanding" is far from simple, as has been shown by studying the reactions of eighteen-week-old infants to situations that are physically "impossible." These studies suggest that human infants are influenced by a heuristic that leads them to expect visual coherence in the physical world. See Plotkin, *Evolution in Mind,* pp. 186–200. For an extended discussion implying other cognitive and perceptual operations of a primary heuristic—in this case, the innate sense of number—see Stanislaus Dehaene, *The Number Sense: How the Mind Creates Mathematics* (New York/Oxford: Oxford University Press, 1997), especially chapter 2, "Babies Who Count."

iment" under a new set of conditions, which, over time, favored a left-right complementarity of function in brain *and* hand. The structural changes in the brain that underlie its current functional complementarity were not merely initiated by but also profoundly *shaped* by survival strategies based on a division of labor between the two upper limbs. Throwing, as originally proposed by anthropologist Ralph Holloway, may have been the first step.[3] Toolmaking, if not the *first* step, was also among the earliest of human-like behaviors, whose consequences can now be seen not only in shifting hominid-environment relationships but in the workings of the modern human brain itself. Co-evolution, in other words, implies more than what we recognize as the multilevel interrelatedness of complex ecological systems. On their own evolutionary time scale, biologic systems can and will modify each other and themselves, and they can do so at any anatomical, functional, or hierarchical level. It is this open-ended, sometimes rapid, sometimes glacially protracted, experience-driven process of recursive molding and remodeling of organs, organisms, and organic processes that is meant by the term "co-evolution."

Professor Greenfield's work reminds us that any credible explanation for human intelligence and language must fit both the evolutionary and neurologic facts as we know them. She seeks to meet that test in two ways: first, by identifying brain mechanisms common to human language and skilled tool use and then arguing that this association is not accidental; second, by finding procedural analogies in human language and skilled tool use and then arguing that evolution has created in the human brain an organ powerfully predisposed to generate rules that treat nouns as if they were stones and verbs as if they were levers or pulleys. We will look at the neurologic side of this question in the next chapter; in the remainder of this chapter we will consider the suggestion that we humans are instructed (or constrained) by our genes to build sentences the way we build huts and villages. We will ask, to borrow the classic phrase of Ernst Haeckel, whether in human language and tool use "ontogeny recapitulates phylogeny."*

---

* Embryologist Ernst Haeckel was the author of one of the most fruitful of all hunches ever pursued in biology: that ancestral forms of a species leave their mark on the developmental processes of living descendants.

Although Haeckel had biological structure in mind, the ontogeny of *function* can look very much as if it, too, "recapitulates phylogeny" (as Greenfield and countless others have noted). When we observe a three-year-old child solving a puzzle "like a chimp," we are amused by the coincidence ("analogy"), yet she may in fact be engaged in a highly truncated but nonetheless authentic rehearsal of evolutionary precedent ("homology"). In this imagined progression of cognitive recapitulation, the same child, as a six-year-old, visits the cognitive world and capacities of Lucy; as a seven-year-old she manifests the wit and skill of early *H. erectus;* and at eight years she is perhaps blasting through the genius level for *erectus.* The cognitive evolution of a single human mind certainly does not follow such a literal reenactment of its ancestry, but that does not mean that the idea of such a developmental sequence is a fallacy.[4]

The anthropologist Peter Reynolds has also looked at this problem—the historical, or *phylogenetic,* relation of object combination to language and thought—as has Greenfield, and has constructed a tantalizing scenario for the early kindling and subsequent ignition of human constructional, intellectual, and cultural co-evolution. Like Dunbar, he credits social behavior (especially grooming) with providing not only the incentive but a stable rudimentary context for the exploration and enhancement of symbolic interpersonal communication. Unlike Dunbar, who associates brain and cultural advance with increases in social group size, Reynolds believes the *character* of the interactions taking place in groups was the critical stimulus to the emergence of human language and intelligence, and that small, cooperative work groups provided the context within which linguistic conventions were devised and perfected. The character of interactions is most closely reflected in the types of cooperation found among animals whose survival depends upon social organization.

In *symmetric* cooperation (Reynolds's term), group members share a task (usually hunting) and may commonly interchange essentially undifferentiated roles in pursuit of a common goal. Hunting packs do this, but assignments tend to be flexible, and once the prey is down these animals (unless they have an extremely rigid command structure) normally *stop* cooperating and fight over the spoils.

Reynolds believes that from the earliest periods of tool manufacture, beginning with stone implements, cooperation could not remain *ad hoc*, or "symmetric." He proposes that tool manufacture must actually have evolved as a social phenomenon. In his view, it would normally have been carried out in small face-to-face groups where the full assembly process depended on distinct contributions of the members, each of whom understood and could anticipate the contributions of the others. Reynolds calls this *heterotechnic* cooperation, and offers in support of the idea his own years of direct observation of chimpanzees, of New Guinea villagers, and of Australian Aborigines; the latter, over a span possibly measuring in the thousands of years, have not changed their system for manufacture of a "simple" stone knife.

> Each individual would anticipate what the other was about to do and facilitate it by performing the complementary action. In preparation of the thermoplastic to be used as the handles for the stone knives, one of the men emptied the spinifex seed from a modern plastic bag into a traditional wooden dish . . . and then smoothed out the seed to an even depth. [Then] the two of them lowered it onto the fire to heat. . . . One man handed a stick to the other, who in turn raked the coals. Then the other picked up a stick and joined in. [Later] one of the men takes the warm, viscous liquid and molds it into a ball. . . . Then the two men divide the labor into two tasks, with one of them molding the thermoplastic into roughly shaped knife handles, while the other gives them their final form and attaches them to the stone knife blades that had been prepared earlier.[5]

Aborigines are not simply "jacks-of-all-trades with crude stone tools." In deceptively primitive working conditions they actually demonstrate all of the basic principles of modern manufacturing: "task specialization, symbolic coordination, social cooperation, role complementarity, collective goals, the logical sequencing of operations, and the assembly of separately manufactured parts."[6] It is precisely this type of socially organized and cooperative division of specialized labor for the achievement of a specific goal that is *never* seen except in human tool manufacture and use. No matter how clever chimps may be—or how much they watch and learn or interact with one another—they do not cooperate in the construction or use of tools. Ever.

Reynolds believes that the social habits of nonhuman precursors help explain the development of cooperative labor. Specifically, attachment and exploratory role-playing (and swapping) establish the conditions for the more deliberate and more highly organized manufacturing behavior of adult humans. His own experience working in the computer industry reinforced his convictions: Silicon Valley designers and engineers conceptualize, design, test, and perfect electronic machines in a process that is indistinguishable from that seen in Aborigines in stone tool manufacture. It is in our nature to put our heads and hands together in this way. But there is more to the idea of heterotechnic cooperation than the transformation of longstanding social practices into a plan for cooperative community work strategies. To understand this thesis and its implications, we must also understand how the work products of humans, under this process, differ from those of nonhuman, tool-using animals.

Two terms make the distinction clear. The first term, "lith," refers to any subunit of a tool (or any other manufactured object), called a "polylith," whose components are mechanically joined. The second term, "pod," refers to the subunits of a multiple-unit object or structure, called a "polypod," which depends on gravity for its stability.

Both humans and nonhuman primates can make polypods, which can be small and simple structures: a stack of two stones; or large and highly complex: the gate of Mycenae. They can also be temporary and improvised. Reynolds gives the example of one of his own guides in New Guinea crossing a lagoon by turning two small canoes, a plank, and *himself* into a sort of makeshift catamaran, which he used to ferry the party and their luggage across the water. This *ad hoc* flotation device, as ingenious as it was, qualified as a polypod.

A polylith, by contrast, is an object consisting of any number of joined units, or subassemblies, that can be freely rotated without disturbing the structural or functional integrity of the object. Any hafted tool—a hammer, for example, which doesn't depend on gravity for its stability—meets this definition. Polyliths can withstand spatial rotations because the individual pieces have been attached to one another by what he calls a "join." If a rock is lashed to a stick to make an axe, it is by definition a polylith. It can be used as a hammer, or it can itself become the subunit of a larger construction. An Erector set, by this definition, could be used to produce polypods or polyliths, but a deck of cards by itself could never produce more than a polypod (i.e., a house of cards).

According to Reynolds, nonhumans may make tools and they may use them in sophisticated and ingenious ways, but neither the tools themselves nor any construction resulting from their use are ever polyliths. Chimpanzees may make quite elaborate *polypod* platforms, and indeed can be *taught* to manufacture simple *polyliths,* but are never observed to do the latter without some kind of interference or prompting from humans.

In what he calls a theory of complementation, Reynolds unites the principle of heterotechnic cooperation with the polypod-polylith distinction, and concludes that face-to-face social interactions essential to complex tool manufacture demand "the cognitive skills of causal inference and logical deduction."[7] To make that assertion real and up-to-date, we need to get over to Jack Schafer's garage.

Jack Schafer is a friendly man who owned, not so long ago, a picture-framing shop in Walnut Creek, California. Several years ago I wandered into Jack's shop to discuss a photograph I wanted him to frame. It was a photograph of Charles F. Bretzman, my mother's father. Just a century ago Mr. Bretzman, not yet twenty years old, had come to the United States from Peine, a little town in northern Germany near Hannover. He had traveled around the American West, doing odd jobs and some photography, and had finally settled in Indianapolis. Eventually Bretzman, a pioneer in the use of Cirkut photography, became the first official photographer of the Indianapolis Speedway and built a successful career as a portrait photographer in Indianapolis.[8] The photograph of Bretzman, a self-portrait, had prompted us to talk about photography and about car racing. Car racing, it turned out, was Jack's other passion. In fact, the walls of his frame shop were jammed with huge, stunning photographs of drag racers; while I was still staring at them, Jack invited me to his home to see a car he was working on.

About a week after receiving his invitation, I pulled into his driveway, stepped out of my car, and *fell* back into my own teens. Crouched right

FIG. 9.1 A complex polylith (1949 Mercury) draws the attention of American youth in a highly focused act of heterotechnic cooperation. This classic American photograph was taken for *Life Magazine* by A. Y. Owen in 1957. (Copyright © Time Inc.)

there in his garage, low, the panther of our youth, was a '49 Mercury. Bright yellow-orange painted flames lapped up over the front fenders. The beautifully rolled, heart-stopping chrome grille smiled its perfect, cool smile. Next to the open hood, resting on a stand and so brightly polished you had to squint to see it, was the engine. It would have been at home in Tiffany's window. I walked around the car slowly, conscious that the level of awe I was feeling had already shot well past what I once felt standing at Napoleon's tomb in Paris.

This car, it seems, has a history that is tightly bound up with Jack's own:

> I grew up in Red Bluff, California, and graduated from high school there in 1958. That whole time for me was just cars. In Red Bluff, as you know, there wasn't much to do, so at night you would go down and hang out at the gas stations downtown where you could watch the trucks go by. Main Street was the highway through town. Of course, everybody who worked

at the gas stations was interested in cars, so it all worked together. The kids at the gas stations had pickups and cars and all the other stuff that went with cars in those days. My father sometimes would let me have a car to work on, but I now realize he knew none of 'em would ever work. But he'd let me spend money and he helped me get parts.

A friend of mine had a pickup—a '51 Ford—and it had two carburetors. I'll never forget them. My friend would come by the house and we'd take the engine all apart and put it back together again just to play with those carburetors. So that was the first of it. In those young days I hung around those guys at the gas stations, and they'd let me do things—the grunt work. But if it had anything to do with those cars I would have paid to do it.

My family had a truck shop with steam cleaners, arc welders, gas

Fig. 9.2 Jack Schafer, working at the age of fifteen under the guidance of his father, lays the foundation for a lifelong love of engines, cars, and racing. Now very close to the age of his father at the time this photograph was taken, Jack runs a thriving business (*Flathead Jack* in Walnut Creek, California) restoring classic American cars.

welders, all kinds of jacks and grinders and equipment to do everything on trucks. And we had staff mechanics who just worked on trucks. I had all that, not to mention all the supplies—anything from bolts and nuts and cotter pins. Anything you could imagine you'd need was there. It was like a dream.

Jack remembers that he was so hungry to hang around the guys and their cars that when he was thirteen years old he taught himself to paint decorative trim on cars; he became a pinstriper: "I never really thought much about that, but it's really true. It was just a way to be around cars."

It was in the first years of high school, around '54, '55—somewhere in there. I remember my claim to fame was a race car from Reno, one of the first dragsters I'd ever seen. I went to the Redding drag strip, had all my paints with me, and I asked Bob Watts if he'd let me stripe his car. I didn't even charge him, I just striped it because I wanted my striping on it. Later on, that car showed up in a magazine with my striping on it. I think that was in 1955. Actually, I always had an eye for art. In grammar school when you were supposed to be paying attention I was always drawing. I drew pictures one after another by the ton. All kinds of stuff. Airplanes, cars, God knows what.

After high school I was in the service, and one of the first things I did after I got back was go to a racetrack to watch car races. That same day I *bought* a race car—spent all the money I had. That was in 1962.

Jack raced his own car for six years, till one day he found someone else, a man named Garvin Tonkin, who worked in a local bank, to drive for him.[9] From that time on he no longer drove in races; he just worked on engines, and the '49 Mercury became his specialty.

As soon as I concentrated just on the cars, everything instantly changed. I started watching what everybody else was doing. Then I started having ideas, like my father used to do, about making things better than they were before. Trying things that nobody else was trying, putting things together better than I had before. As my cars got more sophisticated, we started

Fig. 9.3 The 1941 Willys (*top*) had Jack's rebuilt 322-cubic-inch flathead Iskenderian 404A CAM, and the world record he set with that engine in 1962 has never been broken. At about the same time he started putting the same engine in a more modern dragster; while Jack concentrated on the engine, Garvin Tonkin did the driving.

moving up in classifications, too, to where we were setting world records. Like the engine that's in the car here—I still hold the American Hot Rod Association Drag Racing title with that engine. It was set years ago, and it's never been beaten.

I asked Jack how he looks at his early love of cars now, and whether he attaches any importance to the way his father taught him to use his hands.

> I think coming from the country's got a lot to do with it. In Red Bluff there's not much to do; it's not the intellectual capital of the world, so you do things with your hands. My father could make anything work with his hands. He could take anything that wasn't working and make it work better than it ever did before. He just had a knack.
>
> The racing bug kinda got to him, too. Anything I would do, he would go halfway with me, to give me the incentive to go on with it—I guess it was just one of those things fathers do. Myself, I just loved those cars and the sound of the engines and couldn't wait to get hold of them. I was never really competitive then and I'm still not. I just liked playing with the cars. And don't forget, there was always a crew—there was a male bonding deal there, I suppose—something, I don't know.
>
> Now, it's all different. Kids today are different. Cars are different. Nobody fixes anything anymore. Now you just buy parts and change them. Who wants to just change parts? That's all there is for kids to do now, just change parts. That's all it is. My father would *never* buy a part and change it. He'd just take whatever it was apart and fix it, whatever it was. People in racing today, they're the same way. You *buy* horsepower. They just *buy* these cars and go like hell. People who hot-rod take things that don't work and make them work better. That's what hot-rodding is all about.

Not surprisingly, since he so strongly identified with the kids who were hanging around cars and gas stations, Jack does not associate much of his impressive technical education, or his considerable success in racing, with what went on in school. The one interest that might have led to a connection with school was art, though he took just one art class in high school.

> My family always thought that art was where I ought to be heading. I probably would have done something with drawing but it all stopped when I got to high school and took my first class in art. The art teacher and I didn't get along—we locked horns from day one. She hated everything I did because I wouldn't do it exactly the way she wanted me to do it. Of course, I had to be different. Some of it was far-out stuff and she couldn't handle it. She was very regimented. So she threw me out of that

class and made sure I never got into another art class. But I always had a good eye for things, especially colors, and later on when I got into framing, all that just automatically came together.

While an extremely brief and highly selective review of events in the life of a single individual is certainly anecdotal, it is nevertheless striking how readily Jack Schafer's developmental history not only captures but unfolds and illuminates Reynolds's theory of complementation. As a teenager, Jack fell in love with cars and sought ways to associate himself with other boys who were involved with cars. He used his native artistic ability improvisationally, pinstriping cars well enough to gain attention and money, and, most important to him, to gain access to the world of cars. His artistic inclinations were so strong that they even survived the—what shall we call it?—*pinched* notion of art dispensed in a high school art class whose teacher visited her hostility on him with real energy.

Jack began to grow under the guidance of his father and his uncle, and in the presence of his friends—the "bad boys" who liked cars. It was here that the ingredients called for in the formula for heterotechnic cooperation were put in place. But there was something important at work here above and beyond a formula for the creation of polyliths. Jack did not simply participate in a succession of "face-to-face task groups exhibiting complementary roles instrumental to a common technical goal"; he was formally educated and grew up in these de facto classrooms.

It was nothing but play at first: his father let him work on cars beyond any hope of resuscitation, and at the same time he and his friend with the '51 Ford pickup kept taking apart and rebuilding the engine with the two carburetors, "just like it was a real hot rod." Later, in the real racing world, he formed a productive—literally, a *winning*—partnership with one driver, freeing himself to concentrate on the engines. He became so good at this that the two together won several world championships, and one of his engines still holds an unbroken world title.

The essential truth of these experiences—which failed to materialize within his community's formally constituted learning environment—is that Jack and his family and friends invented their own alma mater, the places and the people (peers and elders) with whom they shared their passion for engines and cars. It was through those experiences that Jack began a lifelong exploration, recapitulation, and refinement of the aptitudes, skills, and feelings that had first surfaced as a teenager's love affair with carburetors and horsepower. His affinities for mechanical things, for the roar and the beauty of an engine, led him to cars, the exemplary polylithic creation of American industrial and pop culture. Now, as an adult who has done many other things, he is engaged in an energetic and energizing rediscovery of what matters to him in life. He is back at the cars again.

Until now, we have not had a theory of any kind to account for the way Jack has lived his life, or to regard it as anything but another story, a human curiosity. I submit that his early and ongoing life tells us something important about humans, and it has to do (just as he says it does) with the hand and with learning. It also *complements* Reynolds's theory of complementation by explaining why heterotechnic groups have the appeal, the generative capacity, and the staying power they seem to.

The evolution of humans *in society* has apparently led to a high probability that there will be Jacks, Garvins, David Halls, and others who will assemble and expand their own elemental propensities in personally distinctive and meaningful ways. Many will be greatly aided in that process through the support they build for themselves in small groups of individuals who work, teach themselves, and socialize together because they care about the same kinds of things. In contrast to the solitary classroom teacher, obliged to present a curriculum whose logic and procedures are *by design* indifferent to the student's special aptitudes or interests (indifferent certainly to their *passions*), the members of these self-constituting groups are particularly alert and responsive to the possibilities for unique, personal contributions from their friends. Indeed, it is the mix of personal differences that afford individual members the incentive to learn and to prove themselves through contributions that are authentically important to the group.

Kathleen Gibson, a biological anthropologist at the University of Texas Health Sciences Center in Houston, commented at a 1990 conference in Portugal (Tools, Language and Intelligence) that "human intelligence results from delayed development and a consequent permanent immaturity of the brain." She did not mean that humans are doomed to an incurably juvenile mental life.[10] This handsomely crafted statement embraces both Stephen Jay Gould's complex schema of neoteny and Henry Plotkin's postulate of secondary heuristics. She also means, for nonacademics, that there is always the possibility that people, no matter how old they may be, may still be working their way closer to what they have in mind for themselves. A year or so after our initial interview, Jack decided that the frame shop was no longer what *he* had in mind for himself. He gave up the picture-framing business and started rebuilding hot rod engines full-time. He told me:

> My whole life is going back to cars now. Just to get my hands in there and go again, and it's all coming back to me. I'm more patient now. I do better work than I used to do. I'm a little older and wiser, and if things don't go right I don't force them; I back up and make them perfect. This car is perfection all the way. That engine's beautiful. Next I'm going to wire it. I hate electricity, but I've been doing a lot of reading on it and now that I've got it figured out, I can't wait to get started on that. Then the car goes to the upholstery shop and it comes out painted. Then I'll start another one.

In 1998 Jack finds himself in the middle of a thriving enterprise he had invented just a few years earlier, helping other hot rod enthusiasts locate one another and the parts and equipment they need to put these classic American cars back on the road. "This probably still wouldn't seem like art to my old high school teacher," he said, "but I think it is."[11]

10
The Articulate Hand

articulate (from the Latin, *articulare:* to divide into sections)

1. *verb:* **to divide** or separate—in sound production, to refine diction in speech, or to emphasize individual notes in the playing of a musical sequence; **to unite** by joins or joints; **to express,** utter, explain, give meaning or clarity to;
2. *adjective:* **fluent, eloquent, intelligible, organized, precise . . .**

In 1861 a French physician named Paul Broca proposed that spoken language depends upon the integrity of a specific region in the brain.* He did this by showing that the brain of a patient left "speechless" by a stroke had sustained damage to a small area of cerebral cortex, the posterior-inferior portion of the left third frontal convolution. Shortly thereafter—in 1874—a German medical student named Karl Wernicke argued in his doctoral thesis that a second kind of language impairment, or aphasia, could be caused by damage to yet another part of the left side of the brain, the posterior aspect of the superior temporal gyrus. With over 120 intervening years of experience in which to compile a catalogue of language impairments caused by brain injury (mainly stroke), neurologists still take it as axiomatic that, for the vast majority of humans, language is critically dependent upon neural activity in two separate but intimately connected areas of cortex in the left hemisphere (Fig. 10.1).

While exceptions to the rule are common (and a source of special delight to aphasiologists), the clinical picture of acquired language loss has changed

* This discovery was contemporaneous with the rapidly spreading interest in both Europe and America in phrenology, a folk "science" (authored by Gall and Spurzheim) that claimed that important human traits were localized in the brain and could be assessed by palpating the surface of the skull. Broca's discovery also followed by only two years the publication of Darwin's *On the Origin of Species by Natural Selection;* and don't forget that Sir Charles Bell's treatise on the hand had been published in 1833. We may have the toys, but *they* had the ideas!

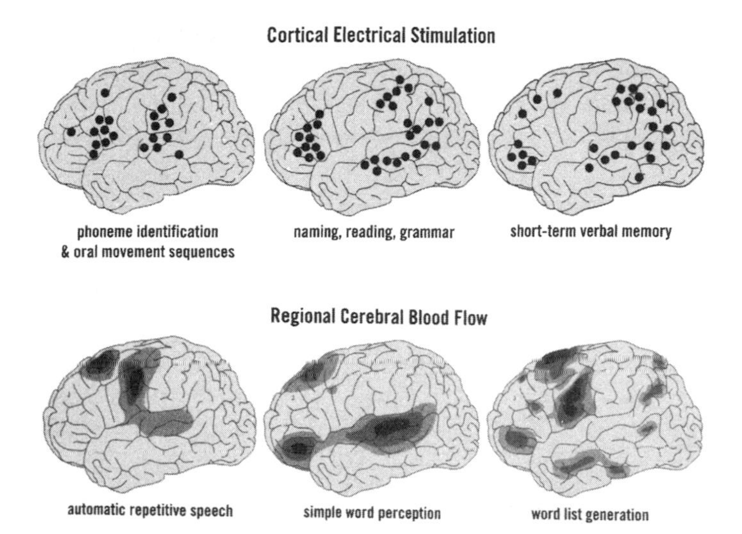

FIG. 10.1 The first suggestions that human language is critically dependent upon specific regions of the left hemisphere of the human brain were made in the mid-nineteenth century and were based on studies of patients who had lost the power of speech (and/or speech comprehension) as a result of brain injury. One hundred years later, sophisticated brain imaging techniques began to demonstrate just how complicated human language is from a neurologic point of view. Basic questions concerning the origins of human language and the relationship of *linguistic* structure to brain function remain unsettled. (The diagrams above are from Terrence Deacon's recent book, *The Symbolic Species,* W.W. Norton, 1997. Reprinted with permission.)

very little in the past century. Loss of tissue (a lesion) in Broca's area produces motor, or *nonfluent,* aphasia (speech and writing are impaired, while comprehension is normal or nearly normal); damage in the posterior portion of the superior temporal gyrus (Wernicke's area) produces sensory, or *fluent,* aphasia (the output of words and writing is normal or nearly normal in quantity, but semantic choice is defective and the words or sounds produced can be so abnormal that speech becomes gibberish, and there is usually loss of language comprehension, or "decoding" ability).[1]

Given the longevity and thoroughness of the clinical record, it is perhaps surprising to discover that there is absolutely no connection whatever between the information contained in this catalogue and the field of linguistic theory. Neurologists have been dutifully keeping the books on aphasia for over a century, and we haven't the first clue why specific lesions in the language areas of the brain produce the language disruptions they do or how these disruptions relate to the generation of language in the brain. This gap simply underscores the depth of *everyone's* ignorance concerning the origins of human language. We do not know, from an evolutionary point of view, how language came into being, and no one has even the rudiments of a theory as to how the *intact* brain in healthy individuals functions in relation to the formal operations of language itself.* Steven Pinker, in his inimitable fashion, puts our knowledge and our ignorance into this nutshell:

> A very gross anatomy of the language sub-organs within the perisylvian might be: front of perisylvian (including Broca's area), grammatical processing; rear of perisylvian (including Wernicke's and the three-lobe junction), the sounds of words, especially nouns, and some aspects of their meanings. Can we zoom in still closer, and locate smaller areas of brain that carry out more circumscribed language tasks? The answer is no and yes. No, there are no smaller patches of brain that one can draw a line around and label as some linguistic module—at least, not today. But yes, there must be portions of cortex that carry out circumscribed tasks, because brain damage can lead to language deficits that are startlingly specific.
>
> There are patients who have trouble with nouns for anything but animals, patients who cannot name body parts, patients who cannot name objects typically found indoors, patients who cannot name colors, and patients who have trouble with proper names. . . . One patient could not name fruits or vegetables: he could name an abacus and a sphinx but not an apple or a peach. . . . Does this mean that the brain has a produce section? No one has found one, nor centers for inflections, traces, phonology, and so on.[2]

---

*Just because damage to Broca's or Wernicke's area or to the connections between them will produce an aphasia does *not* mean that either area is what *produces* language, any more than the fact that a blowout of the left front tire will stop your car means that this tire is what makes your car go.

However poorly we understand its origins and operations, there is no dispute that the left hemisphere (in most cases) is "specialized" for language processing. The nearly universal location of essential language controls in one hemisphere—usually the left—is a central pillar upon which the claim of lateralized cerebral dominance among humans rests. Because of the apparent antiquity of other signs of laterality (manual performance asymmetries), it is assumed that this arrangement has been in place for a very long time, and that a primitive language must have existed and supported the worldwide migration of *Homo erectus.*[3] Hence it is widely believed that the neural infrastructure for language must have been in place by 100,000 years ago, which means that the brain could, or was "preadapted to," accommodate the irreducible minimum of two basic requirements of language; these are, in the official terminology of linguistics: (1) the arbitrariness of sign and (2) a discrete combinatorial system.

Increasingly, linguists are asking whether sign and gesture should not be located on a continuum of communicative movements. For example, David Armstrong and William Stokoe at Gallaudet University and Sherman Wilcox at the University of New Mexico have proposed a "taxonomy of gesture" with four levels: a primate (and primitive) level that would include common animal displays of aggression; a level of universal, transparent iconic gestures, like the "L" shape of the hand standing for a gun; a third level of gestures shared within specific cultures (like David McNeill's "emblem" for gaining the attention of a waiter); finally, a level of gestures used in both speech and ASL that are mutually unintelligible between hearing and deaf communities.[4] These fourth-level gestures meet the formal "arbitrary signs" requirement for language.

In a recent interview, psycholinguist Harlan Lane explained to me how hand movements in American Sign Language (ASL) compare with the common gestures that accompany speech:

> Sign includes movements beyond what is done in the hands—body shifts and limb movements and facial expressions—but still the core of it is in the hands. Yet it's very unlike what we jokingly call (mostly in New Yorkers) "talking with your hands." In fact, one reason it took so long to

discover that sign really is a language is that hearing people who saw it thought it was nothing more than an elaborate form of gesture. But it is unlike gesture.

One of the differences that strikes you right away—there are many—is that ASL (American Sign Language) is a code. That is, the relation between signs and their meanings is mostly opaque; not totally, but mostly, as with English. In English, "boat" doesn't have to mean *boat.* It could mean *boat,* but it could mean *tree,* if we had that convention. Likewise, certain rapid movements of the hands might not mean "weather." As a sign, they could mean *retreat,* or *sleep,* or almost anything. So signs have a certain opaqueness that gestures tend not to have.

Another reflection of the fact that ASL is a code is that everyone says the same thing in more or less the same way. Gestures are idiosyncratic, but (as with mime) normally demonstrative and easily understood.

The economy of ASL comes from its being a code. The basic content of the message comes down to conventional, codified units, so ASL is much more rapid than gesture. Suppose you want to say "scrambled eggs" in *gesture;* you could do it if you're really gifted—you could do a pantomime sketch, but it would take you a long time. In ASL it would take you about a fifth of a second to say "scrambled eggs."[5]

The second requirement of language that had to be met by the evolving brain was provision for a "discrete combinatorial system." It is here, apparently, where chimp language—or at least where the attempt to teach chimps to produce human language—breaks down. As Pinker explains, sentences appear on the surface to be word-chain devices. "Steven went to the store and bought a Rod Stewart record" is a verbal action path involving three nouns—a person, a place, and a thing—connected, or "chained," by two verbs. The sentence works quite well as a simpler chain reduced to just five lexical items: "Steven went store bought record." If you further simplified the chain by removing the tense markers you would have "Steven go store buy record." It doesn't strain credulity to imagine that a few very smart apes, or most recent immigrants to the United States with no previous experience in English, or practically *all* two-year-old children, would get the drift of this condensed sentence and, if they did, could probably reproduce it in some way. The hard part for them would be understanding the code values of the lexical item—what one meaningful combination (morpheme) refers to or what the words mean. The *easy* part would be understanding the straight-line verbal action path that reproduces the chronology of the *event* action path. Even a chimp or a bonobo with language training might get that.

But, as Pinker explains, it was Noam Chomsky's genius to perceive that word chains cannot explain how language really works. Language is built more like a tree, with limbs whose branching patterns support clusters, or subassemblies, of words. Words, therefore, are logically related to one another not on the basis of their order or proximity to one another in a chain, but on the basis of their association with specific branches. Think of a real tree. If you reach up and pull on a particular leaf, can you be sure that the leaf next to it will also move? No, because if the adjacent leaf is on another branch—that is, if it is part of a different subassembly and a different structural component of the tree—it will keep its place. So it is with words in sentences.

It is only because of the existence of this hidden structure of language (Chomsky calls it "innate syntax") that we would be able to generate and comprehend a sentence like this: **Steven,** whose car bore the marks of years of Boston driving and who, because he was looking forward to impressing his date for the evening, hoped to avoid any new dent-producing encounters in or around Mass Ave., **went** to a **store** where he normally did not shop because they never had anything he wanted, but with uncommonly good luck on this occasion—to his amazement—found and **bought** a Rod Stewart **record,** in fact, exactly the one he had wanted.

Steven went store bought record. It is still the same sentence. Indeed, it is still the same *chain,* and, conceivably, on a good day, even the celebrated bonobo Kanzi could separate the wheat from the chaff and get the meaning. But because of the tree structure, which supersedes the confining tendencies of a chaining strategy, we can juggle the order of our five critical words, and produce a shorter, simpler-*looking* sentence that the chimp would never get: The Rod Stewart **record** wasn't in the **store** that **Steve** usually **went** to, so he **bought** it somewhere else. *Record store Steve went bought.* Adios, Kanzi!*

Humans, because they have a language "device" that not only lets them substitute words for individual leaves but *tells them which branch each leaf grows from,* can sort it all out. Children normally can do this by the time they are four years old, give or take a year.

If it is true that clinical disruptions of language (the aphasias) do not conform to the linguist's conceptualizations, does that mean that linguists and neurophysiologists disagree about the nature of language? Not at all. It just means that we do not yet understand the physiologic correlates of language. The brain, in order to produce language as Chomsky characterized it, must first accomplish two purely *in-the-brain* tasks: first, it has to be able to group and label causally related events in the brain—in effect, sift through the noise and compose some of it into signals; second, the brain has to be able to compare those signals, turn them into a code it can read, regenerate, and play with without losing the original correspondence of internal and external events written into the code. This grouping, labeling, and comparing going on in the brain is what neurophysiologists *infer* must be going on in the brain to give it the competence to detect external events ("arbitrary signs"), translate them into equivalent in-the-brain electrochemical "signs," and then treat these internal signs as isolable units of language that will behave like a "discrete combinatorial system." But none of this neurophysiologic activity can be related to real language until it gains access to an input-output channel: in-the-body sensorimotor systems for detecting external events and generating bodily responses.

---

* My sentence example, by the way, illustrates Pinker's comment about the rough but tantalizing correspondence between linguistic and brain operations found in aphasic patients. Recall his informal map: "front of perisylvian (including Broca's area), grammatical processing; rear of perisylvian (including Wernicke's and the three-lobe junction), the sounds of words, especially nouns, and some aspects of their meanings." With that map in mind, let's go back to my sentence example. A patient with damage in Broca's area, losing grammar-processing ability, might be reduced to a very simple chain of the kind we allowed Kanzi to master: "Steven . . . um . . . store . . . um, *record!*" The patient with Wernicke's aphasia, losing sound-meaning coding ability, might say something like this: "He was over there stealing, I mean Steving, you know, I mean storing, which wouldn't be part of it anyway."

The in-the-brain part of this process is not unlike what you do when you go shopping for chairs for your living room. You leave home with an internal chair in your head: it can be a brown or a green chair, a soft or a hard chair, a large chair on its side or small chair standing upright, and it can be all of them together in all possible permutations, even though they are obviously not the same chair or even the same kind of chair; they are linked through a system of internal codes your brain employs to stand for them. Arriving at the store, you retrieve the chair-in-your-head and compare it with the actual chairs now stretched out before you in rows and rows across the showroom floor. Through a visual transformation, while you are still in the store, you can easily place any of these real chairs into various possible locations in your own distant living room. Likewise, you can also go home to your actual living room and "try out" the new candidate chair, which is now in your head. This is an example of an internal representational system, which is fundamental to human cognition. You detect, encode, sort, categorize, compare, store, regenerate, and recombine the actual world in code form inside your head.

Pinker thinks this process is at the heart of what is called "mentalese," or inner thought, and it is based on the *physical symbol system hypothesis* derived from Alan Turing's original description of an all-purpose representational machine.[6] Because I like the sound of it and approve of the inherent joint attribution to Chomsky and Turing, I will exercise poetic license in this chapter and refer to it, whether conceived of as an anatomically dedicated structure for language or as part of the brain's "cognitive architecture" being opportunistically employed for that purpose, as the "ChomTur," which should be given the same spoken cadence, and accorded the same respect, as the Raptor in *Jurassic Park.*

Whatever we call this device, and wherever and however it is constituted in the brain, since you have it, you are a candidate for human language. All you needed to get your ChomTur to work that way was to connect it to an input-output channel. Usually, but not always, that means ears and mouth—hearing and speech. (We will come back to the exception.)

The most persuasive evidence for the existence of a genetically determined or instinctive universal linguistic competence among humans, as Chomsky is famous for having pointed out, is found in language ontogeny.

Young children will acquire speech so long as they begin to *hear* it in early childhood. Formal language instruction is not required to set children irreversibly on this path; they get the idea of speech very quickly, and by around the age of two they begin to produce informative, complex, and grammatically *correct* sentences whether we ask them to or not. Chomsky argued that since speech is launched so readily in children, each child must of necessity have an *innate* capacity for language. This neurologic potential, waiting to detect, record, analyze, and reproduce whatever samples of ambient language the environment will supply, will unveil its full complement of rules and regulations once the child meets a mother tongue.

At this point in the development of our ideas concerning language, we should pause and remind ourselves of the need for caution as we try to examine separately the interwoven strands of language and thought without breaking their connections. At this moment, I think, it will help to reattach language and thought to what is happening to the child's body as the brain is preparing itself to learn about the world and to send and receive coded messages.

The attainment of early language milestones in the child always takes place in company with the attainment of very specific *motor* milestones, which are pursued and met by the child in response to the influence of another powerful developmental blueprint. The motor blueprint, or heuristic, can be tracked to the earliest stages of life, indeed, to events like kicking and thumb-sucking, which occur in intrauterine life. Once born, the new baby reaches for and touches her mother. Almost as soon as she is able to touch, she will begin to grasp. Within a matter of weeks, anything she can hold in her hand she will bring toward her body and will soon try to put in her mouth. If her mother places something in front of her, she will stop whatever she is doing and look straight at it. She will probably smile, coo, or babble, and then look back at her mother.

When she is one year old, when distinctive utterances are becoming more common, she can already hold small objects between her thumb and index finger and move them from one hand to the other. She can also pick up a block and let go when she has finished banging things with it. At this age, when she points or touches toys, pets, and other things, she is beginning to accompany her actions with sounds that quickly come to resemble human speech and that we call "baby talk." At one year she has also been crawling for a few months and is ready to try out her legs for standing and walking. This change in mobility will expand her territory very quickly, and within that territory she will be an insatiable investigator.

As she pushes past old boundaries, she will begin to attach names to all the new things she encounters, and her baby talk will quickly turn into *real* talk. As she expands her vocabulary, she will also expand her *use* of words. Oblivious to any distinction between monologue and dialogue, she will employ words to direct her own attention and the attention of others around her toward whatever interests her. Words will help her to specify what it is she wants to have, or to do, or what she wants her mother or father to do.

Pinker's book *The Language Instinct* follows by exactly sixty years the publication of what may be its single most important predecessor, *Thought and Language* (1934) by the Russian psychologist Lev Vygotsky.* Without the advantage of contemporary investigative neuroscience, Vygotsky and his contemporaries laboriously and meticulously compiled extensive observations of language development in children, from which they inferred the existence of a developmental process essentially identical to that pictured by modern developmental psychologists.†

> This is how [William] Stern translates first words: "The childish *mama,* translated into advanced speech, does not mean the word 'mother' but rather a sentence such as 'Mama, come here,' or 'Mama, give me,' or 'Mama, put me in the chair,' or 'Mama, help me.' "
>
> When we observe the child in action, however, it becomes obvious that it is not only the word *mama* that means "Mama, put me in the chair," but *the child's whole behavior at that moment* (his reaching out toward the chair, trying to hold on to it, etc.) . . . The only correct translation of *mama,* or

---

* The literal translation of his title is "Thought and Speech."

† Vygotsky's work also anticipated the interest of contemporary anthropologists in the relationship of evolving *Homo* tool use to the emergence and structure of modern language. Pinker acknowledges an interest in anthropologic theories of the evolution of language but, like Chomsky, is unwilling to credit them, based on what is now known.

> of any other early words, is the pointing gesture. The word, at first, is a conventional substitute for the gesture; it appears long before the child's crucial "discovery of language" and before he is capable of logical operations.[7]

All of the early observers of the development of language among children (Piaget, Stern, J. B. Watson) regarded the period between one and two years of age as momentous: somewhere between eighteen months and two years, the child makes a life-changing discovery: *anything* can have a sound attached to it. Armed with that magical knowledge, she attaches words to anything and everything, and demands to know which words go with which things. This eruption of mobility, gesture, and verbalism was seen to be an incomparable moment in human cognitive life and in the genesis of human consciousness. It is clear to us now that the unrestrained growth of vocabulary, beginning before the age of two, not only marks the beginning of concept formation but becomes its most powerful catalyst.* Says Vygotsky:

> The active search for words on the part of the child, which has no analogy in the development of "speech" in animals, indicates a new phase in his development. The child's speech ceases to be a signaling, and becomes a signifying function.[8]

The passion for words rapidly increases the child's knowledge of the world outside herself. She sees how many *different* things there are, and sharpens her eye for their distinctive features. Long before she understands what she is building for herself, words will have become more than implements of registration and control. Her words, washing over the world and

* The child behaves as if she has been given a labeling gun whose relentless operation propels her to take possession of the world by putting labels on everything she possibly can. Pinker (*The Language Instinct,* pp. 267–71) validates the bursting quality of this event: from about eighteen months though about eighteen years, a new word is acquired on average (and at a minimum) every two hours.

her own experience of object relations, will soon be leaping through the branches of Chomsky's syntactical tree; her improvisational sentences, like a schoolmarm who just moved into the house and started giving lessons, will rouse and sharpen her brain's capacity for conscious observation, discrimination, and recall.[9]

What is it that transforms the child's compulsive, chaotic sound-mapping of the world into the primary tool of her emergent powers of discrimination and intellect? It is the progressive commingling of speech with the whole constellation of objects, people, and real-life situations encountered in the child's life. It is the increasing influence of the nexus being built among action, thought, and language. For Vygotsky, "well-developed thought" arises as the verbal behavior of the child undergoes a long metamorphosis during which words that were originally object attributes come increasingly to be manipulated and combined, just as *real* objects are manipulated and combined by the child.

As Vygotsky interprets this process, when the brain begins to treat words as if they *were* real objects, it forms them into small groups, or clusters (like little piles of beads, or toy blocks). Speech at this stage has the random character of *ad hoc*, impulsive, or casual collections of whatever-is-around, and he terms this *syncretic* speech; very soon speech is formed into complexes, or chains, that reveal their logical connection in the child's mind—for example, a pile of blocks all the same size or the same color.

> In a complex, individual objects are united in a child's mind not only by his subjective impression but by *bonds actually existing between these objects*. . . . He no longer mistakes connections between his own impressions for connections between things—a decisive step away from syncretism toward objective thinking.

In word complexes, relationships are concrete and factual rather than abstract and logical. For example, while the young child may associate two people simply because they bring him ice cream, he will later associate people because they have the same family name. The relationships possible between words, or the relationships revealed by words, will trigger an advance to abstract (conceptual) thinking. At this stage he will associate two people as members of the same family when he learns that, *despite their differing last names,* he has been told that they have the same mother.

> We may say, therefore, that neither the growth of the number of associations, nor the strengthening of attention, nor the accumulation of images and representations—that none of these processes, however advanced they might be, can lead to concept formation. That is why the central moment in concept formation, and its generative cause, is a specific use of words as functional "tools."[10]

For Karl Bühler, it is the differences in tool use specifically that contain the most critical clues to the different outcomes of the interplay of language and intelligence in apes and in humans:

> It has been said that language is the prelude to the coming of man. That may be, but even before language comes *thinking in terms of tools,* i.e., the realization of mechanical connections and the invention of mechanical means for mechanical ends. To put it briefly, before the advent of speech, action comes to have a *subjective meaning;* i.e., it becomes consciously purposive.[11]

Vygotsky concludes that speech (or language) and thought are of different biologic origins, and that although both are present in other animals, their interaction is unique in humans.

> We find that in the child, the roots and the developmental course of the intellect differ from those of speech—that initially thought is nonverbal and speech nonintellectual. . . . [But, in humans] thought development is determined by language, i.e., by the linguistic tools of thought and by the sociocultural experience of the child. The child's intellectual growth is contingent upon his mastering the social means oı thought, that is, language.[12]

So, thought and language *begin* independently, yet language immediately begins to transform thought and intellectual growth. What, then, were thought and intellect *before* language arrived? As far as we know, or can imagine, thought and intellect are the sum total of the organizing tendency of the child's entire, rapidly expanding collection of passive and active interactions with the world via touch, smell, sight, hearing, and kinesthesis. It is probably impossible to be more specific than this, although perhaps it helps to imagine that during the first year of life the *primary heuristic* (Plotkin) is materializing in the form of a schedule of necessary developmental events in the child's body (which includes the brain), and that the child's experience of that sequence begins in the form of the dawning, here and now, of *episodic awareness* (Donald). At this stage the heuristic and the child's awareness are working at breakneck pace to organize the child's active movements and sensorimotor explorations.

The only additional point I would emphasize concerning this stage when the thought-language nexus is building is that something *very* important is happening in the hand itself. At about the age of one year the child's hands are rapidly becoming manipulative organs with fingers that will soon be able to move independently. The world of objects, and knowledge of the action of those objects, will increase rapidly, and distinctive actions which can be taken with objects in the hand will also increase. In other words, the thought-language nexus is becoming a hand-thought-language nexus. The child learns with real objects, by trial and error, to make constructions that are inevitably composed as discrete events unified through a sequence of actions. Playing with anything to make something is always paralleled in cognition by the creation of a story. Front→middle→back; beginning→middle→end; Steven→store→record.

The brain is not a passive witness to the expansion of these mechanical and sensorimotor explorations and accomplishments. It sweeps the process forward, and as it does so it defines its own procedures for regulating the flow of information generated by all the interactions that are taking place, and it models its processes and its formulations of the world on the narrative principle. It even, I suspect, creates an internal perceptual biological clock scaled to and calibrated by observable changes in the extrapersonal world—which means the brain acquires, and tunes itself to, the rhythms that the legs, shoulders, arms, hands, chest, tongue, and lips invent in their responses to the world. There is in fact strong neurophysiologic evidence that a specific part of the brain's motor system, the *supplementary motor area,* actually performs this function.[13]

This formulation has significant implications for the interpretation of events that will begin to unfold at or near the end of the first year of life. If real language is to develop, it will do so because of the transformation of babble under the influence of a number of processes going on simultaneously, all leading to a personal configuration of the child's "ChomTur" and its linkage to an input-output channel for language. No one knows how the final configurations, calibrations, and hookups necessary for the emergence of language occur, but it is inconceivable that the whole of the child's early sensorimotor experience is not involved. That is, the additional element of unplanned and unplannable experience through which the brain "learns to respond and generate its own reasons for behavior" (Plotkin's *secondary heuristic*) is necessary to the ontogeny of language in the child. If nothing else, the participation of secondary heuristic processes guarantees that a child will not surprise his French-speaking parents by answering them in Japanese sentences or with a Brooklyn accent.

With all these elements in place, incoming sounds that were previously part of background noise can be routed through the ChomTur, where those that keep coming back again and again, producing the same old neural signature (patterns and sequences of impulse) can be grouped and labeled in code and "stored" in some form. The process eventually can be reversed, so that the coded chunks can be retrieved, sorted, compiled, routed back into the input-output channel, and converted to sequences of motor commands whose execution produces sounds that go out to the world and come back in through the input channel for quality control (without which fidelity and continuity would not be achievable). Does language actually come into being and work this way? I don't know, certainly, but why wouldn't it work that way? Since we have no direct access to the ChomTur or its source codes, we will just have to imagine it.

Our next, and final, task in this chapter is to locate the hand *specifically* in the nexus of thought and language in living individuals. I will not slavishly repeat here what by now must be well fixed in your understanding: the hand is involved from the beginning in the baby's construction of visuomotor, kinesthetic, and haptic representations of the world and the objects in it. This is a profoundly important role for the hand, whose importance in both cognitive and emotional ontogeny cannot be overstated, but we have already discussed it. In addition, in the last chapter and earlier in this chapter we reviewed Greenfield's and Vygotsky's observations and conclusions concerning the influence of manual play and object manipulation in the early acquisition of language and cognitive skills in the child.

Against that background, we now take up the hand in relation to language and thought in two new and very different ways. First, we will examine the hand as a recently (re)discovered alternative to the vocal apparatus in language output. Second, we will discuss the hand in relation to what neurologists call "praxis," by which is meant "skilled use." At the conclusion of those two discussions it will be clear to you why this chapter bears the title it does.

Earlier in this chapter I quoted Harlan Lane's explanation of the phrase "the arbitrariness of sign." This was provided in the course of several meetings during which our main topic was not linguistic technicalities but American Sign Language. Lane, a protégé of the celebrated Harvard behaviorist B. F. Skinner, has been interested in the neurologic basis of human language since his earliest days as a graduate student. Strongly influenced by Skinner's work on verbal behavior, Lane founded the Center for Research on Language and Language Behavior at the University of Michigan, in Ann Arbor, and subsequently taught psycholinguistics at the Sorbonne, in Paris. He is now Distinguished Professor in the department of psychology at Northeastern University, in Boston, and the recipient of a MacArthur Foundation Fellowship for his work with the deaf. After he spent the first decade of his academic career working on speech perception and speech production, his ideas about these subjects, and about language and thought, changed suddenly. It happened in the early 1970s:

> While I was in San Diego in 1973 I ran into Ursula Bellugi, a psycholinguist whom I had met before. She took me to her lab, where there were some deaf people signing. While I watched, she translated into English

> what they were saying. It took me some time to absorb what she had shown me; Ursula explained that sign language is not a code on English—she said, "It seems to be a language. There are rules for making up words and rules for making sentences out of the words, but the rules have to do with space and shape—it's an entirely different way of doing language."
>
> I was really stunned. It was like being told there's another ocean that you had never heard of. After a few days of looking into it and digesting it, I began to realize that this meant that language was not about speaking and hearing, which had always been my assumption. It meant that the brain had the capacity for language, and if you can't put it out through the mouth, you put it out through the hands.

This was a revelatory moment of the kind that does not often come so accidentally to people in Lane's circumstances. When Bellugi showed him her deaf signers and shared her opinion that sign, *as they demonstrated it,* had to be a real language, it meant the language generator in the brain must be indifferent to the form and medium through which its messages are transmitted. Are we using Fed Ex? UPS? Who cares? It's just a *messenger service!* Nobody knew or even imagined that this might be the case until 1960, when William Stokoe first proposed that sign language had a grammatical structure in the visual-spatial mode comparable to that of spoken language.*

While he was still in San Diego and unexpectedly eager to learn about the history of the deaf, Lane learned to his amazement that there were no published works on the history of deaf people in the United States. He became very suspicious. "Remember," he told me, "that before the civil rights movement, before Montgomery and Selma, black history in this country was about Abraham Lincoln. I was pretty sure somebody was doing a number on these people."

He decided he would write the missing history, and because the movement to educate the deaf began in France, he and his partner, Frank Philip, returned to Paris. They began by collaborating on a translated collection of

---

* Stokoe's revolutionary insight advanced and prepared the ground for the rediscovery of a similar claim made by Roch-Ambroise Bébian in 1817. See endnotes 16 and 20.

publications authored by the French pioneers in education of the deaf.[14] The giants in this story, who jointly parented a rich apostasy of both theorizing and activism concerning the communicative and mental life of the deaf, were the Abbé de l'Epée, his protégé Roch-Ambroise Sicard, and *his* protégé (named by his parents in honor of Sicard) Roch-Ambroise Bébian. It is a story well worth recounting here in brief.

Epée was not the first to understand that the deaf can communicate effectively with one another through the use of hand signs; he was, however, the first to demonstrate that the deaf could be taught to communicate with the hearing and speaking world by learning to make signs for written French. Sicard took over this teaching method and established a major academy for the development and promulgation of deaf education based on sign. He advanced Epée's initiative with an effective but essentially slavish system for mapping hand gestures onto word equivalents, giving the deaf a signed version of written French. He believed that this technique not only would permit them to make their needs known but would be an essential tool if they were ever to learn how to think and to become educated.

Bébian, the third in this succession of giants, closely followed Sicard's lead but eventually came to an astonishing conclusion about sign language: that it lacks nothing in its power to communicate both concrete and abstract ideas. It was Bébian's 1817 publication *An Essay on the Deaf and Natural Language* that presaged the conflict in which Lane now finds himself embroiled.[15] As Bébian said:

> Sign language is different [from signed French]. In it intellectual ideas are always expressed clearly and easily. This assertion may at first appear incredible, but actually nothing could be easier to understand. Even the simplest external object is still compound; our idea of it is a combination of the qualities through which it can affect our senses. The idea of a peach is composed of the idea of its shape, taste, aroma, color, and even the tree that produces it. All these features provide us with signs that will give a rather good description of the isolated idea of a peach.[16]

The claim that sign is a legitimate language on linguistic-communicative grounds had been argued, documented, and published in France more than

150 years before researchers in modern linguistics were coming to the same conclusion in this country. While Stokoe and the San Diego group may not qualify as the real pioneers in this story, together they initiated the process that has now put the claim on solid neurologic footing. As you remember, in 1861 Broca showed that patients with damage to a particular area of cortex in the left hemisphere are likely to develop a form of speech loss (aphasia). Using the same analytic approach, the San Diego group discovered a nearly complete overlap of abnormalities among hearing aphasics and deaf people who had suffered breakdowns in sign language as a result of brain damage. It was found that ASL equivalents of Broca's and Wernicke's (and other) aphasias exist and that they are produced by nearly the same brain lesions that cause these forms of aphasia in patients with normal speech and hearing.[17]

This is *big* news. If a neurologist were to look at the brain scan of someone who had suffered a stroke and predicted from the location of the damage that the patient would be aphasic, it would make no difference whether the patient was deaf and a signer or previously normal in hearing and speech. This is the finding that left Ursula Bellugi, her colleagues, and other linguists (including Lane) in shock when studies of brain pathology in cases of aphasia in signers were first reported, and it is one that will generate additional research well into the future.

The final proof that sign is a "natural language" comes from the work of Laura Ann Petitto at McGill University in Montreal. In a ground-breaking series of observations and experiments, she has shown that deaf children acquire sign in the same graduated manner, and on precisely the same timetable, as hearing children acquire speech. Furthermore, among bilingual deaf children who are unimodal (two different languages, both in sign) and bimodal (one language signed, the other spoken), the developmental profiles for the two languages are identical. Finally, Petitto found that in the extremely rare instances in which *hearing* children raised by deaf parents are taught sign as a native language, the processes and milestones parallel those observed in both deaf and hearing children.[18]

Neurologic confirmation that the brain is indifferent to the specific input-output channel has both theoretical and practical consequences of profound importance. There are, of course, undisputed advantages to spoken language: even if a gestural form of communication preceded and provided the "innate syntax" for language, humans generally prefer having their hands free for work—even when they are gossiping. But a gestural basis for the deep structure of language would have profound implications for cognitive science—for our theories about how brain activity relates to thought and intelligent behavior—because it would forge a powerful conceptual link between gesture and praxis, and compel consideration of the possibility that these two manual functions are critically linked in the brain itself.

What of the current dominant theory that language represents the marriage of symbolic capacity in the brain to the expanded range of vocalizations that resulted from evolutionary remodeling of the human oropharynx? From about the age of one year, children produce vocalizations that, in very short order, begin to conform so closely to adult *verbalizations* that we deem them to be "first words." By the age of two years, these same children are producing vocalizations that are uniformly understood by adults to be words and *grouping* words into two- or three-word clusters. By the age of four years, words come in longer sequences ordered almost exactly as adults order words in grammatically correct sentences.

Given the phenomenal consistency with which these events occur in children, it is no surprise that language development has until quite recently been attributed to the powerful rewards children receive for their early word-making performances. There is, in fact, good reason to make such an inference. At the level of word-sound similarity, the child behaves like a parrot (or Terrence Deacon's remarkable seal, Hoover) that can detect and imitate speech sounds. So, one might conclude that at the word level, human language originates in the same basic auditory-vocal perception and imitation capacity found elsewhere in the animal world but builds on those elements and multiplies their effects, beginning at a very early age. Human speech, in this model, is seen as growing out of animal calls and communications and erupting into language because of the great size and computing power of the brain. Even if we cannot fully explain human language, we can conceptualize it as growing out of behaviors not unique to humans.

Human language, according to this version of the "continuity" theory of language evolution, generates meaning by a process of associations between sound and experience. A baby hears a cat's meow, begins to imitate that sound, and later, pointing at a cat, makes a meowing sound. Next, playing by itself, the baby picks up a small stuffed animal and again makes a meowing sound. "Meow" for this child has come to represent either the cat or something associated with the cat, and the utterance has become a word, no longer just a random parroting of an environmental sound but a vocalized sign "symbolizing" the cat.

But by the early 1930s, Vygotsky had already realized that a child's ability to speak could not in and of itself account for the origins of *thought.* Recall that Vygotsky said:

> . . . neither the growth of the number of associations, nor the strengthening of attention, nor the accumulation of images and representations—none of these processes, however advanced they might be, can lead to concept formation.

A quarter-century after Vygotsky wrote those words, Chomsky established that the child's long tutorial associating sounds with things and events in the real world could not explain human language. As we have seen, the heart of Chomsky's argument was that sentences in almost every language are built by the same rules, and children learn these rules without ever having to be formally taught them and without being able to explain why they follow them.* These rules are innate. So Chomsky took the Vygotsky claim one step further: sound-experience associations not only fail to explain the origins of thought but also fail to explain the origins of language.

If neither anthropology nor human ontogeny can expose an evolution-

* Another stunning accomplishment of an innate capacity to generate sentences is the provision of rules for severely limiting the possible combinations of sequences of words in a chain. Suppose I asked you to play a game called "Make a Sentence" and gave you the following eleven words: *daughter, son, I, my, my, dance, march, want, to, to, and.* There are nine different words in this sequence and two repeats. The number of possible sequences that could be generated from these words is about ten million, according to a mathematician of my acquaintance. Yet neither you nor I nor anyone else would be likely to generate more than about ten acceptable *sentences* from this sequence.

ary path to language via modifications in the human airway, why would linguists not propose the *hand* and its specialized use for communication as a prime mover in the evolution of language? There may be several reasons: first, very few linguists have personally adopted sign as a mode of communication, hence they cannot bring experiential knowledge to their conclusions about gestural theories of language origins;* second (as has been suggested by linguists who *do* sign), the prevailing assumptions in linguistics concerning the aural/oral origins of language have severely impeded the development of alternative explanations. Notwithstanding these impediments, however, interest in the possible gestural origins of language has recently been gaining serious support.[19]

Armstrong, Stokoe, and Wilcox, prominent linguists whose proposal for a "taxonomy of gesture" we encountered earlier, have presented a comprehensive case for the gestural origins of language in their new book, *Gesture and the Nature of Language.*[20] While their case cannot be reviewed in detail here, its most provocative assertions are important to note:

- the term "gesture" includes vocal-articulatory as well as limb movements, and is defined as "an equivalence class of coordinated movements that achieve some end";
- the visual channel is superior to the auditory channel for processing object relations in relation to a body schema and physical interactions with external objects;†
- syntax and semantics cannot be separated, and "embryo sentences are already inherent in simple visible gestures."[21]

---

* The anthropologists Nicholas Toth and Peter Reynolds both have strongly encouraged other anthropologic researchers to learn how to manufacture and use stone implements, since doing so can radically change one's view of the process and its origins.

† This component of the complete argument is attibuted to the work of George Lakoff and P. D. Deane, who have asserted that "grammar is ultimately spatial"; to the findings of a Scottish group (D. J. Perrett and colleagues) concerning single neurons in temporal cortex that are selectively activated by visual detection of movements associated with specific object-related tasks; and the work by A. Marshack relating language development to "the two-handed, vision-mediated problem solving capacity of a bipedal hominid." See *Gesture,* chapter 2.

These authors conclude:

> Hominid hands did shape tools for striking and piercing and cutting, did ignite and control fire, did fashion clothing and habitation, and did domesticate animals and cultivate plants. But with their hands and developed brain and greatly increased eye-brain-hand neural circuitry, hominids may well have invented language—not just expanding the naming function that some animals possess but finding true language, with syntax as well as vocabulary, in gestural activity.[22]

I welcome this assertion but would extend it in the following way: *praxis* refers to a heterogeneous class of movements in humans that, *by virtue of motivated planning and rehearsal,* exploit novel biomechanical (structural) modifications in the human hand in order to gain precise and extended control of external objects. Because of their intentionality and the precision (or stereotypy) that arises through rehearsal, these movements become iconic. *The praxic movement, in other words, is perforce a sign for the act which it accomplishes, irrespective of the communicative intent of the doer.* Indeed, as any successful basketball or tennis player will attest, high-level skill in reading the intentions of an opponent from his or her practiced moves is indispensable in a close contest, as is the ability of a player to *conceal* from an opponent the onset of one's own strategically critical moves.

This point brings us to the final piece of the complex mosaic that has been emerging in this discussion of the relation between the hand, thought, and language. It has long been known that aphasic patients frequently have difficulty with certain facial, trunk, or limb movements that were fully automatic prior to the occurrence of the neurologic event that caused the language loss. Such patients, for example, may have trouble buttoning buttons, lighting or blowing out a match, combing their hair, or hopping. The name for this disturbance of movement control is *apraxia:* the word means "loss of skilled movement," and it can accompany an aphasic syndrome or occur independently.

For a number of reasons, a neurologist may decide not to call an acquired impairment of motor skill an apraxia. The reason, almost always, can be traced to difficulties that arise when the attempt is made to separate the linguistic from the the gestural or praxic consequences of damage to the linguistically dominant hemisphere of the brain.

First, an aphasia can mimic an apraxia: the patient does not understand the neurologist's verbal request to perform a specific movement. If the patient has no idea what you mean when you say, "Strike a match," you cannot claim that he has lost the skill of striking a match. The rule is, you cannot diagnose an apraxia if the patient has impaired language comprehension. Second, if there is extensive damage to the brain, the hand most often used for doing the task may have become weak or may be paralyzed. In this case, it would be erroneous to conclude that the *internal idea of the movement* has been lost, any more than would be the case if the arm had been cut off. So here is another exclusionary rule: loss of skilled movement because of weakness or paralysis is not apraxia. In fact, the formal definition of apraxia has, from the beginning, been rife with exclusionary criteria: "a disorder of complex higher motor behavior, characterized by the inability to perform purposeful skilled movements in the absence of elementary motor deficits (weakness, akinesia, abnormal posture or tone), abnormalities of sensation, or impaired comprehension or memory."[23]

In their recent elaborated definition of apraxia, Geschwind and Damasio added a further qualification that is actually quite remarkable: "[The definition] implies that the movements are learned. Thus defined, apraxia is a disorder of the high level control of the execution of some classes of movement, *as requested by an observer in the verbal or visual mode.* [italics added] The movements fail to be carried out entirely or are performed defectively."[24] In other words, apraxia is *not* a symptom, a complaint, or a neurologic disorder. It is the failure to pass a test, just like failing to pass a driving test. The neurologist's test, by the way, is not nearly as complicated as a driving test. It is a test of your ability to use a comb, a toothbrush, to imitate blowing out a match.

Now, suppose you are a neurologist and your new patient is a chef whose stroke has left him unable to assemble the ingredients needed to make a cake, and unable to manipulate measuring cups, spoons, or baking pans, though he is not aphasic, weak, or paralyzed. If this chef did not respond to your verbal request that he strike a match, you could hand him a match to see what he would do with it. If he did nothing, you could demonstrate with the match to see if he would imitate you. If he still did nothing, you would write on your note that he was apraxic. Suppose, instead, he did *all* of the things you asked him to do on your battery of praxis tests yet was still unable to manipulate materials relating to baking. You would be obliged to say, perhaps, that he has a "skill-specific apraxia" and no other abnormalities. No one reading your chart would know what you were talking about.

Let's go back and alter the situation, such that when you examine him, the chef does *not* pass your battery of praxis tests. Your curiosity and persistence, however, lead you to improvise an extension of this battery. Now you place a birthday cake with unlit candles in front of him, *then* give him the match, and he rewards you by lighting the candles on the cake. *Now* what would you say? I think you would be obliged to infer that you had found a way around, or through, at least *part* of "apraxia" by tapping into a domain of his knowledge and experience beyond that of ordinary praxis or language—that you had unmasked the *chef* in your patient. With one notable exception, which we shall now consider, neurology has never formally explored these links among cognition, skill, and communication.[25]

It has long been agreed that there are two qualitatively different forms of skilled, representational movement, best exemplified by the differences between gesture and sign. But no one has ever seriously suggested that there might be a whole other class of representational movements—those imbedded in highly practiced movements which are *meaningful* but *neither gestures nor signs.* The one exception and serious candidate for this distinction, musical skill, is treated as if it were, like language, an isolated skill, instinct, or "intelligence." Occasionally it is referred to as "the language of the emotions."

But consider the following: almost anyone can *imitate* piano playing, as in the game of charades, by adopting the classic pianist's posture, then waving the arms and wiggling the fingers. Indeed, this kind of imitation has been so convincingly done in both filmed and live performances that the charade has gone undetected. But the genuine article—real expertise, musical skill—cannot be faked: what is communicated, and the feelings generated by the listener's understanding and active participation in communication, are unmistakable. If a neurologist included "show me piano playing" as part of his apraxia battery, he would be unable to distinguish a nonmusician's correct response from that of a trained pianist, since pantomime is all that is called for.[26]

Musical skill provides the clearest example and the cleanest proof of the existence of a whole class of self-defined, personally distinctive motor skills with an extended training and experience base, strong ties to the individual's emotional and cognitive development, strong communicative intent, and very high performance standards. Musical skill, in other words, is more than simply *praxis,* ordinary manual dexterity, or expertness in pantomime. Despite a budding interest in expert performance, arising within both neurology and cognitive science, there is still no real inclination to see musical skill as evidence of the existence of a whole *family* of physically and cognitively demanding, hierarchically structured, creatively rich human skills that (like sign) have communicative content and are "put out through the hand."[27] To the cognitive scientist and the neurologist alike, the highly trained, creative hand remains almost entirely unseen, and unrepresented in the thinking of clinical and theoretical neurology.*

Oliver Sacks, in *Seeing Voices,* agrees with Poizner, Klima, and Bellugi that signers may develop visual-cognitive "enhancements" in the brain, and asks if their experience does not lead them to "linguisticize" space or to spatialize language. What about that same enhancement in *anyone* (signer or not) who has devoted attention to and sought to improve the fluency with which he or she exercises a specific ("constrained") cognitive or sensorimotor skill? Would not, as I suggested earlier, carpenters, jugglers, surgeons, magicians, and so on, *necessarily* have developed such enhancements in the course of advancing their fluency in execution of a cognitively demanding sensorimotor skill? What do we imagine actually happens in the brain dur-

* The important exception to this generalization, of course, is found in the highly evolved investigative world of sports psychology. I make a distinction between the expert performance of a runner or a high jumper, however, and that of a painter, sculptor (for example), or dancer because of the extreme subjectiveness of performance measures for the latter activities.

ing a painter's lengthy apprenticeship and during her subsequent period of development? Why do we fail to see that any advanced physical skill has its own internal "logic" with specific minimal operational characteristics (limited input sources and output channels, a limited range and type of timing operations and calculating operations) and the freedom to be modified over time based on experience? Language and music fit but are not the only examples, and are not (in my view) diminished by this expanded description.

Sacks, in his typical renegade fashion (for a neurologist) began his recent extraordinary paper "Neurology and the Soul," thus:

> There has always, seemingly, been a split between science and life, between the apparent poverty of scientific formulation and the manifest richness of phenomenal experience.

In the body of his paper, he continues:

> Implied in all this is the necessity for an adequate concept of the individual and of mind, a concept of how individual persons grow and become, and how their growing and becoming are correlated with their physical bodies.
>
> The infant immediately starts exploring the world, looking, feeling, touching, smelling, as all higher animals do, from the moment of birth. Sensation alone is not enough; it must be combined with movement, with emotion, with action. Movement and sensation together become the antecedent of meaning.
>
> This evolution of self, this active growth and learning and becoming of the individual, is made possible by "selection," the strengthening of connections within neuronal groups in accordance with the individual's experiences (and needs and beliefs and desires). This process of selection cannot arise, cannot even start, unless there is movement—it is movement that makes possible all perceptual categorization.[28]

We have now come to a point where we can more fully sense the convergence of the neurologic, linguistic, developmental, and anthropologic perspectives in our search for an understanding of the role of the hand in human life. This merged perspective prepares us to examine more closely the role of the hand within contemporary social and cultural contexts. More particularly, it prepares us to consider how we do, or how we might, develop our own unseen, dormant, *knowing, praxic, inventive* hand and put it to our own personal and expressive use—to make it speak, and ourselves more articulate.

11
In Tune and Evolving *Prestissimo*

WHEN I WAS A STUDENT AT JUILLIARD, WE ALL PRACTICED FOURTEEN HOURS A DAY, AND WE KNEW THAT ANY TIME SPENT AWAY FROM THE PIANO WAS A WASTE OF TIME.

—Misha Dichter, *pianist*

GIVEN ALL WE HAVE SAID UP TO THIS POINT concerning the ancient origins of human hand structure and function, musical skill presents evolutionary theorists with a real puzzle. It is not difficult to invent odd and colorful ways whereby a musical performance—whether the blunt wielding of a musical instrument or an extreme accomplishment in digital dexterity—might have improved the odds for the birth of a musical child; but the question of music's seeming lack of biologic utility is both too interesting and too important to treat as a parlor game.

Musical skill is obviously highly correlated with, and dependent upon, refinement of hearing, and the links between auditory perception and human speech and language are so strong that the Darwinian brief for a "musical ear" could easily be argued by an eighth-grade science student. Music, of course, is *not* the same as speech or language, even though both are routed through the same input-output channels. Justine Sergeant, of the Cognitive Neuroscience Laboratory at the Montreal Neurological Institute at McGill University, put the music-language overlap issue this way:

> Music and speech have certain aspects in common: Both are used expressively and receptively; both involve sequential motor activity for their production; both are constructed of perceptually discrete sounds that can be

> represented in a writing system. . . . One can speak of a musical grammar in the mind of the composer, performer, and listener that in many respects parallels the grammar of language. . . . However, speech and music also differ in important aspects: A musical phrase does not convey the same sort of information that a verbal sentence does; it evokes feelings or emotions—patterns of body tension and release—rather than *referring* to specific ideas or objects.[1] (Italics added)

Perhaps musical behavior is an independent instinct but one not yet so widely distributed as language among humans—a complex adaptation with its own underlying ensemble of computational circuitry. As we have just seen, the possession of a specialized control mechanism for language—the ChomTur—just means that a set of disparate neurologic capacities (which, in hindsight, we can now label "behavioral antecedents of language") coalesced over time into a heritable communicative trait. Perhaps musical skill has a comparable evolutionary history: while no one was watching, Mother Nature kept selecting for improved versions of the "musical brain" until it was an adaptation, or (to exercise your growing vocabulary) a novel human heuristic.

But why would this have happened? Why would the brains of certain privileged subsidiary lines of *Homo sapiens sapiens* now come equipped not only with a ChomTur but with a BachTur as well? Where do we find the *selection pressure* that explains the descent of musicians? The real difficulty posed by that question, if we are to call music an instinct, is this: How do we explain the gifted musical hand? What in our ancestral environment ever demanded anything remotely approaching the speed, accuracy, and exquisite force and timing control exhibited in upper-limb (and particularly finger) movements essential to expert musical performance? As a musician might ask: Where did those *chops* come from? Without some nonmusical need or benefit deriving from this degree of manual proficiency, musical skill just looks like a bizarre accident, as if humans had inadvertently picked up hummingbird or centipede genes somewhere along the way. Or our ancestors really did need music to survive, and we have simply missed all the evidence for that.

Dr. Sergeant, who loved music as much as anyone, was stumped by this question and found herself constrained *as a scientist* to assign music to a category of behaviorally valueless, hence evolutionarily inexplicable, behavior:

> Music . . . is a creation of the human brain that made use of structures it inherited from evolution, and that were designed to serve biologically relevant functions, in order to develop and sustain a domain of activity as yet unheard of and *of no direct biologically adaptive value.*[2] (Italics added)

In the years since I first wondered about my daughter's piano playing: "How does she make her fingers go so fast?" I have had the opportunity to reconsider this question from the vantage point of a rapidly growing clinical experience with injured musicians, augmented by contacts with an extremely diverse mix of musicians and educators representing a wide cultural, stylistic, geographic, and experiential range of musical activities. Taken together, my own clinical work and the ideas gleaned from these many contacts have yielded a single revelatory lesson with implications reaching well beyond the world of music: "music," musical skill, and the musician cannot be understood without reference to the cultural elements that nurture individual musical development and shape musical interactions among people. Until I became familiar with the work of Henry Plotkin and Merlin Donald I did not understand why this was the case.

Any claim from biology that music has no survival or adaptive value is a claim easily disposed of by reference to Plotkin's theory of primary and subsidiary heuristics. Recall that a secondary heuristic is the product of a heritable capacity to generate real-world, life-term survival strategies; the specific strategies embodied in the heuristic are not themselves subject to genetic transmission, even though once in place in an individual they will act with the same force and stability as if they *had* been inherited. For humans living in society, it is contact with other humans that usually has the greatest impact on the establishment of secondary heuristics and their specific perceptual-motor domains. The individual is likely to accept a society's presets—its current tables of weights and measures for a given heuristic—or autonomously generate his or her own. A child growing up in France will normally speak French as his first language but would speak

Japanese (or ASL) first if that was the language his parents first exposed him to; he could acquire other languages later if he chose to.

Human language and music would both be examples of secondary heuristics according to Plotkin's conceptualization. Both are ubiquitous in human society, both have considerable utility to humans who adopt them, and both have pronounced effects on psychomotor development. Neither occurs in children denied exposure to others who demonstrate and use them, and survival into the childbearing years is possible without them. *However* . . . when developed as behavioral specializations, both can offer the individual considerable nonheritable *local niche* advantages, including the ability to preserve and enhance life by extending the working years of the individual beyond that brief period, beginning in puberty, when the primary heuristic is broadcasting just one message and jamming all the other channels.

When Sergeant said she could see no biologic relevance to music, she was speaking in classical Darwinian language. But if you put neo-Darwinian spectacles on and keep them on long enough to adjust to the look of things, the musical career and the mystery of musical ability make perfect sense. The concept of secondary heuristics adapts the classical Darwinian engine to the short-term scale of human life by allowing survival to be defined in purely cultural and personal terms. The human no longer has to depend on the judgment of biology to declare what has survival value, but by intelligent choice can calculate how best to take advantage of his or her own resources to contend with the present environment and circumstances. Plotkin's classification of intelligence as a secondary heuristic makes it clear why individuals interact so quickly and so powerfully with what is present in the culture, not simply in music but in almost any goal-oriented or culturally rewarded activity.

With respect to music itself, it is obvious that in our society there are two entirely separate musical cultures, the professional and the amateur, and that musical activities associated with each operate under very different sets of qualifications, expectations, and rewards. The professional culture, which is the primary focus of this chapter, has every characteristic of a Darwinian ecosystem, but the survival game is being played *prestissimo.* As a result, music technology, musical training, musical virtuosity—*everything*—is being pushed rapidly along channels defined by the culture at a very high rate of speed. The musical career of the classical musician places a premium on musical intelligence, biomechanics of the upper limb, and pedagogy. A hopeful musician must be guided to superior teaching at the earliest possible age and the high standards of professional musicianship, founded on heuristics locked from childhood in the fast-forward position, continue to be sustained by every musician's need to win and retain the approval and following of critics, audiences, and agents. In addition, every prominent musician must compete for attention not only with other performers, but with the wholesale *misrepresentations* of performance presented in commercial recordings, where every effort is made in the editing process to mask or deny the musician's susceptibility during performance to unscheduled deviations from plan.

Franz Liszt may or may not have launched the powerful fixations on performers imbedded in our own cultural biases about professional music, but he certainly did his part to assist: to Liszt is attributed the remark, "Le concert, c'est moi!"—*I* am the concert!* Liszt, of course, did not create the human brain or its motor system, human emotions, our affection for theater, or mob psychology. He merely noticed how easily these elements could be orchestrated to his own considerable advantage. He realized that the concert stage was ripe for conversion into a platform for the display of human prowess, and correctly guessed that people would come to concerts to see such a display as eagerly as Romans flocking to the Coliseum. Liszt, in other words, threw his hat—and the career of the musician along with it—into the Darwinian universe.

---

* The Canadian pianist Glenn Gould adamantly rejected this idea. A mere seven years after an extremely promising performance career began, Gould announced his departure from the concert stage, and for the remainder of his life presented his work in recordings that were openly (and heavily) edited. He maintained that live performance was little more than grandstanding, and seemed to be saying that the only honest way to be a musician was to stay away from audiences entirely and to strive for recordings that faithfully represented the composer's intentions. Although Gould died in 1982, his recordings continue to outsell those of any other classical pianist.

It is ironic that Liszt was practically an exact contemporary of Charles Darwin—Darwin was born in 1809 and died in 1882; Liszt was born in 1811 and died in 1886. The parallels in their respective effects on human thought and life are genuinely uncanny: each produced a radical reframing of existing descriptions and understanding of specific environments. Darwin addressed himself to the implications of competition in the entire world of biology; Liszt to a particular, uniquely human subset of that same world. Liszt recognized that music had already changed from an amateur pastime, done for the love of it, to a *profession.* The serious musician needed a lengthy training process and faced rising performance standards and audience expectations, but could look forward to at least some prospect of a living wage.

Although musicians had always spoken to and for people's feelings, the Lisztian declaration—a new formula for survival—profoundly transformed the relationship between musicians and audiences, and in so doing transformed the definition of musical skill. Classical performers now had to be capable of virtuosic performance because that was what audiences demanded. But audiences and other musicians were just part of the early Darwinian landscape of professional music, and many of the other elements had come into place two centuries before the celebrated Viennese competition of Mozart and Salieri in the late 1780s. Even then, music was being written specifically to evoke and prove the musician's virtuosity, and musical instruments were being modified to create and project dense and thrilling cascades of notes, rhythms, and harmonies.

> At the gigantic concert given in 1615 in Dresden by command of the Elector of Saxony, a certain Riposky from Cracow had a contrabass over 24 feet high carried on a wagon pulled by eight mules. A small ladder was built so that the neck could be reached, and on the strings of this monstrous instrument one ran an enormous bow with all the might of one's arm. But this contraption did not seem enough; later came the grandiose idea improvising a contrabass with the help of a windmill from which big cables were stretched; four men were needed to set them in vibration by means of a large piece of indented wood. On one side of the orchestra

> there was a great organ on which Father Serapion exerted himself with hands and feet; a few cannon in battery replaced the kettle drums. The performance was worthy of these preparations; the *prima donna* Bigozzi from Milano sang so well and so long that she died three days later.[3]

This vignette by the harpsichordist Wanda Landowska makes it clear that the professional musical culture created its own archaeological context for defining and then reciprocally *redefining* the fitness of the musician. Musical instruments were no longer merely the source of a particular expressive voice chosen by composers to accommodate the requirements of musical text, or visual ornaments adding to the aesthetic appeal of the music. Once Liszt proclaimed the beginning of the era of the soloist, pianos, violins, guitars, and other instruments would rapidly become potent selectors for physically adept musicians and a display-oriented musical repertoire.*

Predictably, the skills and the lives of individual concert artists came to be deeply imbedded in musical iconography. As audiences in increasing numbers heard the legendary artists—pianists like Liszt and Mendelssohn, the Schumanns, Rachmaninoff, violinists like Paganini, Kreisler, and Heifetz—these artists and the audiences who surrounded them forged the powerful mythology that now attaches to the classical soloist and engulfs those who seek to master any solo instrument.

It is remarkable how easy it has been to disguise the workings (and the accomplishments!) of the Darwinian engine on the musical stage. The musician, at once the principal agent of selection *and* its main target, bears no resemblance to a hunter or a warrior. On the contrary: the swaddling clothes of the music profession were laid on in the churches, and the musician-descendants of that heritage have continued to associate themselves

* The modern concert grand piano, for example, has its origins in the fortepiano, a harpsichord-like instrument that substituted a hammer that strikes the strings for quills that pluck them. The release action for the hammer had given the artist the means to control loudness ("dynamics") through finger action, or "touch," and of course demanded skill in doing so in order to control the loudness and the harmonic complexity of played notes. Subsequent changes in construction of the piano frame made it possible for the sound of the instrument to be projected greater distances, which meant bigger halls, larger audiences, and a bigger "gate."

and their music with the aesthetics of church and court, and with the self-effacing manners required of those in service to high estate.*

A very different situation exists in the culture of amateur and folk music. The noncommercial environment does not preclude serious work, competition, or high standards, but there is no Darwinian engine at work thinning the herd. Evolution still occurs, but it is a different process when the survival value of musicianship has been replaced by the musician's personal search for a musicianship that adds to the value of survival. The amateur musician of mature years may no longer be of use to the gene pool, but the meaning of life within piano, mandolin, and drumming circles does not end at that point.

After more than a decade of substantial involvement with musicians who have been sidelined by physical problems related to performance, it is now clear to me why so many individual musicians are in trouble: they are in trouble because the Darwinian world, even when its public goals and the private goals of its practitioners are loving and spiritual, is by definition completely impersonal and systematically ruthless. Musicians hoping for a successful career will push themselves to the limits of physical and mental endurance, and more often than not will work under conditions of extreme psychological adversity.†

It is a sign of our times and our culture that classical pianist Misha Dichter, participating a few years ago as a panelist at a summer conference in Aspen, Colorado, was offering his prescription for pianistic perfection

---

* Thinking about the situation in this way has tempted me to conclude that Victor Borge's enormous popularity stems in part from the fact that in his performances he is poking fun at more than longhair music. These shows are replete with slapstick routines in which the piano itself is the butt of the joke: Borge, in other words, assails not just the sanctity of the high priests of music but that of the altar itself.

† Younger musicians are rarely prepared for the realities of niche warfare left in the wake of Liszt's declaration. The benevolent disguise of the musical career continues to deceive both young musicians and the public at large. It is not difficult to see why the bloom has never gone off this particular rose. Music itself is a universal anodyne to life's harsh realities: it beckons us to courting and worship; it enlivens our anniversaries, cushions us in times of grief, and rallies us to displays of civic pride and unity. Music unites us, instructs us, and affirms our detachment from the impersonality of the biologic world.

to an audience composed mainly of doctors. His audience was beginning to realize, as growing numbers of both physicians and musicians now acknowledge, that the musical career is hazardous; that only a few musicians ever make it to the top; and that among the very best of the top performers a high percentage will experience significant and often disabling symptoms and injuries associated with playing. The more carefully we study the histories of the early giants, the more we can see how commonly such problems have occurred from the very beginning: this could have been the case with Liszt himself, who stopped giving solo piano performances in his thirties; Chopin rarely played the piano in public; Robert Schumann's hand, Glenn Gould's hand, Gary Graffman's hand, Leon Fleischer's hand, possibly even Horowitz's hand, all failed when these phenomenally gifted and successful pianists were at the height of their careers.

Several people living and working both in the world of musical performance and in the "performing arts medicine" world have helped me to understand more clearly the application of the Darwinian equation to the special situation of the professional musician. Two in particular have had a profound and continuing influence: Dorothy Taubman and Patrick O'Brien. Dorothy has been a powerful and controversial national figure in piano pedagogy for some years. Patrick is just as well known in the guitar world, although his work has not aroused professional feelings as Dorothy's has. Perhaps that is because Dorothy's domain is the grand piano and the grand platform—the sounds, the ceremonies, and the politics of the cathedral; Patrick's is the simpler habitat of the guitar and the lute—vespers and the gentle and reflective intimacy of the chapel.

I have spoken with both Dorothy and Pat often and relied on their advice about patients many times over a fifteen-year period. Dorothy proved more reluctant than Patrick to discuss the connections between her personal life and her professional work. Pat, however, was quite willing to be public about the roots of his lifelong relationship with music and with musicians, both as a performer himself and as a teacher. He has been thinking about the relationship between his own hands and his technique for a very long time; he has elaborated a remarkable approach to the instrument and acquired strong opinions about the cultivation of the musical hand and its surprising vulnerability to injury. His experience and insights seem uniquely appropriate to the questions raised here about the paradoxically difficult lives of professional musicians in our own culture, so I will quote him at some length:

> When I was in the third grade I had the measles. It was the first time and the only time I was ever out of school for three whole days. Someone gave me one of those puzzles of holes in a wooden board the shape of a cross. There were pegs to be moved around: all of the pegs were red except one, which was blue, and you had to figure out how to move them so that the blue one would end up in the middle. I spent about two and a half days solving that puzzle. When I was a little older, I played billiards. I was fascinated by puzzles and games that require repetition and careful development and a certain kind of concentration, and drawn to any skill I could slowly think out and work out. I wasn't always sure I could handle the world improvisationally, and working on problems of this kind created a safe place for me to be.
>
> Even as an adult the problems and puzzles I enjoy always seem to relate to things I can analyze and work out through repetition—something I can gradually come to control. Juggling is an example, and so is sleight-of-hand. There is a particular emotional element involved in this kind of work, a way of perceiving things and a need to predict what's going to happen next. It's a very little world—in your hands. Whatever you can do with your hands gives you a small world that you can actually cope with, as opposed to the big world, where perhaps you can't. I think this interest in small things explains my habit of observing people in a certain way. I am sure my sensitivity to outcomes has to do with being very insecure and easily intimidated as a kid. That feeling leads you to watch other people's faces carefully as you speak to them.
>
> My father was a machinist who worked for Sperry Rand making parts for gyroscopes. He valued precision both in process and in making tools, and I'm sure I have something of a sense of craft from his outlook. He was very laconic and, in the way of the Irish, completely undemonstrative. My mother was born here and was a mix of Italian, Sicilian, and Neapolitan. She was very expressive, very immediate in her reactions, very emotional; she was the antithesis of my father.

As a child I heard music constantly. My first contact with an instrument was when I was a tiny kid in school; we played little plastic flutes, but from the beginning I was determined to play correctly. One cousin of mine—we were almost exact contemporaries and were always together as little kids—saw me playing a plastic flute in first grade and immediately fell all over himself laughing at me. I addressed myself physically to this toy instrument as though I were a professional, and that was hilarious to him.

I began playing guitar sometime thereafter. I was self-taught and don't really remember the chronology very well. Somewhere around the age of seven I had saved some money delivering papers for my brother, and used it to buy a cheap guitar. I found books that taught chords and I copied things off records. I was doing this for a long time before I learned to read music. I actually was playing for a long, long time professionally—I made a good deal of money at it—before I read music.

In high school I bought a classical guitar because I was interested in South American jazz—bossa nova, originally a highly political idiom and politically a protest music that I was interested in. So I bought some books, thinking I would learn something about playing with my fingers—before that I had played with a pick, mostly playing chords.

The transition from pick to fingers is not a small thing, and it is a process I began to understand much better after I began to play the lute. When you play the lute, you don't use your nails. In fact, you still want to use gut strings if you can. The whole process involves getting closer physically to the control of the string without any intermediary. Not the hard, stiff insensitivity of a pick, not the steel strings, not an amplifier, not even your fingernails. As I was getting closer to a feeling of touch I found a particular satisfaction in being closer to the string, feeling more input from the string, being able to articulate more and more carefully from the right side of the instrument.

The first and only teacher I ever had was a well-known player I went to when I was going to Brooklyn College. He himself had studied with an old Russian named Alexander Bellow, who had been a pianist, an arranger, and a conductor. My teacher had a specific way of going at the strings, which didn't work very well for me—it just wasn't a very good technique for what I wanted to do. Eventually I had to leave the technique behind, but I did so in bits and pieces. Some of the things he had given me simply blew up in my hand in the late sixties. That was when I began to play more and more adventurous modern literature, which often is not composed for the guitar and doesn't fit on it very gracefully.

As I started to push my playing, my hand started to hurt constantly, particularly what we call the "A" finger—the ring finger of my right hand. Rather quickly I developed a *gigantic* tendinitis. So there I was with my hand problem. I began to ask people questions; I called up doctors, therapists, chiropractors, all the musicians I could think of, and no one seemed to know what this was. In fact, no one even said the word "tendinitis" to me.

A few people said to stop playing for a while, and I tried that. It didn't help. I just went through an awful time, thinking I wasn't going to be able to play or teach anymore. Eventually I began to see a way out of it. I did work my way out of it, but it took me a couple of years to do.

That whole time was strangely like being eight years old again, sitting in bed with the measles, fiddling with that peg game day after day. I just worked it out, alone in a room, quietly by myself. Finally I found a way of playing which made my hand hurt less. And the change gave me a better tone. It gave me, eventually, more speed, more power, and much more independence between the different fingers in plucking. The repair job gradually coalesced into a systematic way of approaching the instrument, but I have only recently had any idea what's underneath the change: it comes down to avoiding unnecessary distal flexion in the fingers. That was a habit which I had gotten into, and it was a costly one because of the pattern of sustained repetitive cocontraction I had fallen into. I still just barely have a sense of what *that's* about.

I think that the predisposition to handle the problem as I did comes in part out of my background. I was brought up in the Roman Catholic Church, and I had a father who would say "allergies and divorces are for rich people," and we couldn't afford them. What he meant was that you had to solve your own problems. You had to analyze what was going wrong yourself, then figure out how to deal with it, or just cope with it. Every week when you're a little altar boy in the church, in the days when I went to the church, you said: "Mea culpa, mea culpa, mea maxima culpa." You really assimilated the idea of personal responsibility; *you* had done something wrong and there were consequences. So I was ready, when I began to see my hand hurting, to accept that I had done something wrong, and that was why it hurt.

The pain actually helped me in an odd way. I could *feel* every day whether I was doing the right thing or not, because my hand would feel better or worse. Merely stopping playing didn't help. I stopped for a month; I didn't use my hand for *anything* for a couple months at one point, and it still hurt like hell. When I did finally solve the problem I had no idea how I had done it—I hadn't even a theory. Of course, theories are *all* I have now, but for a long time I had only the most naïve concept of what I was doing.

A considerable part of my work now involves helping injured musicians. The players with the most severe problems usually require changes of technique, changes in the way individual fingers are used, and a very slow, careful imprinting of an entirely new pattern of motion in the hand. In the process of working *that* slowly, evaluating everything *that* carefully, listening to the tone of each individual note, just watching the finger move slowly, the injured player will have to begin to change *as a person.*

They have to let go of a certain kind of goal orientation which has always told them to grip and grab everything as hard as they can, to get ahead in their career. It is as if a misguided life metaphor is visited on them in their hand. And they *have* to change. Whereas previously perhaps they dashed their way through things—approximating—they now have to begin to think about *every* motion. And they must develop a very, *very* refined sensitivity to what their body feels every time they move it. They have to learn to completely relax between notes, which of course they may never have done before in their lives. Perhaps they have never done that with *any* part of their personality, in anything they do. They move as fast as they can, play as fast as they can, and turn the metronome up a notch arbitrarily, once every day, force-feeding themselves the instrument.

> They have to let all of that go. They may even have to say, "I don't care if I never play a concert again; I just want to be able to *play* again." There's something really profound about that, forsaking the goal of winning a competition, getting into a certain conservatory, playing better than the guy down the street, whatever it is. They have to be able to say, "I want to do it, even if only privately, for myself. I'd like to be able to play something simple." These are almost universally people who would *never* have said that before. They have to stop *driving*. Some people just can't make that change.
>
> Having acquired a particular combination of insights—the ability to see into problems in technique—would not necessarily give you the empathy you need to work with other musicians with problems. I went through so much pain in my own situation, such a terrible crisis, wondering whether I was going to be able to play again—what would I do with my life if I couldn't play? There was nothing else I wanted to do as much as I wanted to play the guitar. That gave me considerable empathy for other injured musicians.[4]

Patrick O'Brien makes it clear how powerfully musicians in the professional culture are driven to develop and maintain the skills they need in order to secure a place in that world, and how costly their participation in the competition can be. We will return to his observations about the educational process below, in chapter 15.

Now, let's see how the argument for a musical heuristic, including the BachTur, behaves when you look at a few simple facts:

- Great pianists did not exist before the piano existed. We would be unaware of this "natural" ability in humans were it not for the existence of the piano and pianists to demonstrate its use.
- Few great pianists exist, but the piano, plus heightened public interest in (and rewards for) great piano playing, has created a search engine for discovery of pianistic talent. It has also created a demand to "raise the bar" in solo piano performance, a goal achieved through the separate but complementary processes of musical composition, modification of

> musical instruments, early identification of "musically gifted" children and the creation of special training environments for them, and the institution of serious musical competitions for children.*

Very early during the comparatively short life of this cultural experiment with keyboard playing it was apparent that people like Bach, Mozart, and Beethoven had *something* that other people—including other musicians—lacked, or at least had considerably less of. And while it is customary to assume that "inborn musical talent" or "high musical intelligence" is the consequence of a specific genetic tendency—an instinctive, intentional, a priori, predisposition to be a musician—the claim becomes untenable when you consider musical experience in the world at large. Even if one were to confine the search for a musical gene, or the "essence of the musical germ," to our own society, one would quickly discover that high levels of musical achievement are far too broadly based in both makeup and experience to be attributable to a genetic tendency any narrower or more specific than that controlling human language, which (as we know) is universal and expressible in the face of considerable variability at the level of the input-output channel.

What we are left with when we seek to explain musical talent on a biologic basis seems best characterized as an assembly of neurologic and behavioral potentials that arise from within and are uniquely defined by specific cultures. The "computational" requirement for musical achievement is the ability to perceive, understand, retain, reformulate, and interpret (or generate and regenerate) musical ideas and information. This aptitude is not identical but highly congruent with the aptitudes underlying spoken language, and, like spoken language among the hearing, it depends upon a highly organized and elaborated variant of human receptiveness to sound. The argument is commonly made that auditory processing in speech

---

* It is extremely interesting to consider the impact of the musical prodigy on the rapid evolution of musical skill within the professional musical culture. Despite the obvious charismatic influence of such skill on the evolution of professional standards, the mechanism of transmission is cultural rather than genetic. Apart from J. S. Bach himself, there is no evidence that true genetic prodigies breed more prodigiously than other people, and in fact—for highly individual and usually complex reasons—comparatively few true musical prodigies pursue professional musical careers.

behavior and musical behavior cannot be the same because imaging tests show nonoverlapping metabolic patterns for them in the brain; but this is like arguing that track racing and street driving are "different processes" because infrared pictures taken from high altitude show localized, circular patterns for the former and linear, diffuse patterns for the latter. True enough, but what does that tell us about the internal combustion engine or driving skill? If there *is* a BachTur in the brain of a musician, its operations are linked to those of the ChomTur: if a musician becomes aphasic as a result of damage to the ChomTur, as Ravel did, the musician's ability to represent and generate musical ideas is virtually always impaired.*

The upper-limb (or "output") requirements for an instrumentalist are not unique either; they depend upon the possession of arms, fingers, and thumbs, specific but idiosyncratic limits on the range of motion at the shoulder, elbow, wrist, hand, and finger joints, variable abilities to achieve repetition rates and forces with specific digital configurations in sequence at multiple contact points on a sound-making device, and so on.[5] Peculiarities in the physical configuration and movement capabilities of the musician's limbs can be an advantage or a disadvantage but are reflected in (and in adverse cases can be overcome through) instrument design: How wide can you make the neck of a guitar? How far apart should the keys be on a piano? Where should the keys be placed on a flute—in general? and for Susan and for Peter?

I cannot say if this is good news or bad news for musicians, but I see no reasonable alternative to the conclusion that the *music* heuristic (or musical talent) is *invented* and fashioned in the brain and body of a child early in life in response to what the specific culture offers that particular child by way of musical examples, opportunities, and rewards. The earlier and the better the fit between the peculiarities of the child's mind and body and the cultural offering, the greater the likelihood that the child will be labeled as musically talented and offered (as bees offer their queens) a diet of experiential royal jelly. One *must* step outside one's own culture to see how strikingly different and culturally determined our definitions of musical talent are.[6]

---

* The retention of "procedural" musical skill (being able to learn to play notes in sequence, or to play music already memorized) is not necessarily excluded among aphasics, however.

John Blacking, a social anthropologist and a trained musician, came to just this conclusion after living with the Venda, a traditionally nomadic tribe who live in the Transvaal in South Africa. The Venda are apparently unique in the world in that music has long been the foundation of their social life, and from the time of early infancy every member of the society is immersed in musical activities that repeatedly intersect, moderate, and inform the daily lives of the people. Initially intending to teach "advanced" European music to Venda children, Blacking discovered that children of preschool age were better able to perform and to *explain* and teach their own music than anyone except professional musicians in the world he came from. Writing about his experience, Blacking observed:

> We talk freely of musical genius, but we do not know what qualities of genius are restricted to music and whether or not they might find expression in another medium. Nor do we know to what extent these qualities may be latent in all men. It may well be that the social and cultural inhibitions that prevent the flowering of musical genius are more significant than any individual ability that may seem to promote it.[7]

Blacking's discovery of the Venda musical culture, set against his own very different education and experience, led him, as an anthropologist, to question the segregation of the two favorite debates of modern cognitive psychology: *nature vs. nurture* and *single vs. multiple intelligences.* His insight anticipated the reformulation of these two questions by Henry Plotkin, who declares that the nature vs. nurture debate is moot: the primary heuristic *constrains* the secondary heuristic, hence there cannot be a *general* intelligence in the sense of a capacity to learn everything about everything. And although the primary heuristic both creates and sets limits to multiple intelligences, they cannot be predefined in *cultural* terms.[8]

The work of Harvard University's renowned professor of education Howard Gardner, beginning with *Frames of Mind: The Theory of Multiple Intelligences,* has had a profound impact on the thinking of the professional educational community and has created a healthy and vigorous advocacy for diversification of the educational process in ways that acknowledge the diversity of human potential. As Gardner says in his introduction to the tenth-anniversary edition of his book:

> While I do not question that some people may have the potential to excel in more than one sphere, I strongly challenge the notion of large general powers. To my way of thinking, the mind has the potential to deal with several different kinds of *content,* but a person's facility with one content has little predictive power about his or her facility with other kinds. In other words, genius (and *a fortiori,* normal performance) is likely to be specific to particular contents: human beings have evolved to exhibit several intelligences and not to draw variously upon one flexible intelligence.[9]

If one considers the example of "musical intelligence" from a cultural perspective, and the division of musical experience into professional and amateur cultures in our own society—noting particularly the distinctive learners, forms of learning, and social and behavioral outcomes found in these environments—one begins to see what intelligence is, in both Henry Plotkin's and Howard Gardner's terms, and how a culture can define intelligence in individuals without *redefining* it in the *species.*

One could easily argue—and I do—that the modern professional musician illustrates the de facto evolution of intelligence taking place right under our noses. This evolution does not represent a change in biologic intelligence, but the establishment of a culturally defined and valued form of intelligent behavior through early and intense educational manipulation and the subsequent rewarding of musicianship, both with special incentives for success and severe penalties for failure. Within the space of just a few hundred years there has been such a radical transformation of music, and of the nature of the musician's life and work, that present-day musicians can in a certain sense be compared to the specially adapted species Darwin found on Galápagos. Their musical development is subject to circumstances so unusual and extreme that professional musicians are actually evolving *as we watch.* It is a *virtual* new species, however, because the information controlling the new musical intelligence and skill is imbedded in musical institutions and has no effect on the genetic composition of living individuals.

Hence "musical intelligence" as we know it would disappear within a single generation if all newly born children were denied exposure to music and access to musical information. Human language would vanish just as quickly and completely, absent childhood exposure to it, despite its being an "instinct."

Nothing could be more clear than the fact that "musical intelligence" is highly rewarded and highly developed within both traditional Venda and modern Western societies. In both cultures the lives of these children will be increasingly filled with musical experiences and relationships, and their musical vocabulary and expertise will grow as they enter adult life. Venda children will draw upon their intelligence to cultivate personal and musical lives in ways that are very different from those among *our* "musically gifted" children; Venda children simply expand upon musical activities within a social context, doing more or less music, *as experts,* depending upon opportunities presented to them and their own inclinations.

By contrast, our culture has established specialized educational practices for the specific purpose of producing "great musicians," so we are inclined to isolate the most musically responsive children as soon as possible in preprofessional learning situations to prepare them to spend their adult lives immersed in the communities, protocols, and agendas of people regarded as musical thoroughbreds. As a result, we have the pleasure of seeing the benefits of a high-pressure survivalist educational experience in music. We also have the *obligation* to care for those who cannot keep up, as a variety of health professionals are now doing with younger and younger musicians.

It is the dark side of music's educational success, in fact, that leads me to suggest that it may be premature for us to crow so loudly over the slaying of the dragon of General Intelligence. Obviously there is something drastically wrong with a notion of General Intelligence that promotes and preserves artificial ranking of human potential while failing to account for the diversity of human achievement and expertise. But a theory of *multiple* intelligences is hardly an improvement if it merely replaces one grand, cultural horse race with *multiple* cultural horse races. If we assume that we know *in advance* what a certain child is cut out to do, we do violence to a process that is innately engineered to contend with unseen possibilities. Evolution, it has been said, is smarter than we are.

Human intelligence can, and almost always *will*, be directed and "divided" by the culture (which is to say, captured and programmed by selective reinforcements and inhibitions) before the child has any real opportunity to discover or define constitutional affinities through self-initiated interactions with the world, and to explore, shape, and refine those affinities on the basis of experience. This process begins so early in life and proceeds so rapidly that we have no practical way of discovering in a particular individual what the genetic and experiential roots of his or her "talent" actually are.

So, in an odd and paradoxical way, Gardner's "realism" concerning the heterogeneity of human intelligence could in the long run actually help perpetuate the misdemeanors of oppressive educational policy by supporting a culturally partitioned view of human potential. This is an issue to which we will return in chapter 15.

Obviously, the more we understand about human learning, the more truly rational our provision of formal education can be. In that regard, it is worth considering Merlin Donald's conclusion in *The Origins of the Modern Mind:*

> As we develop new external symbolic configurations and modalities, we reconfigure our own mental architecture in nontrivial ways. The transition (to mythic and theoretic cultures) has led to one of the greatest reconfigurations of cognitive structure in mammalian history, *without genetic change.* In principle, this process could continue, and we may not yet have seen the final modular configuration of the modern human mind. Theories of human evolution must be expanded and modified to accommodate this possibility.[10]

Donald was not really thinking about musicians in this rather more edgy than euphoric close to his book. Without putting words into his mouth, I take his concern to be that a human brain whose development is driven by a narrowly defined culture is far more likely to be our downfall than the runaway computer we have all been worrying about since we watched Stanley Kubrick's filmed version of Arthur C. Clarke's apocalyptic fable, *2001*. One of the truly reassuring aspects of the musician story is that it shows us what can happen when the brain evolves as far as it can *with the body acting as a brake* on the operation of secondary heuristics. What might happen if the body no longer were to define, limit, or even help "calibrate" the brain's continuing experiment to expand its reach? Would it surprise you to learn that we are already beginning to find out? We will play with this idea in the Epilogue.

12
Lucy to LuLu to Rose*

EVERYONE WAS A LITTLE TENSE AS THEY SAT DOWN AT THE TABLE, BUT THAT'S AS FAR AS IT WENT UNTIL THE QUAIL WERE SERVED. IT WASN'T ENOUGH HE'D MADE HIS WIFE JEALOUS EARLIER, FOR WHEN PEDRO TASTED HIS MOUTHFUL, HE COULDN'T HELP CLOSING HIS EYES IN VOLUPTUOUS DELIGHT AND EXCLAIMING:

"IT IS A DISH FOR THE GODS!"

—Laura Esquivel,

Like Water for Chocolate[1]

THE ENHANCEMENT OF APPETITE WITH AESTHETICS is a defining trait of humans, and the habit must have started when primates moved into the trees. So far as we know, tree climbing became an important survival

* The title is a paean to Daniel Dennett, the author of *Darwin's Dangerous Idea* (1995) and the man who singlehandedly cracked the mystery of Richard Dawkins' "meme" for me. From Dennett's suitcase bulging with memes, I selected the famous "baseball meme" for an explosive triple play: "Tinker to Evers to Chance" (p. 299). In homage, and for the memologists, I offer a culinary triple play. It contains a variation on the theme of explosive developments, and one that should appeal to evolution's punctuated equilibrists. "Lucy to LuLu to Rose" stands for the cultural ignition of an evolutionary development that changes everything overnight. Lucy (*Australopithecus afarensis*) to LuLu (the restaurant that launched a young chef's career in San Francisco, also symbolizing the arrival of the period when mealtime for *Homo sapiens* had become an art form in every city of any size) took 3.25 million years to happen. LuLu to Rose (the name of the third successful restaurant opened by the same chef in San Francisco) stands for the speed with which a cultural niche can be not only defined but stormed by someone with the right stuff (3.25 years in this case).

strategy nearly 65 million years ago, when changes in global climate produced a proliferation of flowering and fruit-bearing trees—this was an ecosystem made in heaven. The insects and birds assured the free circulation of plant genetic material, and their presence in the vicinity of all the blossoms and fruits must have made mealtime in the treetops a feast for all the senses.[2]

Drawn by the natural appeal of the menu and the setting, primates brought with them some rather special equipment that would have nudged them very strongly in the direction of a connoisseur's attitude toward food. Consider that ground-dwelling mammals typically have a rather pragmatic approach toward nutrition, just as ground-dwelling reptiles before them always did. A meal is either overrun and literally *mowed down* (the monotonous vegetarian life of elephants, rhinos, and cattle); subdued, then propelled into the gut either for temporary storage or for digestion (the athletic carnivorous life of the cats and dogs); or scavenged (the thieving life of the hyena).

Cats, bears, and other large mammals routinely nest in the large, lower branches of trees; but to get up to where the picnic was, up at the level of the canopy, mammals needed rather specific changes in upper-limb structure (you might want to flip back to chapter 1, "Dawn," to refresh yourself on this point). Here, in the opening days of the primates, the kernel of an idea (or a heuristic) for human aesthetics seems to have been inserted into the human genome; the time had come for mealtime to be transformed to a dining experience. Reasonably secure and unlikely to be disturbed, the tentatively hedonistic primate could relax, cool off in the light breeze, enjoy the vista, the smell of the flowers, and the birdsong that came with his beetle surprise with fresh fruit topping. This habit apparently survived the transition back to ground living, and we have been buffing and polishing it (along with the silver) ever since.

The descendants of arboreal primates are very different creatures, and of course much of what had happened to the body by the time Lucy was heading out across the open savannah was necessarily utilitarian; the australopithecines could ill afford any bodily frills if they were to survive in their new and unfriendly environment. But selection, as we know, produces its effects through serendipity as easily as it does through adversity: an adaptation for warfare may eventually prove its continued worth in peacetime simply by the reframing of circumstances. Professional cooking is a case in point: the warrior's body is admirably suited to the arena of a modern restaurant kitchen, where a chef will find occasion not only to lift skillets and pots but to hurl them. When you consider the other physical demands of this job—chopping, slicing, pounding, and kneading, for example—you may well wonder how anyone without military instincts and experience could hope to survive at it.

The other essential requirement for any chef is the aesthetic one—a highly refined appreciation for the subtleties of aroma, taste, and texture. Bring all that together, harness the requisite neuromuscular virtuosity of *Homo sapiens sapiens* to a lively palate and a passion for food and its creation, and you have the possibility of a new menu, maybe even a new cuisine. There is no end to the elaboration of themes and the ingenuity with which the eating experience has been enlarged through the imaginative mixing of appetites and aesthetics. The Indians can now scarcely eat their curries without ragas; a French meal would not be offered unaccompanied by candlelight and flowers; and the Japanese, who remain faithful to the oldest traditions of their culture, find it natural to mesmerize the diner with a marriage of marination and the martial arts.

Chefs in the modern world have to be both strong and imaginative: kitchen work requires herculean physical endurance and absolute psychological indomitability; second, survival demands a killer instinct in the business world (the ability to get more than turnip juice out of a turnip, in other words); third—and this is the only attribute the public consistently pays for—a winning presentation demands a keen eye and a solid feeling for theater and ritual. The slightest deficiency in any of these three "true-life" categories means certain doom within months, if not weeks, in the contemporary world of the culinary arts. It is not a life for the frail constitution or the mundane personality; it is a *jungle.*

With some of the above considerations (and universal Darwinism) in mind, several years ago I read a short article in a San Francisco newspaper about Reed Hearon, a recent arrival and highly regarded contributor to the internationally renowned restaurant scene in San Francisco. What attracted my eye and riveted my attention were the following lines:

> Reed Hearon is hotter than a *habanero* these days. He's just opened LuLu, a chic-but-casual Mediterranean restaurant on Folsom Street; his new Cafe Marimba is due to open on Chestnut Street in April; a line of Mexican spices and salsas is on its way to market; and his "Salsa" cookbook will be published in July. Not bad for a math/philosophy honors student who worked his way through the University of Chicago and the University of Texas at Austin with restaurant jobs.[3]

This clearly was someone to meet. Honors in math and philosophy; worked his way through college (and spent time in Chicago, Santa Fe, and Aspen, all places where food is taken seriously); and in his first year he sounds like a Disney syndicate, with two restaurants and a book about stuff he's also marketing in his own catalogue store.

The first of several meetings I had with Reed took place at LuLu; he has since gone on to other enterprises, and now runs Rose Pistola, an Italian restaurant on Columbus Avenue. So busy and so successful has he been that Rose was reported to be the best Italian restaurant in town almost before it opened.

Before we met, I had tried out LuLu with friends who have considerable experience in the Bay Area restaurant world, and we were not disappointed. Of the meal, I vividly remember roast quail and corn on the cob, both unlike anything I had ever tasted before—I felt, as Pedro did of Tita's quail with rose petal sauce (in *Like Water for Chocolate*), that it was a dish fit for the gods. As promised—and quite uncustomary for San Francisco—the dinner was served "family style." My hunch, obviously, had been right, and after a few phone calls Chef Hearon and I sat across from each other to delve into his work and his ideas.

Reed Hearon is, first of all, visually striking. He is tall and handsome, with an unforced and reflective smile and, despite the mayhem that prevails in any of his restaurants, a completely calm disposition. His voice is soft, evoking southern Texas charm and gentility. At our first meeting in 1993 I asked him about his beginnings in the restaurant business.

> I started in Austin in 1980 or 1981. I ran a restaurant there for two or three years and later discovered that what we'd been doing was called "south-

western food." I didn't know that at the time. After doing this for a while, I felt I had hit a dead end. I'd learned an awful lot from books but hadn't had any formal training in cooking and really wanted to work with some people who were good.

I ended up working at the Rattlesnake Club in Denver in 1984—I had just packed all my belongings in a car and when I moved out there I took any job I could get in a kitchen. I ended up being one of several people who had the title of "chef" there, in an example of what I call the "King of England School of Management"—you call people something important and that makes them happy, even though there's nothing behind the title.

After a year I went to Santa Fe, helped Mark Miller open the Coyote Cafe, and also helped him write *The Coyote Cafe Cookbook.* I worked there for a couple of years, went back to Texas just in time for the collapse of the Texas banking industry, and ended up coming out here and redoing the Corona Bar and Grill. Then came Aspen and the Caribou Club, and when that was done I returned to open LuLu. Now I'm working with my partner on something called Cafe Marimba, which will open in the summer.[4]

Meanwhile, my book *Salsa* just came out, and there's another manuscript due a month ago that's still three or four hard months away from being complete. It's a larger piece on some historical trends in Mexican cooking.

Over the course of several visits to LuLu I learned a number of things; two subjects we covered seem to me of particular relevance to this book. First, and most obvious, is the matter of cooking as a specific skill involving the hands.

I think everybody who cooks enjoys working with their hands; there's an immediate sense of accomplishment. A big issue in the kitchen certainly is "hand speed," which is just the amount of time it takes to chop things, or to complete a given task. Some people are intrinsically very quick and some people are slow as molasses and are never going to change. Some folks do very well in an academic environment but not terribly well in a restaurant. Let's take as an example the Culinary Institute of America in New York—a graduate toward the top of their class will have a fairly wide exposure to a range of fundamental techniques. They may or may not remember how to make a particular sauce but they have probably been exposed to it. They may know at least the fundamentals of making stocks and sauces, they know how to butcher things, they know how to open oysters, they know the fundamentals of grilling or sautéing, they know the names of things.

Professional cooking is very much a system of guilds and formal schools. I actually don't have any formal schooling, but probably most of the people in the kitchen here do. The culinary schools are getting to be fairly popular around the country and they turn out a healthy number of graduates with at least some rudimentary skills. Generally the graduates haven't much idea about how those skills are applied in the real world until they've had restaurant experience. I suppose that's part of the reason I really tend to think of cooking as much more a craft than an art. I picked up most of what I know just by experience and reading. I'm a firm believer in the idea that if you read two books on a subject written by knowledgeable people you will know more than ninety-five percent of the people in the entire world know about that subject. To learn enough to equal the other five percent will take you the rest of your life.

We bring people into our kitchen and give them a few years of working as a line cook, which means that they may work in the pantry doing cold appetizers and salads or they may work the grill. That's probably the most grueling part of learning to be a chef; there's a rotation through all the stations, arranged so that each station gets progressively a little harder in terms of the skill levels necessary to execute things well.

With greater responsibility and, depending on the person, after anywhere from six months to a few years of this, our cooks are ready for the next step. Here, that means being put on salary, which in turn means they're privileged to work sixty hours a week for not much more money than they were making working forty hours a week. With the promotion they get some oversight responsibilities, and they begin to learn about the business side of things: how you order things, how you figure out what to pay for them, when you need to buy them. The more highly skilled people handle food products which are more expensive.

> That's a complicated problem for restaurants because not many other businesses use inventories which are perishable in a matter of a day or two. We have upwards of three or four hundred items in inventory in the kitchen, and probably half of them are perishable—they last three or four days. So, the ordering is a substantial task to learn. Our best guys are getting to be very good cooks, they're learning how to be creative and focused at the same time, honing their cooking skills, and getting a taste of managing people and products. Pretty soon—sometimes all of a sudden, with no more background than what we give them—they may want to go out on their own as a chef. In that case what they really should have is an MBA and some real estate background.

It was at this point that the conversation moved in an entirely unexpected direction. We had talked about the interesting but still relatively predictable transition between a basic interest and the skills required to pursue it, focused and honed under the powerful pressure to succeed. We even talked a little about the unlikely connection of theoretical mathematics and the restaurant business. It was his casual mention of his book *Salsa* and his book in progress on early traditions in Mexican cooking that eventually led to what follows. In *Salsa,* he writes that he cannot imagine anyone but the Mexicans inventing salsa. Says Hearon,

> I think my interest in Mexican food has parallels in the nouvelle cuisine in France in the seventies or the new American cuisine in the United States in the eighties, all of them really involving rejection of a European aesthetic in food. The food prior to the Conquest of Mexico was very highly evolved, very elaborate and very ritualized. Much of what people associate with Mexican food—being greasy or cheesy or made with a lot of pork—in fact comes from Spain. Pre-Conquest Mexican food was made without pork or beef, many of the nuts, or any of the usual cooking oils. The only cooking oils available to any degree in pre-Conquest Mexico were armadillo fat and pumpkin seed oil, and there's not a lot of fat in either one of those. Early Mexican cuisine involves a lot more toasting things on clay *comals*—these are ceramic griddles which, I suppose, are the historical suc-

cessors to hot rocks next to a fire. Tortillas were combined with chilies and spices that had been toasted and then ground together to make elaborate sauces, or were made into tamales or stuffed in leaves and steamed.

My earliest experiences in Mexico were like those of a lot of *gringos* at a certain time: I was a small child being appalled at the images of poverty, and apprehensive I might get sick and die from eating the food. Other than that, my responses as a child were not much different from what they might have been had I been in Paris or Bombay: it was just a different culture.

But then I started to feel the warmth of the Mexican people. I think this was something that came with teaching myself Spanish. And as I made the effort to appreciate the Mexican culture, that culture proved itself to be an enormously warm and open one, and a natural one for sensing the importance of both ritual and dialogue.

I see much of cooking as a dialogue. The dialogue may be very simple: an individual in need of nourishment is being fed—that's one kind of basic dialogue in life that goes on between people and the land they live on. Even though food is a basic necessity, that dialogue can become very stylized, depending on where you are. The dialogue can also take place on a cross-cultural level. At the close of the twentieth century in North America we are seeing an enormous influence from Mexican culture on United States culture—did you know that salsa outsells tomato catsup in the U.S. now? Thirty years ago that would have been unthinkable.

The issue of culture and emotion has a lot to do with what interests me in food—and food is *very* emotional. Your first experiences with food were at your mother's breast; food *is* love at that level. I find a strong emotional language or connection with food. There are dishes that we all like, but only because of the connection we made with them as children, or through people who are important to us. We all remember some favorite dish our mother cooked for us. And there are meals that are highly ritualized, that carry an emotional or social message—such as Thanksgiving—but which have no other real place in our diet or dining practices.

The warmth in Mexico is a very special thing for me, particularly in remote parts of Mexico where the pre-Conquest cultures have survived in some form. Mexico seems like the place that would necessarily have invented salsa because salsa is a small gesture to improve an ordinary experience—eating—made by people who are for the most part of very limited means in what they are able to eat. Its effect is utterly transforming of the food *and* of the experience.

You can appreciate what this means by thinking of the heritage of much of North American cooking. We have a very strong Protestant heritage here, and well into this century it was thought to be somewhat sinful if food was overly embellished, simply because it was believed wrong to excite the senses. Which is what a salsa *does.*

Salsas seem to root very deeply into the notion of the world as a magical place. This is a view that was held by the native Mexicans prior to the Conquest and even seems to be retained in their peculiar version of Catholicism. In Mexican Catholicism, as I understand it at least, the saints became replacements for an ancient pantheon of gods, and Mexicans' relationship with the saints actually reflects and reinforces the old theology and the old culture. There, the gods, who are sometimes in conflict with one another and who need to be appeased, represent a variety of natural forces at work. The pre-Conquest view of the world was that it was a place where people are not in control; they are at the mercy of great, powerful forces whose power they address through rituals. Food was part of that ritual, because the food itself comes out of and reflects the operation of uncontrollable forces—whether the rains come, whether the corn grows, whether things ferment or not to create beverages.

That certainly isn't *our* philosophy. Our view is that as human beings we are capable of conquering any or all of nature. Nineteenth-century Puritanism in the United States, despite its great repressions, was fundamentally an optimistic religion, holding that man could redeem himself through work. Exciting the senses was dangerous, because it tended to push you off this path, to distract you from doing God's work.

I think Mexico is fundamentally a very pessimistic country. There, the view is much more, well, do your good works and a rock *still* may fall out of the sky and hit you on the head tomorrow. So you might as well enjoy the family, which is the central context of life, and pursue the simpler pleasures of eating, and existence, and love. Those are precisely the pursuits that tend to get pushed aside in a culture that preaches disaster prevention through work.

To go back to what I was saying earlier, I am very interested in the issue of dialogue. This touches a notion in mathematics that's been very powerful in the twentieth century, that of the map from one set to another. For any two objects, a mapping or a dialogue just refers to the relationship the two have to one another, and it may be a relationship that operates on more than one kind of set. So the *emotional* map between an individual and an individual dish can be the same map or dialogue that exists between one culture and another culture, or between each culture's dishes.

As abstract as that sounds, I think it does help explain what ritual is, because the profound thing about ritual is its ability to be moved around and not to be dependent on individual places and times. If the catechism means something to you and you say it now or said it a hundred years ago, or if you say it with irony or say it with heartfelt feelings, it is *still* a ritual and it is still having its effect on you by connecting you with another object or set of feelings. So as we explore rituals, or traditions, in food from pre-Conquest Mexico; on the emotional level we may find that they have the same kind of impact now as they originally did on people.

It is hard to know how much of Reed's philosophy and his delight in creating foods and environments have become a factor in his commercial success. One could harbor the cynical thought that he has simply hit on a formula that brings people in (in droves), but I believe otherwise. I believe he has designed his restaurants to be places where the sort of people whose company *he* would enjoy might come to sit around the table and share a meal and one another's company, just as he learned to do a long time ago, where salsa comes from.

During one of our conversations, Reed called cooking a "lunatic job." Maybe so, but he has done so well that it has been almost impossible to keep up with his culinary march. The "Datebook" section of the *San Francisco Chronicle* of December 26, 1993, carries the following opening paragraph in a year's-end piece on dining in San Francisco.

> It seems an anomaly: the economy is weak but the number of destination restaurants that opened in 1993 probably exceeded any in the decade. Bradley Ogden came back to the city to open One Market. Pat Kuleto and Nancy Oakes teamed up for Boulevard. Jeremiah Tower reached to the Napa Valley with Stars Oakville Cafe. However, the grand prize winner is Reed Hearon, a veteran of Corona Bar and Grill, who blazed onto the dining scene with *three* winners: LuLu, LuLu Bis and Cafe Marimba. ("A Feast of New Restaurants," p. 17)

Reed also received the 1996 San Francisco restaurant critics' coveted Chef of the Year award,[5] and Rose Pistola received the James Beard 1996 award as Best New Restaurant to Open in North America. Quick hands, a mind disciplined by the study of mathematics, a personality warmly disposed toward a celebration of life, and a commitment to cooking traditions where "food is love," have combined in Reed to produce a solid equation for success, and what appears to be a growth industry that knows no bounds. Yet another restaurant—The Black Cat, described as a "late-night restaurant and bar"—is scheduled to open in San Francisco's North Beach just as this book goes to press. And despite the distinctly Mediterranean direction of his restaurants, he continues to write about Mexican cooking: *La Parilla: The Mexican Grill* is a collection of recipes from the countryside of northern Mexico (Chronicle Books, 1996), and *Bocaditos: The Little Dishes of Mexico* is a collection of tapas-like dishes (Chronicle Books, 1997).

Reed's extraordinary success contains an extremely important object lesson concerning the interplay of "ordinary physical skill" with a host of other factors that influence the paths people choose to follow (or that they invent) as they move to integrate their experiences and formal education into adult life and work. In the book's final chapter we will consider how well educational theory corresponds to, or addresses, the interaction of the biologic and social processes through which unusually successful adults appear to have capitalized on the opportunities to attain distinctive forms and high levels of personal skill, intelligence, and independence, *beginning from the time they were still children and innocent of adult needs or ambitions.*

13
Tough, Tender, and Tenacious

THE EXECUTION OF AN ACTION BY NO MEANS PROVES THAT WE KNOW, EVEN SUPERFICIALLY, WHAT WE ARE DOING OR HOW WE ARE DOING IT. IF WE ATTEMPT TO CARRY OUT AN ACTION WITH AWARENESS—THAT IS, TO FOLLOW IT IN DETAIL—WE SOON DISCOVER THAT EVEN THE SIMPLEST AND MOST COMMON OF ACTIONS, SUCH AS GETTING UP FROM A CHAIR, IS A MYSTERY, AND THAT WE HAVE NO IDEA AT ALL HOW IT IS DONE.

—Moshe Feldenkrais[1]

"*FELDENKRAIS* THERAPY? Hmm . . . sounds interesting. What exactly is it?"

I was sitting in my office taking a history from Margaret Bemin, a high school language teacher who had suffered a stroke several years earlier. She had walked into the office supporting herself with a Canadian crutch held by her right arm. The left arm was flexed at the elbow and hung close by her side, the hand a tight fist. A metal brace supported her left ankle.

She explained to me that her working life had abruptly come to an end the day her left side stopped moving. She had been hospitalized, the usual tests had been done, and she had then gone to a rehabilitation hospital to work on regaining the use of her left side in hopes of learning to walk again. Her initial efforts at rehabilitation had been a partial success, insofar as she had discovered that life *is* possible in a wheelchair, especially when you have the kind of family encouragement and support she had; she had not, however, been able to return to her job teaching high school Spanish. And since teaching was the real love of her life, no rehab program could have been counted a success if it did not allow her to get back to a classroom blackboard.

Someone told her about a young man in Berkeley named David Bersin, that he did something called "Feldenkrais." She went to see him, and he began working on her in sessions doing slow, easy movements, guiding her body with his hands. Eventually her left side started to move again, she learned how to walk with a brace and a cane, and she got back to the blackboard and back to her life.

"You should meet David," she said innocently.

I did go to meet David—I'm not certain precisely when I did, but it was over ten years ago. Calm, friendly, casual, and assured, David told me that Moshe Feldenkrais was an Israeli physicist interested in the mechanics and control of bodily movement. He had developed a series of "movement lessons" that seemed to change the fundamental organization and character of body movement. He had used these at first to help friends with aches and pains, a dance teacher solve problems with students, a child with cerebral palsy, an adult who had had a stroke.

"How is what you do different from physical therapy?" I asked David when I met him.

I wasn't tape-recording our conversation, but I recall his explanation very well. The gist was that Feldenkrais had discovered that certain movements, done gently and in particular sequences, would lead the person doing them to a heightened sense of what the body was doing, the mechanics of getting from one place to another. Eventually, with practice (by doing movement "lessons"), the person could learn to move more smoothly, efficiently, and, indeed, pleasurably.

The underlying idea had occurred to Feldenkrais over a period of time just by observing people, watching them with a physicist's eye and thinking how difficult it would be to get a motorized vehicle to move as the human body does. The more he thought about walking, the more he marveled at the sophistication of the brain as a controller of movement. Somehow athletes could perform, dancers could dance fluently, sometimes magnificently, and the incomprehensibly complicated physics, biomechanics, and physiology of the body—the underlying ballet of bones and joints and muscles—would take care of itself completely outside of awareness, confidently, silently, and reliably orchestrated by the brain.

But Feldenkrais saw a paradox: most people *don't* move as dancers and athletes move. Most people slouch, tilt, shuffle, twist, stumble, and hobble along. Why should that be? Was there something wrong with their brains? After considering what dancers and musicians go through to improve control of their movements, he guessed that people must either be ignorant of the possibilities or refuse to act on them. So they just heave themselves around, lurching from parking place to office to parking place, utterly oblivious to what they are doing, to their appearance, and even to the sensations that arise from bodily movement. He suspected that people just lose contact with their own bodies. If and when they do notice, it is because they are so stiff that they can't get out of bed or are in so much pain that they can barely get out of a chair. *Then* they start noticing.

Feldenkrais wondered if refinement in movement might be restored to the person who slows down to pay attention to the feel of the body—someone who stops rushing, pulls over, and takes time to pay attention. He tried his idea out on some friends and quickly found that his own hands could assist another person's sensitivity to even the smallest movement. By guiding their movements, using the pressure of his hand itself to heighten sensitivity, he seemed able to get people to pay attention to their own movements.

What he was doing did not seem complicated. The goal of the guided movements was not to learn how to move, in the sense of learning to do a new dance step. The goal was not to stretch ligaments or muscles. It was not to increase strength. The goal, as he saw it, was to get the *messages moving* again and to encourage the brain to pay attention to them.

Feldenkrais eventually wrote a book called *Awareness Through Movement.* The title expresses his increasing conviction that the basis for disturbed or degraded bodily movement was, as he had originally supposed, inattention, ignorance, or laziness. But the book suggests that sometimes there was something *really wrong* with the person. He had *learned* to move abnormally. The cause might be obvious—a stroke, a head injury, or a pulled muscle—but more often there was a hidden reason whose discovery could come as a complete surprise. Sometimes (in fact quite often) it seemed that the body in its posture and movement had become a walking metaphor for a whole life out of balance, or for a life distorted by the need to mask or forget pain. Body language was not necessarily waving your arms at the grocer for selling you a bad cut of meat; body language could be the silent and disguised voice of a traumatized life.

The more time Feldenkrais spent with people, the more often he discovered that by initiating the correction of movement he could unlock memories of old, forgotten events and buried feelings; more often than not these memories turned out to be disagreeable. Changing a posture or a movement in and of itself had an uncanny tendency to flush out stories of disappointment, of loss, or of some personal derailment. Working to improve the movement—without ever suggesting such a purpose—seemed to be all that was necessary to initiate unintended and nonverbal psychotherapy.

Feldenkrais had not originally planned to teach others to be movement trainers. Inevitably, though, people who watched him work, or who had achieved personal relief as a result of their experience with him, wanted to learn how to "do Feldenkrais" themselves. One of those people was a woman who, as a young child, had watched Feldenkrais giving his lessons to a small group of people in her own living room near Haifa, Israel. She didn't take the lessons herself then. It wasn't until she was a graduate student that she recalled having seen this man (whom she had in the meantime forgotten) some years before, doing his peculiar gymnastics with people in her parents' home.

As a teenager she had taken a few Feldenkrais lessons from her ballet teacher but remembered them later only as a pleasant add-on to the ballet lessons; her impression then was simply that the lessons seemed to make it much easier for her to dance. The full realization of who Feldenkrais was and what he was doing did not come until much later—quite suddenly, during her graduate training as a psychologist—when she heard about him and went to watch him give a lesson to a group of people in a hall in Tel Aviv. In one of those epiphanies one hears about—for her, a jolting surprise—she *unforgot* him, realized what he was doing, and decided that it was her job to become his student.

Anat Baniel is small, lithe, dark, and very attractive. You cannot be in the same room with her and be unaware of her presence. This would be true if she were hiding under a table with a blanket thrown over her, such is her native charisma. I would say she is catlike except that cats are blasé, aloof, habitually in repose. Anat is restless—either pouncing or ready to pounce. I have wondered if her demeanor now is as it has always been, or if she may have picked it up as a quirk of her association with Feldenkrais, the antimissile defense scientist. Anat is all energy—husbanded, poised, and targeted.

Before I met Anat I had already come to appreciate how potent an ally a Feldenkrais practitioner can be to a neurologist. After meeting David I found many opportunities to send him patients who had become stranded in the shambles of a musculoskeletal disorder. Some had had strokes, some suffered from Parkinsonism, some had multiple sclerosis. Always, when walking was a problem, Feldenkrais lessons improved control, balance, and endurance.

Because of this experience, I had already come to accept the potential value of Feldenkrais training when I learned that Anat had recently moved to northern California from New York. Her advance billing was impressive—she was "something special." My first opportunity to see her work came at an informal workshop I had organized for the Health Program for Performing Artists at the University of California at San Francisco. The San Francisco Bay Area, as everyone knows, has become something of a cornucopia (or a street fair) of life- and body-changing philosophies and practices. It offers every imaginable variety of massage, breathing, and herbal therapies, acupuncture, hypnotherapy, mud baths and compost packs, and, of course, Feldenkrais lessons—these are the conservative choices on the list. Wanting to respond more effectively to the unusual needs of some of my musician patients, I decided to invite a mixed group of "alternative therapies" practitioners to meet with me and one of my patients. I would describe his problem, let him discuss it, and ask for their help. My challenge to them was to demonstrate, in an open session, how they would deal with the problem: If I were to refer my patient to you for help, how would you see it, and what would you do? The patient was a young guitar player plagued by wrist and arm pain who had agreed to a public discussion of his problem. It was an eye-opening event for all of us.

Of the many interesting exchanges that took place that afternoon, the one I recall most clearly was that between the guitarist and Anat. Unlike most of the others, when it was her turn she made no move to approach the chair where this young man was seated. She chatted with him about music, about playing the guitar, and asked him specifically how he had learned to play the instrument. She asked about other physical activities he had taken part in during high school, particularly around the time the guitar had, for all practical purposes, become his life. He laughed, saying that the happiest day in his four years of high school was when he learned that studying the guitar would fulfill his requirement for a physical education course. He hated "jocks" and couldn't believe his luck at being able to spend his gym time playing the music he wanted to play (rock, of course). Anat, although her turn had come near the end of our session, was the first therapist to ask him to play the guitar.

After watching him play for a few minutes, Anat became very serious, very quiet. Then she spoke:

> Well, let me put it this way. The guitar means everything to him. He has spent all of these years since high school doing nothing but thinking about the guitar and loving it with his fingers. If you watch his hands, you will see a thirty-year-old man, very talented, very musical. However, as he has told us, the rest of the body is of no interest to him. And you can see that. If you look at him sit, if you watch him turn his head or shrug his shoulders, you will see an eight-year-old boy. And that's the problem. Those hands need the body and the experience of a thirty-year-old man to support them. If I were to work with him, I would try to get them back together. I think it would take a while, but it could be done.

And that was it. No one said anything. Her analysis registered with all of us as being transparently and distinctly correct.

Several years later I asked Anat to see an extremely difficult and, for me, upsetting case. A young woman training for a triathlon had suddenly been stricken with uncontrollable twisting and pulling movements of her neck. The condition, which is called torsion dystonia, can be, in its chronic form, one of the most devastating of all neurologic disorders. Young people in their prime have been for all practical purposes completely demolished by this nightmarish movement disorder. The head, pulled by powerful, writhing contractions of the neck and shoulder muscle, tilts severely backward and to the side, and may be pulled and twisted in slow rhythmic movements. Often, at the beginning, these movements can be suppressed or modified by tricks that the patient invents—holding the head back against a wall; resting the chin against the palm of the hand; touching a spot on the back of the head with one finger. But the situation always gets worse.

The prospects for improvement are now much better than they were when I first saw the young woman owing to the availability of Botulinum toxin, a drug that interferes with muscle contraction. It can be a truly miraculous treatment. However, this drug was not available for use in the United States when I first saw the young athlete. At that time, many people with dystonia were being advised to travel to Canada, where the drug was available. She had decided not to go. With no real options, I asked Anat if she thought she might help. She said she would try.

To this day I cannot account for what happened. The young woman had already given up her job and was housebound and immobilized. But after her third Feldenkrais "lesson" (the first of which I personally videotaped) she went out dancing with her husband. Her friends and family were so shocked by the change that they wondered if she hadn't been faking the problem from the start. Unfortunately, after the treatments stopped, the movements returned very quickly. Anat and her associate, Mary Spire (another Feldenkrais therapist whom I knew and trusted) agreed to continue the lessons, about thirty in all, which spanned nearly a year. The last time I saw her in my office, after my year in Düsseldorf, there was no trace of the problem and she had been free of symptoms for more than two years. Curiously, she was not comfortable discussing her recovery with me, as though just thinking or talking about the dystonia might bring it back. It was as if her therapy had only subdued an extremely dangerous beast that was still inside her. She was "cured," but she did not want to tempt fate.

As you read the following interview with Anat, it will help if you have a picture of her working. The room and the atmosphere are quiet; it is intimate. The "client" lies on a table or on a mat; Anat dresses simply, wearing a short-sleeved blouse, dark slacks, and sweat socks. She talks softly but moves quickly from one side of the table to the other. She touches, then reflects. She takes an ankle in her hand, lifts it, then slowly moves the whole leg until half of the body is stretched, the other half in a fetal position. She coaxes the trunk of the person over the side of the table, cradles the head, moves it in a gentle rocking motion, and sometimes offers reassuring words, as a mother might to soothe a child who needs to be held. She asks how something feels. She laughs.

Born near Haifa, Anat is the daughter of a research biochemist and inventor. Her mother is both a homemaker and an artist, and her older brother is a playwright and director, also living in Israel. Anat herself works and teaches comfortably in a realm that technologically advanced medicine tends to discount or to consider a historical curiosity. But Anat knows what she is doing, and there is no doubt Feldenkrais knew what *he* was doing when he decided to let Anat come for lessons. What I have seen Anat accomplish, and the method by which she obtains results, puts me in mind of the obsolete meaning of the word "doctor," which comes from the Latin word *docere,* meaning "to teach."

When I was three we moved to Haifa, to the top of a mountain, and I became maybe the youngest hiker in the world on my own. I would leave home and just walk, and somehow my mother didn't seem to be concerned. Miraculously, I never got lost. When I was a little older, I used to hike from the top all the way down to the seashore and then back up.

I was a very poor student; I really didn't like school, it didn't feel right, the teacher didn't feel nice. I was also a daydreamer. I remember around the fourth grade my father was driving me somewhere in his car and he asked me, "Do you like to daydream?" And I said I did; I guess the teachers had talked to him. So he asked, "Do you like to daydream *a lot*?" I said, "Yes." And he said, "How much?" And I said, "All the time."

I had a very vivid inner life, extremely so. And I was lucky that my father didn't know what to do about it and so decided not to interfere. He didn't want to spoil it. Somebody told me that one of the best things that happened to me as a child was that I was left alone, because I turned out to be a pretty unregimented person. It was years before I realized that most women are raised to *feel* that they're not as smart, or they shouldn't *appear* as smart, or they should have different *kinds* of intelligence than men. I absolutely was not informed about that.

Feldenkrais himself always liked and had women friends who were either creative or very intelligent. Some were very well known artists in Israel. My father placed a very high value on education, and he said, "You must do what you love." When my brother decided to go into the theater and my mother worried about it, my father said, "If that's what he loves, that's what he should do. People should do what they love, because they have a lifetime to do it."

Music and dance were my passions as a child, but when I went to high school I only wanted to be popular with the boys. I became the top student in my class, but my reason to do it had nothing to do with school—it was just a way to be popular with boys. My passions were music and dance, hiking, and my friends.

Once, when my father went to Paris, he took us with him; that was where I saw Nureyev for the first time. It was a life-changing experience—he was so phenomenal that the audience wouldn't let him stop dancing. I don't think it was so much the height he jumped or the acrobatics, it was something of the quality, it was a feeling, in how he moved. I don't know if anybody can talk about it directly, but the artistic quality, the passion, the man—I wouldn't have been able to say it in words at age eleven, but I was grabbed by it as soon as he walked on stage. We were all mesmerized like people in a church; it was like listening to Casals play the cello. This quality I also found in Feldenkrais.

I had met Moshe when I was very young. My father would bring researchers and once a week Moshe would come from Tel Aviv and give a lesson at our house, on the floor. I was three or four and I used to watch it. Much later, as a college student, I had been considering both dance and medicine; I tried to visualize myself in a hospital and was sure I couldn't function in that environment. At that point in my life I even didn't remember Feldenkrais or what I had learned of his work through my ballet teacher. It just wasn't in my consciousness. It was only after I began trying to find ways to use what I knew that I suddenly remembered Feldenkrais—my "lessons" as a child—and I went looking for him. I knew *nothing* about him. My father told me he was in Tel Aviv, and said, "Look in the phone book and call him up."

Anat calls her discovery of Feldenkrais "a guru story." She began attending his lessons, just as an observer. "He gave me my first lesson, and after that he couldn't peel me off. That was it. I never left."

> It took me seven years before I could start explaining what was going on. But what I *said* was: "This man opened the closed box in my soul that I didn't know I had." That was my verbal translation of my experience. It was so delicate, it was so . . . *human!* It was true contact, a feeling of the other person, true communication in the person's reality. I knew that instantaneously; I didn't need more than one lesson to absolutely know that. I didn't know whether I could *learn* how to do it. I didn't know whether I could ever understand it. But I had no question about what he was doing.
>
> I asked only one question in the first four years, near the beginning, and he didn't like it. So after that I just watched him work. I would come in and say, "Hello," then go sit in the back on a small stool—just disappear into the background. I never looked at a watch at a lesson, but when it came to the end, or almost the end, I would start crying. You know, sometimes when you hear beautiful music, you cry. It happens to me. Every one of his lessons made me cry like that, but always at the moment of the end of the lesson. Without understanding *cognitively* what he was doing, I knew when the lesson got to its resolution point. And I spontaneously responded by crying.
>
> I joked about this later: I was working as a civilian psychologist for the Israeli army while I was going to see Moshe work, and one week I couldn't come because of my work schedule. When I came the following week, he said, "Where were you?" I was shocked! At that time I wasn't even sure he even knew I was coming because we had no specific interaction—I would just say, "Good morning," and "Thank you, goodbye." I was like this orphan he let in and out. When I realized that he noticed that I hadn't come, I wondered why. I finally had the idea that when he looked at me and saw me cry, he realized it was the end of the lesson. Years later he said to me, "You were like a newborn child with no judgment and no idea. You just absorbed and absorbed. You learned from me like a newborn infant."

He was the one who started me with the children, without ever telling me or explaining that he was going to do it. He just said, "There is a child I don't have time to see, and I told them to see you." That's how it started. And so I said, "Okay, I'll go do it." I think working with children has given me this idea, which isn't often discussed in medicine: a lot of disease—medical disease and emotional "dis-ease"—is an outcome of a lack of full development. It's not something we can get to just by removing a psychological block. There actually *are* no blocks in that respect, but there *is* the block caused by lack of learning and development. In that sense, healing is a process of continuing development and learning. It's not a single, miraculous event that people imagine, a catharsis or something like that. So when I look at people with problems, more and more I ask: "What have they not learned? *What in their development have they missed?*"

Of course there are problems due to traumatic events in childhood, or disease—you name it. Feldenkrais said that ideal development would happen if the child was not opposed by a force too big for its strength. When you say to a small child, "*Don't touch that, it's dangerous!*" you create such a forceful inhibition that you actually distort the child's movement, and growth, in a certain way.

Feldenkrais taught us to look for what *isn't* there. Why doesn't movement happen the way that it should, given gravity, given the structure of the body, given the brain? For all of us there is a sort of sphere, or range, of movement that should be possible. Some people get only five or ten percent of that sphere, and you have to ask, "What explains the difference between those who get very little and those who get a lot?" Feldenkrais said that the difference is that in the process of development, the body encountered forces that were disproportionate to what the nervous system could absorb without becoming overinhibited—or overly excited, which is a manifestation of the same thing.

It is important to understand that the major effect of trauma doesn't come from the trauma itself, especially in early childhood. The trauma per se, if the system kept moving, growing, responding evenly in all directions, would just be another thing that happened and eventually become nothing. But it was just a brief event, so why does it take away so much? The answer is that *violence* distorts functioning in some way, I guess in the brain, and spontaneous growth and learning in a certain direction may just stop.

But it's not a block in the sense of an obstruction; you can't just remove it and expect everything to be normal after that. The person will need to go through a process of apprenticeship, of learning *how to be otherwise,* by developing a repertoire. I think the failure to understand this idea explains one of the potential weaknesses of verbal therapy, because you can become aware of what happened to you and why you are the way you are and so on, and gain a certain amount of freedom from it and more options, and relief from anxiety and guilt and so on. *However,* you still need to learn to do what you never learned.

Child abuse is an example; I think it always affects the breathing patterns. I work with somebody who was abused. I start communicating through touch, gently. I explore, I try to get them to feel in their body and their breathing that there is an alternative to their way of breathing. The moment the breathing starts to change, there is a very strong tendency for the memory to come to the foreground. Very often people remember the event associated with the pattern before it was changed. Moshe used to say that if the lesson is done well, and the movements done gradually, a bad memory is not experienced as a terribly traumatic thing. Sexual abuse is so violent that the memory always brings a certain amount of anxiety to the surface, yes, but somehow the nervous system seems to be able to take it.

I think what Moshe gave us was the way to access what's in the person, not only in terms of their potential ability but also their inclination and need to learn to do it. We *need* to grow, we *need* to develop; the psychologists say we need to grow, and they're actually right. People need to continue growing. However, growth is a very concrete event. It's a very *physical* event, even if it comes through talking. Talking is very physical. Emotions are very physical. Thoughts are very physical.

And the hands, the beauty of the hands, is that they speak a universal primary language. It is the language of newborns. That's the language before words. Everybody had that language or we wouldn't be alive. People who do the research say that touching premature babies fifteen minutes a day increases their growth rate by forty-five percent. I don't remember the exact figure, but it's something pretty stunning.

When I was still working in New York, I saw a child from Ann Arbor diagnosed with cerebral palsy. When he finally started going to school, he had a hard time with arithmetic. The child brought me his homework. I looked at his paper and I asked him, "How much is two plus two?" And he said, "Five." And I said, "Right! And how much is six plus one?" And he said, "Let's see . . . nine." I said, "That's right, and how much is . . . ?" And the mother, who came all the way from Ann Arbor, was about to *faint!* I said "right" even though he was making mistakes!

Of *course* he's going to make mistakes. That's why they've come! He obviously doesn't know how to do it. But I was looking to see what he *does;* I was looking to see what his nervous system is doing with these questions. How else can I find out what he needs? And I found out, by using his mistakes to guide my questions. I think it took about ten minutes. The more I said, "Right," and "Go on," and "That's fine," the more he was ready to give me information, to *answer.*

So I found out what the problem was. He had *no idea* what numbers mean. He had no feeling or understanding of what counting is. He had just tried to memorize the right answers. I spent the rest of that time and the following time with him, actually giving him the feeling of numbers, which means *distinct* events or *things.* And where's the easiest place to find that out? It's your own body. As soon as we did it with movements, he understood it. We did it up to a hundred. That was all it took; he had it after that.

I asked Anat to comment about the learning that goes on formally in schools. What can possibly be done in schools to give children a better chance to learn?

That's a big one—actually it's vital. It's not by chance that there's so much breakdown. The question is, where to start? There's a lot of talk about class size; no matter how young or old the children are, some of them need personal attention, but class size isn't the only problem.

Without going into contents or subject, I think a big problem is when the teaching is done independent of the child's subjective reality. Somebody walks into the room to teach something without taking into account in a real way the students who are there. For me, it's a little bit like what's wrong with classical physical therapy. I want this child to crawl, so I'm going to put him through these exercises, *one-two-three, one-two-three.* Well it's a great idea, but *one-two-three* not only doesn't get every child to crawl, but very often it induces a traumatic state, with dissociation, more self-hatred. Since it doesn't bring into account and connect with where the child really is and how he or she is actually operating at the time. The child needs more than *one-two-three,* sometimes many other things, in order to be able to crawl.

Let's take teenagers, later—teenagers, whose hormones are running mad in their bodies, boys and girls checking one another out and maybe yesterday they had a party and God knows what happened there and you're sitting there and guiding them through history or mathematics or German. *You're not connecting with anything that matters to them.* Our understanding is that in this way we connect to their brain.

In that sense there needs to be a revolution in comprehending what works in learning. Feldenkrais used to say over and over again, and initially I didn't understand what he meant—he said: "Teaching and learning are two independent processes, and usually they do not correlate."

Take, for example, teaching a child to read or write. There is a way to work with a nonreading child so that by the time they finally do it, it can be almost a nonevent. It happens very quickly. Or you can teach it and teach it and teach it and it's hard for them and hard for them and hard for them. And what do they learn? They learn that it's hard. You know, we learn *everything,* we don't just learn what we're supposed to learn, or what the teacher believes is being taught. So I learn that I feel horrible, I learn I don't know how to do it, I learn it's difficult, I learn my mother gets angry—you understand? I learn *everything.*

Another thing goes hand in hand with this. Both independent of whatever you teach them and in terms of the specific content of the teaching, it is vitally important for people to continue to develop their feeling world—their kinesthetic and perceptual world. It's very unfortunate when a child is put in the classroom and made to look outside of himself to learn the things "out there" and as a result needs to dissociate from himself to do it.

It is very beneficial when the focus is on the children. You can do arithmetic around the parts of the body, you can relate almost every subject through the self, you can make learning very egocentric in that sense, but you will create *much* less egocentric people in the long run, people who can really function far better. In order to do that, it is necessary to *detect* learning when it is happening; learning is usually well under way long before the outcome you're after is there. I often work with children in my practice who have severe motor problems, and I have to get them started with whatever it is they need to learn, and I have to know that what I am doing is working even though what I'm after might be two years away. I don't ever *talk* about walking or reading or writing before the child is actually ready to do it.

With Anat, it seems to me, we see again (as we saw with Reed Hearon) how the hands can bring an individual not only into a distinctive kind of work but into transforming relationships with people and ideas. As in many such cases, the hand as an instrument of action and contact may become, or seem to be, merely incidental to a more complex process or activity. But even when the hand eventually yields the stage to other skills (more true of Reed than of Anat), its historic role in the acquisition of knowledge and skill during the apprenticeship remains in the foundations, continuing to feed the dynamic processes of the imagination. In a recent report on computers in education published in the *Atlantic Monthly,* Todd Oppenheimer offers his own version of this point:

Kris Meisling, a senior geographical-research designer for Mobil Oil . . . still works regularly with a pencil and paper—tools that, ironically, he considers more interactive than the computer, because they force him to think implications through.

A spokeswoman for Hewlett-Packard, the giant California computer-products company, told me the company rarely hires people who are predominantly computer experts, favoring instead those who have a talent for teamwork and are flexible and innovative. Hewlett-Packard is such a believer in hands-on experience that since 1992 it has spent $2.6 million helping forty-five school districts build math and science skills the old-fashioned way—using real materials, such as dirt, seeds, water, glass vials, and magnets. Much the same perspective came from several recruiters in film and computer-game animation. In work by artists who have spent a lot of time on computers, "you'll see a stiffness or a flatness, a lack of richness and depth," Karen Chelini, the director of human resources for LucasArts Entertainment told me. "With traditional art training, you train the eye to pay attention to body movement. You learn attitude, feeling, expression. The ones who are good are those who as kids couldn't be without their sketchbook."[2]

Anat's experience raises a number of additional issues about human learning, to at least two of which we shall return in chapter 15: how experience *with the body,* as Vygotsky and others have argued, establishes (or, contrarily, can distort, or even block) meaning or intellectual understanding; and why learning is impeded in the absence of personal interest. Anat also raised the subject of a mentor's power to catalyze the integration of skill and intention into the creation of original and completely personalized work; this issue, which *must* be of paramount interest to teachers, is inextricably bound up with the other two.

"I only draw with software."

FIG. 13.1 Just where, we may wonder, is the computer revolution taking our children? We already know where it has taken *us!* (Reprinted from *The New Yorker,* with permission.)

14
Hidden in the Hand

NOW A SURGEON SHOULD BE YOUTHFUL OR AT ANY RATE NEARER YOUTH THAN AGE; WITH A STRONG AND STEADY HAND THAT NEVER TREMBLES, AND READY TO USE THE LEFT HAND AS WELL AS THE RIGHT; WITH VISION SHARP AND CLEAR AND SPIRIT UNDAUNTED.

—Celsus[1]

THE HEALING ARTS HAVE ALWAYS BEEN CONNECTED with the hands; sometimes—as Moshe Feldenkrais learned and then demonstrated to Anat Baniel and others—the skilled hand guiding another person's relaxed movements can induce not only more comfortable and graceful movement but memories attached to forgotten postures or physical responses to severe stress or injury. Countless variations of "hands-on" therapy have been devised and practiced in every culture we know of, all honoring and elaborating the timeless traditions of primate grooming. The ubiquity, antiquity, and variety of these practices can mean only one thing: back-scratching is more than politics. We all *need* to be touched.

The modern doctor is still very much involved in physical contact with his or her patient, but most often with a tool of some kind interposed. A veritable avalanche of gifts from science and technology seems to have irreversibly transformed the mythic office of medicine, investing in the doctor the means to preserve life and restore health in situations where that outcome would have been unthinkable a mere decade ago. And what has this transformation done to the mythic office of the hand? When Sir Charles Bell carried out his surgeries in London two hundred years ago, he was everything Celsus said a surgeon had to be. The removal of a bladder stone took three minutes of his time from start to finish—"from skin to skin," as the surgeons say. There was no anesthesia, no string quartet playing Mozart over in the corner.[2]

Technology has not rendered the surgeon's hand obsolete. My good friend and colleague Leonard Gordon began his career as an orthopedic surgeon, operating on hands, and he is now a microsurgeon. Using the kind of equipment it takes to run the space shuttle, he reattaches a baby's finger to its hand and stitches together tiny arteries vital to the survival of a nerve that is, itself, so small the naked eye can barely see it. Another colleague and good friend is hand surgeon Robert Markison, who can maneuver a fiber-optic cutting tool through a tiny incision into the wrist of a patient with carpal tunnel syndrome, inspecting the underside of the transverse carpal ligament before deciding where and how much to open.* Researchers at the National Laboratories are working on a robotic and telecommunications technology that will permit a military surgeon to operate on a soldier hospitalized on another continent.

It takes a very long time to become a surgeon, so today's surgeon is just barely "nearer youth than age" when he begins. What about the other traditional qualifications—the steady hand, the ambidexterity, the keen eye, and the undaunted spirit? Do they still count? The question is a serious one and of intense interest to the people responsible for training surgeons because the stakes are high for everyone involved: the candidates, the people who design and run the programs, and of course the patients who will eventually come under the care (or the knife!) of surgeons whose competence they must trust. Competition for available training positions is fierce, the training is many years in length (premedical training in college, four years of medical school, and *at least* five years of postgraduate residency training), and the expense is far beyond any figure you are likely to imagine. Understandably, evaluating candidates for this marathon, as with evaluating candidates to train for space flight, cannot be done casually. What do training institutions look for? How are prospective surgeons chosen?

---

* Both surgeons are avid nonsurgical, off-hours hand users. Gordon is an active tennis player—as a teenager he was the national junior singles tennis champion of South Africa. Markison is San Francisco's undisputed renaissance man of the hand. He is an active jazz musician (saxophone and flute, mainly), a serious amateur painter, a tailor (he makes his own ties and shirts), and he cobbles his own shoes! Hands down, he has the most interesting medical office in the world.

Arthur Schueneman and Jack Pickleman, at Loyola University Medical Center, have been interested in the predictive value of standard measures used by surgeon-professors to monitor and guide the progress of surgeons-in-training. Can the old boys really pick the future stars? To test the testers (and their own tests), these researchers decided to pit the judgment of seasoned surgeons against a battery of standardized psychological tests, each focusing on specific attributes deemed by the profession to constitute the "stuff" of a successful surgeon. They looked at manual speed, fine motor coordination, and bimanual sequencing; they looked at visual perception, including the ability to see important patterns buried in visual clutter and the ability to solve maze problems; they tested spatial memory; they tested the ability to perform under stress. The tests were subdivided into three major headings: psychomotor ability, complex visuospatial organization, and stress tolerance. (The Celsus formula has not changed in two thousand years: it's still hand, eye, and nerves.) They also grouped the apprentice surgeons by age, gender, and handedness, to see if there were measurable differences between the groups in performance of these subsets.*

After munching and crunching their data, they looked to see what the highest achievers in the program had that lesser surgeons lacked, or had less of, based on the psychological tests. Guess what! The eyes have it!

> In evaluating these findings, it appears that, contrary to surgical folklore, pure "psychomotor skill" (manual dexterity) is not the major dimension distinguishing the proficient surgical performance from the mediocre. Rather, perceptual abilities involving the capacity to rapidly analyze and organize perceptions based on multisensory information and the ability to distinguish essential from nonessential detail, particularly when the "signal to noise ratio" is high [*sic*], appear to be the essential precursors. . . . This is not to imply that manual dexterity and verbal abilities are not important to surgeons in performing their activities—they obviously are quite significant; but that the distinguishing features of the superior practitioner are his/her ability to "see" the relevant anatomy of the operative

* This tactic has another benefit: it can unmask a systematic teacher-evaluator bias based on those factors.

> site, even when this might not be immediately visible; to quickly identify important "landmarks" in the incision; and to mentally organize multisensory data and actions at any given point of the procedure so as to allow a smooth and efficient sequence of responses.

At the close of their paper, which, up to that point, I had been entirely satisfied with, Schueneman and Pickleman concluded:

> Pure motor ability is not the critical factor in surgical proficiency. Rather, relatively innate, nonverbal, perceptually based cognitions about complex spatial information appear to play a more central role in the operating room. . . . The results of our multimethod-multitrait approach to the study of these phenomena as they interact with the surgical environment clearly suggest that these abilities are the product of higher cortical, integrative activity of the brain—not the hands.[3]

I ask you to keep those final three words—*not the hands*—in mind as you read through the rest of this chapter. At the conclusion, you and I shall return to them; I will hand you my scalpel so that you may have the pleasure of performing the modest amputation that will make a merely beautiful analysis absolutely perfect. You will see where the problem comes from and will know what to do about it.

Long before I read the paper from Loyola, I had decided I wanted to meet a particular surgeon to see what he might have to say about his own hands—or about the healing hand in general—and to hear any predictions he might have about the importance of the hand to future doctors, given the enormity of change medicine is currently undergoing in this country.

Surgeon with a Magician's Hand

Robert Albo is not just a surgeon. He is also a magician and a preeminent historian and collector of magic. Before meeting him for the first time I already guessed that he would have a special take on the question of quick hands and a sharp eye, but I hoped he might also indulge my curiosity about the connections between medicine and the *other* ancient profession known for hands that produce magical outcomes. Perhaps he would talk about the links, old or new, between those professions, and say something about their operation in his own life.

Dr. Albo's childhood was spent in a poor neighborhood in West Berkeley, California, where, as he puts it, "every day was a fight for existence." Even as a small boy, though, both magic and medicine were in his head. When he was about nine, his parents spent one dollar to buy him a Mysto Magic Set, and he quickly learned that performing magic tricks was a foolproof way to keep the class bullies at bay: once he became a magician they stopped picking on him. During the first summer of his youthful career, he took himself to the neighborhood library, read all the books there on magic, and then, at the suggestion of the librarian, began regular treks to the *main* library in Berkeley to take on the collection of magic books there.

His parents had told him when he was a young boy that they wanted him to become a doctor, and he never resisted their prompting: "I think the only reason I was a good student was because I wanted to get into medical school. I always understood that my grades had to be good enough for medical school, and I studied with only that thought in mind." But the real passion he harbored was never eclipsed:

> I didn't pursue anything else as much as I pursued magic. I knew before I was ten years old that I was going to be a doctor, but the impetus to my life was the magic. Magic was almost a personality crutch—it opened every door in the world to me. It didn't matter where I went; as soon as people found I did magic, suddenly I was everyone's friend. When I was a little boy, ten years old, adults were practically falling all over me.

Magic proved both socially valuable and financially indispensable, especially during college and medical school years. His love of performance helped define what later became the specialty for which he is now renowned in the world of professional magicians: a collector and historian of magic. "Over the years it just went on and on and on, until I had one of the largest private collections of magic apparatus in the world."[4]

I asked him to speculate about the connections between medicine and magic; he replied that not long ago he would have seen very little connection at all; his interest in the history of magic, because it was primarily an interest in performance, had not until recently reached farther back than the time of Robert-Houdin. But when he began his research for a series of lectures he planned to give on the ancient history of magic, he became aware that his two great interests were not as separate as he had kept them in his own life.

> Illusions were being used to control people five thousand years ago. Certainly the priests and the magi did.* They would light an urn in the front of the temple to cause the door to open by simple pneumatics and water—heating water to create steam to turn a wheel that would open a door. To ignorant people, that was a miracle—God himself must have been opening that door. An important distinction from the earliest days was between white magic—performed by the good priest—and *black* magic, which was associated with the sorcerer, the witch, and the witch doctor.
>
> Magic as we know it today arose from the good side, when magic pulled away from the priests and became strictly entertainment. Magicians were circus performers who entertained for a price, without attempting to influence the mind. Eventually, Robert-Houdin took magic out of the circus, made it an art form, and put it into a small theater. That change has always been the basis of my interest.

Albo explains that the illusions in performance magic are based on deliberate and carefully rehearsed deceptions—not the bridge to medicine most doctors would be most eager to cite in explaining the kinship of their profession with magic. Of course, medicine as an organized profession specifically and aggressively seeks to distance itself from deceptive practices; medicine officially abhors the aroma of snake oil. Despite all this formal protesting, however, medicine recognizes that an effective physician regu-

* The word "magic" is derived from the magi, priests in the ancient Persian religion founded by Zoroaster, who had the ability to control *daevas*, or evil spirits. The older Sanskrit word *devas* simply means "god" without implying good or evil, and it is related to *divyas*, meaning "celestial."

larly uses his or her authority in a charismatic way to reinforce the scientific, or "objective," steps taken as part of any counseling or other treatment. Why? Because it helps.

It is in the use of charismatic persuasion that medicine's oldest links with magic are apparent. After all, magicians (or sorcerers) have always relied on their personal magnetism as a means of manipulating the attention of subjects and audiences. For them, doing so was an effective way of demonstrating the possession of unusual powers (occult or religious). Having demonstrated their ability to perform astonishing feats, magicians could attribute their powers to "sacred" or "occult" knowledge or empowerment, and thus influence others to think or behave in a desired way.

Evil or benevolent, the old magician always knew that what he was doing was fakery and that it was the subject's credulousness and distracted attention that provided what was necessary to produce the illusion. The principle was innovative, and it has always been useful to those who understood it. The existence of a psychologically based influence on the receiver of magic anticipates the modern idea of a *legitimate* charismatic use of personality and authority by physicians, placing the cherished bedside manner in the category of "white magic." Doctors regularly try to "charm" their patients into following instructions (to take a specific medication or to stop smoking) or motivate the patient psychologically as part of a treatment.[5] Lewis Thomas, who was dean of both the New York University and Yale schools of medicine, called the doctor's charm "an inborn capacity for affection, hard to come by but indispensable for a good doctor."

But must the doctor literally trick the patient into believing that a truly effective treatment will work? I asked Albo to compare the role of persuasion and deception in medicine and magic—how information is used or withheld to ensure a desired outcome.

> Well, as you know, there is more than one type of patient, and there is more than one type of audience. There are the people who cannot stand to be fooled and who want to know how everything is done. They are distressed and disturbed over seeing a trick that they can't explain. They want to have *some* explanation, and they're far happier watching the performance even if the explanation they have is a faulty one. There are other

Hidden in the Hand

FIG. 14.1 The ancient ties between medicine and magic arise naturally from human responsiveness to illusion and the impulse to seek assistance from supernatural forces. This drawing with our chapter title replacing the original caption is from Robert J. Albo's *Further Classic Magic with Apparatus* (1979), volume 4 of his long-running encyclopedic history of magic, the ninth volume of which is nearing publication. (Reprinted with permission.)

> people who go into a performance who really don't care to know how anything is done. They're there to be entertained, and it doesn't faze them in the slightest that you are doing a trick. They are perfectly content to sit there, to watch and be entertained. They have no concept of how the trick is done and they don't care.
>
> The same is true in medicine. There are those patients who come to you, and you say, "Mrs. Jones, you have a cancer in the breast," and she doesn't care to go into every detail. "You take care of it, doctor, that's fine" is her response. But there are other patients who want the knowledge, the insight—every little detail, all the options—everything laid out in front of them so that they can retain control. The ones who want *you* to have control over it say, "That's fine with me; you tell me what to do and I'll do it." They're wonderful, they're intelligent and they're educated, but that's as far as they want to go.

Traditional "healers" in almost every culture have attained a high level of respect on the strength of their natural gifts and their knowledge of certain, shall we say, trade secrets. But we now live in an era of considerable edginess over any diagnostic or therapeutic method that has not been scientifically validated, and we have a strong tendency to consider older treatments outmoded or ineffective simply because they are older or came into use as folk remedies. We think of truly ancient medical practices as offering no more than mild effects (as with herbs, massage, relaxation, and so on), or as working only through the power of suggestion. If treatments seem to be based on outright deception, we call them quackery or a sham and try to outlaw them. Albo comments:

> I do think there is a correlation between the shaman—the healer and the magician—and the sham, the fooler.* I suspect that the early healers were no better or worse at healing than perhaps real doctors were in the early days. It's only in recent years that we've become adept at healing. In the

---

* Actually, the words "sham" and "shaman" have no semantic relationship, although we seem to use them as though they did. The origin of the word "sham" is unknown; "shaman" is of Siberian origin.

early days of medicine, bedside manner counted for more because there really wasn't much else a doctor could do. The early doctors were guilty of offering sham treatments, and the movement in this century to professionalize medicine, to place it on a scientific footing, certainly came about because of that problem.[6]

Our conversation shifted from the historic overlap of thematic boundaries between healers and magicians to the common need for a deft hand. Albo commented on this parallel between the work of surgeons and magicians, in the process bringing up the work of America's reigning master of prestidigitation, Ricky Jay.

Prestidigitation is from a French word meaning "quick fingers." A *prestidigitateur* was a person who performed magic. The person who coined the word did so because he did not like the connotations of the word "magic." He was a French nobleman who lived in the 1840s, and he used that word to refer to himself in an aristocratic way. He didn't want to be associated with anything as low-life as the circus or the charlatans who went around pretending to change base metals to gold and that sort of thing, which was very popular at the time. A *prestidigitateur* was still a magician, but a much higher class magician.* The best now, of course, is Ricky Jay. He is a magician's magician. I've watched Ricky many, many times and he's a *wonderful* performer. Ricky's real gift is finger sleights and flipping—that kind of thing—and he just amazes you with his skill.[7]

In surgery, as far as the hand itself is concerned, of course dexterity is essential. Hand-eye coordination has always been one of my assets, and that has obviously been important in my interest in sports, and it may be part of why magic came so easily to me. I think I also owe a great deal to the fact that I could read a description of a trick in a book and make sense of it. I could always extrapolate from written instructions to my hands without any real difficulty. It was easy. Once I had it, I had it, and I never lost it.

---

* The word "legerdemain" is the English form of the French *leger de main* ("lightness of hand") and, like the word "prestidigitation," means "performing tricks."

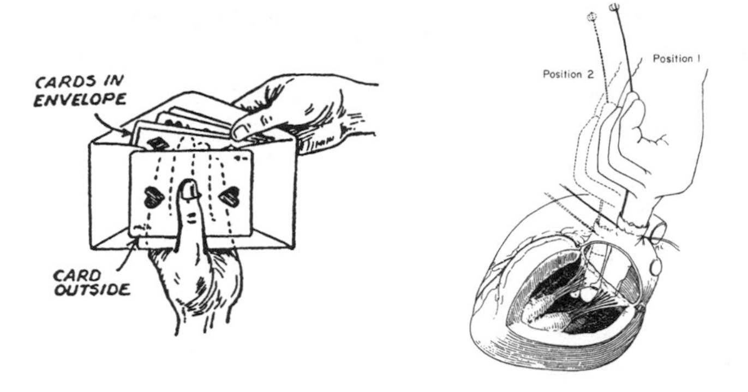

FIG. 14.2 Robert Albo's early interest in magic was in part supported by a native ability to visualize from verbal descriptions the complex maneuvers used in magical illusions. The same ability has served him well as a practicing surgeon. Edward Tufte called attention to the parallels in visual descriptions of magic and surgery in his latest book, *Visual Explanations: Images and Quantities, Evidence and Narrative* (Cheshire, Conn. Graphics Press, 1997). The card illusion is from Jean Hugard and Frederick Braué, *The Royal Road to Card Magic* (London/New York: Faber and Faber, 1948): the cardiac surgery illustration is from Dwight Harken, "Surgery of the Mitral Valve," in Dwight Harken, ed., *Cardiac Surgery* (Philadelphia, Pa: F. A. Davis, 1971). Both illustrations reprinted with permission.

> The same thing happened in surgery. When I was taught in the early years how to tie knots, they only had to show me once. Then I had it. I never had to labor with it. And I've always been one of the fastest surgeons in this area. I'm able to dissect, to operate, more easily than I think most other surgeons that I work with. It just comes easily to me.

Before commenting on the conversation with Dr. Albo, and indicating why I believe his experience has implications well beyond its corroborative value to the conventional view of the hand-eye skills of the surgeon, I want to turn the tables on the medicine-magic story. To do this, I introduce into our discussion a second individual, Mark Mitton, a former student of Quintino Marucci (a.k.a. Slydini), one of the most revered "close magic" artists of our time. Although Mark has no formal connections with the medical world, he recently discovered that even in a modern hospital the juices can be made to flow in the deep roots connecting magic and medicine. Working strictly in the capacity of an entertainer (but acting out of his intuitive response to a hospitalized child), he used his skills as a magician with an overtly medical objective, and with a stunning result. As he described to me what he did, I heard the clear voice and the confident understanding of a physician; he had acted because he saw the problem and knew what had to be done.

MAGICIAN WITH A GRACED HAND

Mark Mitton and I met in his apartment in New York shortly after he and his wife had moved there. We sat at the table Slydini used for his close magic, and which Mark now owns. Mark projects an intense, open friendliness that alerts you to the possibility that he might not be a native New Yorker. He is, in fact, a Canadian by birth, although he has lived in the United States much of his life. Like Robert Albo (whom he knows), Mark was smitten by the magic bug when he was nine years old. That was when he was given a Chex Magic Kit, and the year he saw his first magic show.

By the time Mark was a teenager living in Superior, Minnesota (near Duluth), he was hosting a children's talent show on a local television station and performing sleight-of-hand magic. He saw his first tape of Slydini when he was sixteen and says it inspired him to work even harder on sleight-of-hand technique. But he was equally interested in the psychological side of magic, particularly "psychics and con games," and he suspects this was because he grew up in a very conservative Baptist family. He attended college in the United States, first at American University in Washington, D.C. and then at Haverford College in Pennsylvania. One of his friends at Haverford was a physics major, and the two of them had frequent conversations about the overlap in their interests; they even drew meaningful connections between Max Malini's *Book of Magic* and the physics lectures of Richard Feynman.

After graduation, when his college interests in economics and politics failed to produce a job, he lived with his parents for a while. During that time, he studied karate and "took some magic lessons from a con man." Although he had not anticipated it, karate later turned out to be useful, when he became more serious about magic, because it taught him how to mirror someone else's movements.

Before long he moved to New York, where he began working with a street magician. He decided it was time to meet Slydini, who by then was eighty-three.

> One Saturday I met Jim Sullivan, who worked under the name Cellini and who had been a student of Slydini. We had a long conversation, and all he could talk about was Slydini. He told me what Slydini *really* knew, not what the magic fraternity *thought* he knew. He said, "Every day I thank God that I got to study with Slydini," and he told me what I'd have to say to get Slydini to take me as a student.
>
> So I met Slydini. He was very formal. Always in the lesson he would say, "Do what I do." I had to replicate his movements, and the mirroring made it easier. Mirroring takes a certain skill, and once you know it, once you're in rhythm, it's very fast and it's beautiful. Slydini had a wild technique, but he understood perception in a way I think no other magician ever has. He approached it scientifically, really broke it down. For example, there is a vanishing paper-ball trick based on a principle he called "the coordination," and it involves something called an "erasing movement." When this trick is done by people who didn't study with Slydini, it's actually sort of vulgar—you see a guy chucking paper over someone's head. But in Slydini's hands it was beautiful, and part of it was his letting you see how the person was being deceived. It wasn't just a cheap trick, because Slydini was a real *illusionist.* That's what he taught his students to be.

Mark has high praise for practice but, like juggler Serge Percelly, believes that performing is the secret to learning.

> You can learn from books or from a teacher, but you have to practice, and you have to perform. And nothing else helps you as much as performing. Doing a watch steal is a great example of what I mean. Part of what we do is sleight of hand, and part of what we do is out-and-out deception. If a person takes your watch, you know that a guy's taken the watch off your wrist—it's not really a secret.
>
> The reason so few magicians do the watch steal is that to learn it *you*

must be caught doing it. How do you make that acceptable? Either you find someone who knows you're going to take their watch, or you have a joke to cover you if you do get caught. And that's only the beginning, because in addition to the technical skill and the deception, you need a presentation. Did *you* take the watch, or do you make it look like someone else took it? All this has to fit together. Slydini had very, very strong feelings that it was the overall effect that creates what people experience as real magic. He said, "Listen—I teach you the trick, the trick teaches you the principle, and the principle teaches you magic."

What he meant was this: the only way to teach was through very specific routines. That's what he did. As a student, only by learning and *doing* the routines and by performing them could you start to understand the vocabulary. Anyone who studied with him for less than two years would do everything exactly as he did. It wasn't until later that he would teach you to mold the trick around yourself.

After we had discussed the trick and I had been treated to a demonstration, Mark agreed to talk about his hospital work.

You can hardly imagine the environment kids face in a big city hospital. Maybe it doesn't sound like much, but a magician doing tricks for the doctor, the nurse, the child, his parents, and sometimes even the janitor can get them all together. That was most of what I did, but I was involved in one incident that was different—we helped a boy who couldn't walk even though he was *supposed* to be able to. This boy was there because of a serious kidney condition, and there were a number of factors in play, but still, he was supposed to be able to walk but they couldn't get him out of his wheelchair. After I had spent some time with him, I started thinking about where *I* was at age fourteen. That's not hard: I was a kid and I wanted a job. Carmelo was constantly hustling people for candy, perks, anything he could get—he was basically a hustler. My idea was to find a way to get him to hustle *us* for money. To me, maneuvering him into trying to get a job out of us was the same thing as lifting a watch.

So one day I did a trick we call a card read. You take a card, look at it, and I look in your eyes and tell you the card. It's very intimate; it elicits an involuntary eye-movement whenever you see the card you picked. It's very easy to spot.* I worked out a way for us to do the trick together. I would actually do it, but put myself *way* in the background—like I was doing nothing, which is a big Slydini thing. In exactly the way you eliminate a *movement* from a person's mind, you can also eliminate a person's presence from someone's mind. You're just playing with their perception. So I did the trick and Carmelo got the credit.

One day after we did the trick I congratulated him by saying, "You are an *amazing* actor!" I *meant* it, by the way—and he said, "You know something, you're right. I *am* an amazing actor." And he went on and on about how he was a wonderful actor. I told him he was good enough to join us, and if he did we'd pay him. This was just a *hustle;* it was a *straight* hustle. It was very much the way you take somebody's watch. At least that's how I was thinking about it.

Mark found his chance to draw Carmelo into a confidence game in which Carmelo was the unknowing target.

That moment came three months later, when he decided he wanted to work with us. Since he'd been so sick, and there was no family, he didn't have any chance to make his own way and he was excited about the idea of making money. Even being ill and stuck in the hospital, he still had all the impulses of a fourteen-year-old New York kid who liked to hustle people. At least, I was betting on that, and that's what gave me the chance to pull the little twist.

Three months later, when he was ready, I came in and I said, "I'm really sorry, but it looks like we can't use you after all. I had everything all ready—the money, everything, but we can't use you because I wasn't thinking about your wheelchair. I mean, I didn't know that you can't walk." I told him—truthfully—"We work four floors a day and we use the stairs.

---

* Mark demonstrated the trick to me. I knew exactly what he was doing and still could not defeat him. It was over in ten seconds.

FIG. 14.3 What would John Napier have called this grip? Mark Mitton, an up-and-coming master of "close magic," demonstrates the technical side of a famous sleight-of-hand illusion. As he masks his own extra-prehensile manipulation of cards, coins, and handkerchiefs, Mark unmasks the shortcomings of conventional descriptions of the movements of the human hand. However, as he tells us both in his discussion of Slydini and in his story of Carmelo, a great deal more than unconventional dexterity is needed to produce real magic. (Photograph copyright © John J. Pavlik.)

If we have to wait for you—wait for these elevators—it's going to throw off the whole schedule, and we won't be able to do our job."

This is another common technique in magic. You take something that's very straight, that's *true,* you take hard information and you *twist* it. I took real information—the slow elevators, which he certainly knew about—and just said, "I'm sorry." It was *very* traumatic, and he was crying, but when I went to the door he stopped me and said, "But, I *can* walk." It was just like in the movies. I pretended like I didn't believe it. It was straight hustling—well, worse than that, it was straight ball-busting.

And it did work. That weekend, his nurse told me he'd made a deal with *her:* "I tell you what—if you get me a pair of sneakers, really *nice* tennis shoes, I'll walk." He'd hustled *them* every chance he could up to that point, so that was no surprise. She obliged, got sneakers for him, and he walked. It took us some time, but he got out of that chair. I don't want to pat myself on the back. To me, it was just a very straightforward endeavor. It was dangerous, potentially, and I never wanted to forget that risk. We had quite a long talk about the reaction of the hospital staff to this event. I think it really led to a closer working relationship between our group and the staff.

Reflecting on what happened, Mark commented:

> All of this, of course, is *not* an appropriate activity for clowns in hospitals, but as a person with a strong religious background who was raised on the idea that it's not enough to *be* good, you've got to *do* good, I decided it was appropriate in this instance. In a way, I think the impulse to do this came from reading Plato's *Republic.* The first words, "He went down"—the allegory of the cave—was what I was thinking about with this kid. That's what Slydini did with me. That's what my karate instructor did with me. Each of them saw where I was, met me, and then we went somewhere else together. They were able to meet me because they could remember the time in their life when they were where I was. My fears were real to them. And my lack of fear was also real to them. I could see both of those in Carmelo.

Mark's experience suggests that medicine can sometimes benefit from some of the *old-fashioned* brand of magic, the kind that needs the expert surgeon's "complex visuospatial perceptual ability," but with something added to it, something we might, just for fun, call *psychospatial* perception. Actually, Lewis Thomas had the right word for that something extra: empathy. For Mark, it was "seeing where the other person is, to understand his fear and his *lack* of fear."

Unfortunately, Carmelo did not survive his illness, but genuine "healing" did take place, and it was set in motion by a young magician who met a gravely ill city kid head-to-head, hustle-to-hustle. Mark *was* a physician here, determined to learn the truth about Carmelo and to use that knowledge to help the boy reclaim not only his lost mobility but his dignity at the end of his life.

Both medicine and magic are built on the willing suspension of disbelief that must precede a temporary surrender of autonomy by one person to another. The patient in a doctor's office or in a hospital and the person in an audience watching a magic show—whether or not they need to know *how* the treatment (or the trick) is accomplished—participate in a ritual shifting of power and responsibility to another. Conceding helplessness, the patient says to the doctor, "I trust you. I know you can heal me." The magician is placed on the same kind of pedestal, even if it is only theater. For just a little while he is clairvoyant, wise, and strong. He contains powerful knowledge and can work magic. Sometimes it really *is* magic.

"Not the Hand"

It is now time to return to Loyola to perform the amputation I mentioned earlier. To refresh your recollection, researchers there completed a study of surgical residents and discovered that what distinguishes the superior surgical resident from the others is higher cortical integrative action, specifically in the domain of "complex visuospatial organization." It is a complicated study but it looks clean to me. In point of fact, having spent my early days in medicine as an emergency room physician, and having in that capacity repaired a good many horrendous lacerations, I can vouch for the surgeon's need to know what he is looking for and to know what he is looking *at,* even when the problem in front of him bears no resemblance to anything in a textbook, anywhere. So I have no quibble with the conclusion that the surgeon's *talented and trained eye* makes all the difference in his ability to get results.

But after authors of the study crossed the finish line, they kept going and threw themselves into a ditch. For reasons they do not explain, they strongly imply that the hand has nothing to do with visual perception. This is incorrect. Every surgeon who ever lived had a life history, beginning at the moment of birth, and that life history records a shared apprenticeship of the body—all of it—and the mind. This is not poetry; it is a declarative statement with a wealth of observation and analysis behind it. During all the years this youngster (whose head at some point was flooded with dreams of becoming a surgeon) was in precandidate status, his or her brain was collecting information through the hands and eyes and ears and tongue and nose, all together, in order to make sense of the world. This surgeon's-brain-to-be did not send messages to the hands and then ignore the messages the hands sent back. The reason messages were going to the hands in the first place was so that the hands would reach for, grasp, touch, turn, weigh, join, separate, bounce, and so on, whatever it was that came into their possession. The hands were moved, among other reasons, to obtain information that could be obtained *only* by acting upon the object being held. The information returned to the brain was written in the tactile and kinesthetic language of manipulation and was compared with information coming from the visual system, as part of a process through which the brain creates visuospatial images. When Robertson Davies said the hand speaks to the brain as surely as the brain speaks to the hand, he was explaining how things work.

Technically, of course, one can say that the researchers at Loyola were right to conclude the article as they did. They cannot alter a surgeon's past; all they can do is assess the current state of the individuals in front of them. How Jane and John Surgical Resident got to Loyola is irrelevant once they are on campus. I concede that point to them.

But I do *not* concede that they have provided an explanation for the differences they have discovered and are describing. Every surgeon I have ever met, including Drs. Albo, Gordon, and Markison, was fascinated by the behavior of intricate and manipulable objects from the time of early childhood. They were not merely fascinated with how things looked; they were fascinated with how they felt and behaved in the hand. It matters not a whit whether they were perceiving tennis balls or Chinese disappearing balls—or hitting a ball or picking cards from a deck with their eyes closed or open. To paraphrase Tim Ingold, perception is not something that goes on inside a processor running inside the brain. There is not, and cannot be, anything called perception—including any kind of visual or visuomotor perception—just as there is not and cannot be anything called intelligence, independent of the behavior of the entire organism, or of its entire and exclusive personal history of interactions with the world.

Selection committees for surgical residencies may not need to know this, but parents and children's teachers do.

15
Head for the Hands

THIS MAY NOT SEEM RELATED TO MATHEMATICAL DEVELOPMENT, BUT IN MY MIND IT IS: I BEGAN BUILDING MODEL AIRPLANES OF BALSA WOOD WHEN I WAS VERY YOUNG. BY FIVE, I WAS ALREADY BUILDING THEM BY MYSELF. I USED TO SEE IN MY MIND WHAT THEY WOULD LOOK LIKE AND HOW THE STRUTS WOULD GO IN AND TURN THE DESIGNS OVER IN MY HEAD. MY FATHER WOULD HELP, BUT SOMEHOW I WANTED TO DO IT ALL MYSELF. (*Mathematician No. 7*)

—from W. Gustin, "The Development of Exceptional Research Mathematicians"[1]

INDEPENDENT OF WHATEVER WE DO, THE CHILD IS A CURIOUS, QUESTION-ASKING ORGANISM, EVEN BEFORE IT HAS LANGUAGE. THE IMPORTANCE OF THIS INTRINSIC CURIOSITY CANNOT BE OVERESTIMATED BECAUSE IT IS SUCH A POWERFUL GOAD TO THINKING AND ACTION. . . .

—Seymour Sarason[2]

THIS BOOK HAS BEEN AN INQUIRY into the premise that the hand is as much at the core of human life as the brain itself. Whether you agree or disagree, the question calls for an answer. The hand *is* involved in human learning. At the outset of the book, I asked: "How does, or should, the education system accommodate the fact that the hand is not merely a metaphor or an icon for humanness, but often the real-life focal point—the lever or the launching pad—of a successful and genuinely fulfilling life?"

For a great number of people, the hand is the focus of years of specialized training and becomes the critical instrument of thought, skill, feeling, and intention for a lifetime of professional work. Can the experiences of such people be applied to *all* learners? Could anything we have learned about the hand be used to improve the teaching of children? The specific educational needs and interests of individual children—*your* children or grandchildren—cannot be accommodated by our educational system; it is a major inconvenience to the design of grand schemes for education that each child is different.

In his new book, *The Educated Mind,* the Canadian educator Kieran Egan suggests that any strategy for educational reform must address the effects of education's long history, during which three educational goals have come into being:

> that we must shape the young to the current norms and conventions of adult society, that we must teach them the knowledge that will ensure their thinking conforms with what is real and true about the world, and that we must encourage the development of each student's individual potential.[3]

Egan agrees that these goals are equally desirable; unfortunately, he says, *as a charge to educational institutions,* they are also mutually incompatible. The oldest idea of all, education as socialization, is essential for the survival of any society, whose young must gain the necessary competence for adult responsibility and participation in the life of the culture. Plato's academy challenged that idea, offering as a radical (and elite) alternative a curriculum whose goal was the cultivation of mental discipline through the systematic search for "truth." Not only did the academy *not* augment or reinforce existing educational thought or its institutions in ancient Greece, it was a pointed rebuke to the training of the young for "mindless" conformity with society.

Plato's educational ideal, however, proved notoriously difficult to realize. As Egan reports, Jean-Jacques Rousseau, with the publication of *Emile* in 1762, exposed the failures and the dark side of the academy by showing it to be a place where (in Egan's words) "dull pedagogues took the forms of knowledge that made up the curriculum, organized them into what seemed the best logical order, then beat them into the students. The result was misery, violence, and frustration." The correction of this abuse advocated by Rousseau, and presumably a "fix" for the failures of the academy, was the adoption by schools of "natural" (that is, developmentally oriented) teaching methods. His revolutionary idea is now strongly associated with the names Piaget, Vygotsky, and Dewey. The incompatibilities among educational objectives were not resolved by Rousseau and his followers, however, because Rousseau's idea is as much a revolt against Plato as Plato's was against the schools *he* sought to replace. A developmentally oriented teaching practice is *not* just a better way to realize Plato's search for truth. As Egan says:

> Rousseau and his modern followers are not simply making methodological or procedural recommendations that might allow us to do the Platonic academic job more efficiently. They are actually recommending a different job. . . . In the Platonic idea, knowledge drives development; in the Rousseauian idea, development drives knowledge.[4]

In a series of recent books, Yale psychologist Seymour Sarason focuses on the third of Egan's educational goals. Referring to himself as a "hedgehog with one big idea," Sarason says that education can succeed only when it understands that "every human being is born with the potential to be creative and artistic." The basic inquisitiveness of the human mind serves the fundamental desire of the human to establish meaningful relations between himself or herself and the world, and (intrinsic to that process) to "put his or her personal stamp on some aspect of the surround."[5] Sarason does not pretend that his prescription for education can or should blend the best of education's great troika of ideas. His mission is to revitalize for both parents and educators John Dewey's understanding of the nature of the artistic impulse and its relation to both ordinary life and human creativity. Dewey, in his book *Art as Experience,* wrote:

> The task is to restore continuity between the refined and intensified forms of experience that are works of art and the everyday events, doings and sufferings that are universally recognized to constitute experience.[6]

Sarason began an energetic advocacy for alternative education years ago, after losing an argument with the superintendent of a school for institutionalized adults. He tells that story in *The Challenge of Art to Psychology,* in which he introduces his readers to the work of Henry Schaefer-Simmern:

> I met Schaefer in 1942 at the Southbury Training School, a residential state institution for mentally defective individuals. These "children" were not supposed to be capable of working willingly and enthusiastically for more than a very short time, but what I saw were people struggling for as much as three hours at a stretch to give form to their internal imagery. And, wonder of wonders, *they* decided what they wanted to do. Schaefer's role was to study with each of them *each* of their products. . . . Henry Schaefer-Simmern *never* told a person what or how to draw. He started with a suggestion: "Draw what you want to see so that you see it the way you want to see it, however simple it may be." Following which he would ask the person, "Do you like what you have done? Study it. Is there anything you want to change? Is there something in the picture that bothers you? Do you want to try again?" As he liked to put it, "When we form something through artistic activity, we are formed and changed in the process, and that spurs the developmental process."[7]

Human learning, Sarason argues, will proceed without respect to any predictions of normative measures of "intelligence" so long as several fundamental and comparatively simple conditions are met. Of these, the most important is that all students learn best and most quickly when self-interest orients and drives the search for information, understanding, and skill. This is equally true of infants, young children, adolescents, and younger and older adults, and it applies equally, without respect to assessments of mental capacity. What this means, if Sarason is right, is that any student who chose *spontaneously and voluntarily, on the basis of his or her own inclinations and experience,* to become a brain surgeon could probably prepare adequately to enter that career by the age of twenty if those in a position to support that ambition provided the necessary assistance or at least did not stand in the way.[8]

There are other essential components of the modernized "child-oriented" approach to education. Of these, the most impressively documented is a study called the Development of Talent Research Project. The project, carried out at the University of Illinois by a group headed by Benjamin Bloom, was a retrospective study of the experiences of approximately 120 adults (based on interviews with each of them, and with their teachers, friends, and family), all of whom had, by the age of thirty, gained national recognition for their success in six very different, highly demanding, and competitive pursuits: concert pianists, sculptors, mathematicians, research neurologists, Olympic swimmers, and professional tennis players. In almost every case, most often beginning well before their teenage years, these children had responded to a strong family work or achievement *ethic* (though almost never defined as a specific career expectation) by investing their time in activities that interested them.[9] Lauren Sosniak, whose primary research involved the pianists, said of these children and their families:

> Over time, it seems they created an interdependent and self-sustaining system of mutual encouragement and support. . . . No one had any idea what they were getting involved in at the start; no idea how long it would take, no idea where it would lead. In fact, I would argue now that if the pianists and parents had striven for huge successes from the start, they would probably have been less successful than they were. It seems that the parents' and teachers' spontaneous responses of approval and delight for the youngsters' work were all the more meaningful because they were unstudied. . . . Because aspirations and expectations of concert soloist status grew with the development of skills and understandings, rather than the far-off goal serving as a stimulus for their development, there was never so much to be done that the task would seem overwhelming. . . . *The pianists learned to work toward more difficult and distant goals as they learned to care about achieving those goals.*[10] (Italics added)

Jeanne Bamberger, another modern-day apostle of John Dewey, started life as a piano prodigy—she was a student of Artur Schnabel. Just over ten years ago, Bamberger threw herself into the trenches, along with teachers, students, and student teachers, at an inner-city school in Cambridge, Massachusetts. This move was prompted by her earlier years of research on the differences in approach taken by prodigies and "unmusical" children in the study of music. Led by those findings to question her own assumptions about the nature of learning—more particularly, about the design and conduct of formal instruction—she decided to look at fast learners and slow learners in an open-school, general-curriculum setting.

Although her initial intention was to focus on the children, she quickly discovered that she could not make sense of the efforts of the children without looking at the teachers as well. Educational researchers have been doing this kind of thing for many years, but Bamberger stumbled onto an entirely new approach to the problem as she kept turning her "microscope" to higher and higher magnifications.

> The *Laboratory for Making Things* was created in an effort to address a poorly understood but well-recognized phenomenon: Children who are most successful, even virtuosos, at using their hands to build and fix complicated things in the everyday world around them . . . are often the same children who are having the most difficulties learning in schools. . . . With the emphasis in schooling on symbolic knowledge, it is not surprising that attention focuses on what these children *cannot* do, and it is also not surprising that the school world sees them not as virtuosos but as "failing to perform." [We started] from a different assumption, namely that "hand knowledge" and "symbolic knowledge" constitute equally powerful but different and not equally appreciated ways of organizing worldly phenomena. . . .[11]

The teachers participating in the program met regularly before the "laboratory" accepted its first students and continued to meet after it had begun operating. These meetings were occasions for discussing the presentation of curriculum material and evaluating what occurred during teacher-student-curriculum interactions. Bamberger recalled for me one such session in which Mary, one of the teachers, described what happened when eight-year-old Jeff, who was building a mobile, tried to balance two objects of unequal weight attached to a stick suspended from its middle. Mary noticed that Jeff seemed to know where to push each of these weights to keep from tipping the stick. When she asked Jeff how he knew what to do, his answer was that he "just knew"; pressed to explain, he said, "I had a feeling of it, like on a teeter-totter." Bamberger's comment:

> In telling the story to the teacher group, Mary asked: "What is it that Jeff 'knows' in being able to do that, how did he learn it, and what did he mean by his comment about the teeter-totter?" The teachers found themselves facing some confusion. Unlike Jeff, they had, of course, learned the formula of "weight times distance," but what they had been taught and what they had learned to *say* seemed disconnected from what they could directly see and feel. . . .[12]

A very significant component of the project, and indeed perhaps its most significant accomplishment, was the exploration between "in the body" knowledge and formal verbal representations (or attempts at representation) of that knowledge, both on the part of the teachers and the students. Bamberger further elaborated on these issues in an interview with me:

> We spent a lot of time in the lab helping kids to move between building things, then making descriptions of what they were doing. For this particular group of kids, the biggest problems were with any kind of symbolic, verbal, or numbered material—getting things on paper—and moving back and forth between action and description. We found kids who'd had trouble learning to read. And kids whose problem was not so much *doing* math but *representing* it. Some of them, for example, were terrific at solving mechanical problems, solving problems in building a gear machine, or figuring out how to connect electric circuits. But—and this is the critical point—when they would make instructions so that *someone else* could build what they had built, or when they tried to describe how it worked—something like that—their descriptions, their drawings, and even their notations might focus on features quite different from those you or I might think were the important ones, or were the ones kids are taught to notice in school.
>
> Sometimes, as they were moving back and forth between doing and explaining, someone would get a stomachache because they didn't understand how this lever works, or why *weight times distance* doesn't make any

sense at all. And it wasn't just the students: the teachers became genuinely intrigued with their own discomfort and confusions. They were intrigued that a particular problem was making them confused and wondered why somebody else *wasn't,* and wanted to know what you could do about it.

The teachers, like most adults, were used to keeping neatly separate their school knowledge and their everyday knowledge. But the moments of confusion helped them recognize and confront the incongruences between these ways of knowing—on the one hand, the know-how, the "hand knowledge," that worked "by the feel of it," and, on the other, the meanings implicit in learned formalisms. They learned how to ask questions that probed these incongruences even when that made them very uncomfortable.

I think it's *critical* for teachers to look at their own understanding of these very ordinary things that everybody teaches. Because if you can begin to remember what it was like before, when you didn't know these things, you can get at it. It's very difficult, though, because you've wiped out the memory of *not* knowing. You can't imagine what it was like when you didn't know how to divide fractions. Even a first-grade teacher has her way of teaching, and if it doesn't work, she doesn't have very much to go on, because *she hasn't ever thought about what she knows already!*

It's critical to find out what you *do* know. And look at it, really examine it, not just to take it for granted, but ask yourself about it. That's just as important as finding out what you *don't* know. In fact they go together. You start looking at what you do know, the nature of your understanding of addition—just something like "Why is it that $5 + 7 = 12$?"—and if you aren't careful you very quickly get to something like "What's a number?"

Teachers look at what we're doing and say, "We can't do this." That's what people say, but it's *not true!* It does take time, but think of the payoff! And what are the alternatives? You just keep laying on methods.

In Bamberger's view, the classroom teacher can remain "*still curious, still learning, and okay about being confused.*" Equally if not more important, Bamberger and her colleagues found a powerful way to explore the interaction of intelligence-as-information (book- and language-based knowledge) and intelligence-as-action. As Bamberger describes this classroom-workshop, an older and "simpler" pedagogy fostered the interplay of adult-child modes of thinking and looking, and continually sought ways to help learners (including the teachers) to clarify the nature of their own understanding and to increase their skills in representing what they knew to those with whom they were working.

Bamberger's experience and the findings in the *Laboratory for Making Things* returns us to Kieran Egan. Egan, you will recall, insists that educational institutions cannot resolve the inherent incompatibility of the three "great ideas" that underpin contemporary educational practice. As a way around the impasse, he suggests that teachers modify their approach to children to conform to what is both a sequence and (in his view) a *hierarchy* of cognitive development and function. Specifically, Egan believes that human evolutionary and cultural history offer a template for encouraging the cognitive development of children. Teachers, he says, can exploit for the young what human culture has accumulated in the form of "kinds of understanding."

Egan believes that intellectual development "requires an understanding of the role played by intellectual tools available in the society into which a person grows." In ascending order, the tools found in every society are *somatic, mythic, romantic, philosophic,* and *ironic* understanding. These different kinds of understanding imply a progression of human capacity for thought. Egan appropriates Merlin Donald's model of episodic to mimetic to mythic to theoretic culture, and consequently he does *not* suggest that we discard a lower form of understanding as we move to the next higher form.

> When we become literate we do not cease to be oral-language users, so when we become oral-language users we do not cease to be prelinguistic sense makers. I do not mean that somatic understanding constitutes some common "human nature," but simply that we also make sense of the world in a distinctively human way that is not linguistic.[13]

However much we may as adults rely on the tools associated with the ironic mode of understanding, we retain contact with and can exhibit all of the forms from which this form of understanding is derived. All forms coexist and continue to influence understanding and behavior. What does this mean for teachers? When children first enter school, they are adept in the use of somatic tools and exhibit somatic understanding; later, in the primary grades, they respond most immediately to stories, the hallmark of mythic understanding. By the time they are in the fourth or fifth grade, they are most attracted to romantic descriptions and best able to use romantic tools of thought:

> The preservation of great events, of the memory of the outstanding, has been central to romantic history writing from Heroditus's time to our own. . . . Why are the contents of *The Guiness Book of Records* so much more engaging than the typical math or geography textbook to the average ten-year-old? One answer is that such facts are more romantic; they tell about the wonders of the world, the most extreme experiences, the limits of reality, the greatest achievements, the most exotic forms of life, the most amazing events.[14]

Let me return now to my own experience. Very gradually, as work progressed on this book, I came to realize that an unexpected dialogue was developing between two sets of information I had been considering. From the personal interviews, I had learned that work that demands skilled use of the hands affords people, in Seymour Sarason's words, the means to put a "personal stamp on some aspect of the surround." On the scientific side, findings from a variety of sources pointed specifically to the evolution of the hand and its control mechanisms as prime movers in the organization of human cognitive architecture and operations.

The dialogue was an exchange between the experiences of individual learners and certain large-scale constructs of cognitive and behavioral science. Serge Percelly's description of the pragmatics of juggling, worked out almost entirely on his own, "speaks to" the findings of a research group in motor control and computer science at MIT. George McLean's insights about bimanual control of tools, gained from his remarkable experience following the loss of four fingers of his dominant hand, echo a research laboratory's hard-won discoveries concerning hemispheric specializations for dominant-nondominant hand control. Dorothy Taubman and Patrick O'Brien *deduce* important principles of action of the small muscles of the hand and challenge neurologists who consider musician's (and writer's) cramp to be a consequence of brain disease. Anton Bachleitner notices an important perceptual change as he gains control of marionettes and, based on that experience, teaches his puppeteers to put their own eyes into the heads of his puppets.*

Obviously there can and should be a bidirectional character to this "dialogue between sets" I have been describing. But it is in fact remarkably difficult to find examples of reinforcement of the small-scale particularities of individual learning from the larger world of science. That is, *it does not seem that we have yet learned how to apply systematically to individuals what we know from biology about the nature of human learning.*

The human "hand-brain complex" came into being in association with bipedal locomotion and the comparatively minor structural changes in the hand. As many eminent evolutionary theorists have pointed out (and as I have tried to show from a variety of perspectives in this book), it would be a serious misreading of evolutionary doctrine to suppose that any kind of structural modification in the hand or arm "put the finishing touches" on what was incomplete prior to that time. What was there was already complete: an australopithecine hand is not an incomplete or imperfect human hand. It is a perfectly complete australopithecine hand. It would be equally mistaken to imagine that once the primate brain laid down some kind of internal circuitry necessary to support language, the perfect brain had now been realized, setting the stage for the full suite of human behavioral potentials to come into play.

What we *can* cautiously postulate is that, under the influence of all it was exposed to beginning about 2 million years ago, this hand with an altered grasping potential may have become part of an unprecedented and uniquely successful survival strategy for at least some of the hominids who had come to possess it on a genetic basis. It seems to have been an unusually propitious time for this hand to appear among hominids, whose fortunes continued to improve as the "hand-brain complex" continued to evolve. Our earliest ancestors, greatly aided by inventive tool use and the exploitation of

---

* He will have to wait for a reply from cognitive science, which lacks both the tools and a theory to explore this phenomenon.

gesture as a means of symbolic representation, instruction, and communication, converted a risky nomadic life into a global migration. Coincident with the worldwide dispersion of *Homo erectus,* and with the co-evolution of hand and brain, was the appearance of modern *Homo sapiens* and what we refer to as human intelligence.

Robin Dunbar has argued that the human neocortex expanded to support not only tool use but the demands of increasing social complexity in migratory hominid society: language (gossip) became a surrogate for the manual grooming that primates had always used to establish and maintain essential liaisons. Merlin Donald associates the expansion of hominid awareness beyond immediate time and place with the manufacture and use of tools and the propagation of information gained from experience through *mimesis,* a novel visuomotor behavior with both intentional and representational uses. He accepts that evolution produced modifications of the vocal apparatus and the neocortex to support speech and language, and that these changes in turn supported a rapid expansion of the cognitive realm of behavior during the transition between *H. erectus* and *H. sapiens,* beginning sometime between 100,000 and 200,000 years ago.

Henry Plotkin, examining essentially the same history from a specifically Darwinian perspective, argues that any successful survival strategy for humans would have required two kinds of "devices" (or strategies) working in parallel. The more stable and conservative of these, which he calls a primary heuristic, comprises the totality of species-specific physical and behavioral traits by which modern *H. sapiens* is uniquely identified. Taken together, these traits constitute the basic survival kit for humans during the transition from infancy to maturity.

Because the environment in which maturation occurs may change unpredictably between the time of birth and sexual maturity, our genetic program also allows for *secondary heuristics:* novel behaviors or adaptive strategies invented to meet unpredictable demands presented by the particular environment encountered by each individual. The relative importance of secondary heuristics to any species will depend upon the risk of exposure to novel threats or opportunities. Such a risk for sharks is not very high; for hominids, who were deploying an experimental body into an almost completely unfamiliar environment, the risk was extremely high. Thus, for the australopithecines and for *Homo* the "uncertain futures problem" has never been a remote or abstract prospect; on the contrary, it has always been the dominant circumstance of life. Humans are born into the world not only obliged to learn and to change, but uniquely equipped to do so.

The heuristics construct has at least two important consequences for those concerned with educational theory and practice. First, it invites us to notice that the hand-brain complex affords each of us an extremely diverse set of old and new skills to master (or invent) and the freedom to meet life's predictable or surprise demands as we choose.* Second, and equally important, this construct exposes the artificiality of all the classic mind-body dichotomies. Knowledge of the physics of light and the flight of birds is reflected in the optics of the human lens, the photosensitivity of the retina, in the muscles that move the eye in the orbit, in single cells in the brain's visual system that detect light oriented or moving in a particular direction on the retina, and in bodily orienting reflexes which cause the head and eyes to turn quickly and accurately toward objects approaching from the air. Nor does the old mind-body separation stand up to careful scrutiny when one considers instances of the most complex forms of culturally derived behavior. High levels of achievement in purely "physical" skills like juggling and competitive athletics depend on a mastery of both procedural and declarative knowledge, and follow the same developmental course observed among highly successful mathematicians, sculptors, and research scientists. The clear message from biology to educators is this: The most effective techniques for cultivating intelligence aim at uniting (not divorcing) mind and body.

There is also a critical lesson from biology to the philosophers of education and to the designers of systems of education. There will continue to be individuals whose accomplishments will astonish the experts by proving to be useful or even critically important despite all predictions to the contrary. We can neither fully anticipate our future needs (what we will regard as tal-

* For example, drawing on an ancient primate trick (the grasp reflex), David Hall improvised a solution to a non-trivial educational goal—helping inner-city kids toward a more robust self-reliance—by confronting them with the physical and emotional challenges of spelunking and rock climbing.

ent or genius) nor fully orchestrate a survival strategy for ourselves. We can, will, and *should* do both of these to the best of our abilities, but we will never beat evolution at the game that created us. Therefore we must make provision for those among us who hear a different beat and are impelled to march to it.

Increasingly in the work of those concerned to illuminate human origins, we find evidence that, from the beginning, the hominid hand and its growing repertoire of movements were integral to what was happening in behavioral, cultural, and cognitive evolution. It therefore seems most likely, as Sherwood Washburn intimated, that the brain elevated the skill of the hand as the hand was writing its burgeoning sensory and motor complexities, and its novel possibilities, into the brain—all this while the canopy was . . . , and while the glaciers were . . . , and while other predators were . . . , and so on, and so on. We, the beneficiaries of this incomprehensibly long process, arrive with a selective but deeply imbedded, widely distributed "knowledge" of our own past and acutely primed to adapt to a future we cannot possibly predict. We arrive with our own secondary heuristics: "intrinsic curiosity," responsiveness to the human and material contexts into which we are born, hands and brain like none other on the planet, and the ability to build trust in our own instincts, skills, and judgment.

In his landmark treatise on the hand, Sir Charles Bell noted that "we can hardly be surprised that some philosophers should have entertained the opinion, with Anaxagoras, that the superiority of man is owing to his hand."* Bell, taking exception to what he judged in Anaxagoras to have been excessive regard for a mere bodily appendage, opined that these hands were *given* man "because he was the wisest creature." The perfect match of hand structure to intelligence, Bell reasoned, evinces—both reveals and proves—God's purpose: the assignment of man to a dominant role in the biologic world.[15]

Intelligence, the capacity for innovative response to the world (a sec-

---

* Anaxagoras (born in 500 B.C.) taught philosophy and science in Athens until his credibility sank in the wake of his claim that the sun and moon were not gods but hot stones circling the earth.

ondary heuristic and the major target of formal educational effort), is also an aspect of the entire organism. As we noted in chapter 7 ("The Twenty-Four-Karat Thumb"), the anatomist Frederick Wood Jones dismissed Bell's sermon, asserting that it was "the whole nervous mechanism" that gave the hand its movements and man his superiority; modern Darwinists of the Dawkins and Dennett school delivered the *coup de grâce* to Bell's "naïve" idea, explaining that evolution is both impartial and blind. Its effects have proven so uniquely advantageous to us that we judge ourselves to be the products of purposeful design.

But we should not, as the saying goes, throw the baby out with the bath-water. If modern evolutionary theory precludes us from agreeing with Sir Charles Bell that the human hand evinces design, anthropology and cognitive science reply that this hand *evokes* design. The more one looks, the more it appears that the revolutionary hand-brain marriage qualifies as one of *the* defining and unifying themes of human paleoanthropology, of developmental and cognitive psychology, and of behavioral neuroscience. Bell found the diamond mine, all right, even if he was mistaken about its origin.

It seems to me that we now have good reason to listen and learn from Anaxagoras, Sir Charles Bell, John Napier, Raoul Tubiana, Mary Marzke, Jeanne Bamberger, Henry Plotkin, Harlan Lane, and many, many others whose collective message (given its most eloquent expression by Robertson Davies) is that the hand speaks to the brain as surely as the brain speaks to the hand. Self-generated movement is the foundation of thought and willed action, the underlying mechanism by which the physical and psychological coordinates of the self come into being. For humans, the hand has a special role and status in the organization of movement and in the evolution of human cognition.[16]

Seymour Sarason has illuminated a crucial link between the hand and intelligent action in his exploration of the meaning of creativity. The creative impulse, which is deeply personal, is a critical element at the core of all learning. It requires that information be gathered, ideas explored and tested, and decisions made so that progress can be made toward a personally valued goal. As Sarason says, if this process succeeds, a personal signature will be affixed to the final product. The "creative work" could be a cabinet, a painting, a hot-rod engine, a poem, a back-flip, a smile on the face of a nursing-home patient, a triangle drawn on a piece of paper, or a horseshoe. No object or action is automatically excluded.[17]

Creative acts also arise from the impulse to alter an inner state or to communicate something to others that might alter *their* inner state. It is foolish to imagine you can stand outside such an initiative and judge its success or failure. It is only through creative activity that any individual can define his own understanding of the world and his place in it, and anyone who wants to do that *can* do it. Nothing but the paralysis of innate curiosity or abject isolation can stifle creativity. It was Jon-Roar Bjørkvold who first called my attention to Johan Huizinga's book *Homo Ludens,* which he cites in *The Muse Within.*[18] I have not read Huizinga's book, but I have been in Bjørkvold's home in Oslo and have heard the family jazz band. They play, and they are playful. *Ludens* means "playful," and it is clear that the spirit of play, of joyful or just curious experimentation and exploration, comes to us, just as the hand itself comes to us, as a powerful organizer of learning and growth. We might consider pushing that side of ourselves a little harder.

One observation made by Kieran Egan strikes me as being particularly worth highlighting and exploring further. There seems to be something quite special about the potential of children around the age of ten for "romantic" awakening. This is the ideal time for apprenticeship: the unique adult-child relationship, usually outside the immediate family, in which the child's imagination attaches to mature goals and to a mentor whose caring both about the child and about the activity can have enormous long-range consequences.

I was struck in a number of the interviews I conducted by the consistency with which some people could recall discovering a strong affinity for a certain skill (or someone who possessed it), sometimes as early as seven years old. This seems far too early for a child to be making career decisions, but why, at that age, should a child *not* be able to sense his or her own nascent gifts? David Hall was not much older than that when the school librarian noticed that he had "arms like Popeye." Almost *all* children who prove to be "successful long-term learners" initiate a series of successful professional apprenticeships before reaching their teens, as when Anton Bachleitner saw his first puppet show, Serge Percelly saw his first juggler, Jack Schafer heard his first big engine, Robert Albo learned his first magic trick, and Pat O'Brien played his first toy musical instrument.

Richard Moore is a blacksmith in Red Bluff, California.* The shop he owns was started by his grandfather; before he started high school, Richard was working alongside men who had been welding for many years. He told me about his first visit as a boy to his father's shop, when he was ten years old:

> I just went in there; Ed Derby always liked me and he kinda took me under his wing and taught me everything. I worked side by side with him from about thirteen. He'd give me a project and I remember he'd make me throw some of it away—just wasn't good enough. Do it again—you know? One time he got on me and he made me so mad that I quit. I just took off. I went home, thought about it and thought about it, and the next morning I went in. I'll never forget it. He said, "Well, you think you can do it right today?" And I said, "Yep." I don't even remember what job it was, but I did it again, and he came over and he complimented me on it: "By gosh, you *can* do it, can't you?" Which I knew I could. But you know how kids are, young people, you know, I thought it was good enough but it didn't suit him.

By the time he entered high school (at the age of fourteen) and took a farm-mechanics class, Richard had already had considerable experience working with machinery.

> The teacher was trying to help these two kids; they had this half-inch hole they were trying to drill and they drilled on that thing for twenty minutes, and he'd take the bit over and sharpen it and it just wouldn't cut. I was watching them and finally I just said, "Hey, would you like me to sharpen that drill for you?" The teacher was mad anyway, so he said, "Here, if you think you can do better!" So I went over and sharpened the bit and it shot

* He and Jack Schafer have been friends since they were boys.

right through there. That was it. He just didn't have the right angle on it. He had the heel higher than the point of the bit and so it would just ride.

They had an old forge there and he didn't know anything about that, so I showed him how to start the forge and make the coke. After that, for the next three years, he made me the shop foreman.

Richard's daughter, Jill, is now an old-timer herself at the blacksmith shop, although there are still some men who are put off by having a woman estimate a welding job.

Jill, my daughter, she's always worked, and in the summertime she'd come down to the shop. She'd have on her bib overalls and her ponytail and her cap, just drove everybody nuts. I think she was about twelve, thirteen, and a guy from Lindy, which builds welders, came in. He had this beautiful machine they were going to demonstrate. So after everybody in the shop tried it, he asked me, "What do you think?" I said, "I got one more person I want to try it." He said, "Who's that?" I hollered, "Jill, go grab a hood!"

So she came out there. This guy kinda *looked* at me, and I said to Jill, "Try this welder, see if you like it." So she went over, started welding with it. I said, "What do you think?" "Oh," she said, "it's all right, but it's not as good as the Mig," which was the welder we had at the time. And this guy damn near fell over. But she could run as good a bead as a couple of the guys I had working for me. She's a certified welder now. But there aren't many girls who do this.

We begin life with our parents as the first teachers, learning through early exposure to toys, language, music, and other children and adults. We perceive change in ourselves through countless interactions, formal and informal, with others to whom we are drawn or driven: teachers, relatives, friends, and rivals. As emerging adults reaching toward the heady goal of independence, we seek to match ourselves to an exemplary archetype, and through emulation of this person (his or her principles and work) we are directed toward a life of productive work, companionship, and reward. The socialization that formal education strives so hard to inculcate is, as Kieran Egan argues, actually built into this process as we increase our familiarity and facility with tools (hand-held and cognitive) offered to us through our contacts with those around us.

No wonder learning is so hard to control, so easy both to direct and to misdirect. It is brain and hand and eye and ear and skin and heart; it is self alone and self-in-community, it is general and specific, large and small. The interaction of brain and hand, and the growth of their collaborative relationship throughout a life of successive relationships with all manner of other selves—musical, building, playing, hiking, cooking, juggling, riding, artistic selves—not only signifies but *proves* that what we call learning is a quintessential mystery of human life. It demands energy but produces more than it consumes. It is enigmatic and enzymatic. It marks the fusion of what is physical, cognitive, emotional, and spiritual in us. Learning is the principal tool through which we exercise our species' compulsion to survive, both individually and as a community.[19]

Both the desire and the capacity to learn are present in all of us, and both are difficult (though not impossible) to extinguish. Both grow, take shape, and continually reinforce one another through the action of the many seen and unseen hands that will touch us, move us, guide us, challenge us, and protect us for as long as we live. The desire to learn is reshaped continuously as brain and hand vitalize one another, and the capacity to learn grows continuously as we fashion our own personal laboratory for making things.

I can think of no better way to close this chapter, and the book, than by quoting Tim Ingold, a social anthropologist at the University of Manchester, England, and an individual who evidently shares my dissatisfaction with the *cephalocentric* view of intelligence.

> Western thought, whose penchant for constructing dichotomies is one of its main defining characteristics, has given us a distinction between intellect (as a property of mind) and behavior (as bodily execution). We may of course describe as "intelligent" an animal whose actions manifest certain sensitivity and responsiveness to the nuances of its relationships with the components of its environment. But it is quite another thing to attribute that quality to the operation of a cognitive device, an "intelligence," which is somehow inside the animal and which, from this privileged site, processes the data of perception and pulls the strings of action. . . . Cogni-

> tion is an accomplishment of the *whole animal,* it is not accomplished by a mechanism interior to the animal and for which it serves as a vehicle. There is therefore no such thing as an "intelligence" apart from the animal itself, and no evolution of intelligence other than the evolution of the animals with their own particular powers of perception and action.[20]

It may be that a greater appreciation by parents and teachers of these facts would bring formal learning and self-directed "life learning" into a more mutually reinforcing relationship. It may also be that the most powerful tactic available to any parent or teacher who hopes to awaken the curiosity of a child, and who seeks to join the child who is ready to learn, is simply to head for the hands.

Epilogue

THIS APPENDAGE to *The Hand* is really a short waltz around the hall, or maybe a polka, with several ideas I tend to think of as evidence of the inherently restless nature of the ideas in this book. I suspect a completed book more often than not will come to haunt its author with a new list of unanswered questions. I now find myself wondering about an obligation to pursue answers to some of the questions raised but not answered here.

The eminent biologist Ernst Mayr points out in his most recent book that the science of biology is more descriptive than explanatory and cannot be expected to duplicate the procedures or the answers associated with the world of physics.[1] While the irrepressible drive of science to get at first causes—to explain everything on the basis of fundamental laws of nature—makes perfect sense in the impersonal world of physics, this ambition is usually frustrated when exported to the world of the neurosciences and the cognitive sciences. When someone claims that "brain science indicates you should do *a, b,* or *c* in order to achieve *x, y,* or *z,*" the appropriate response is skepticism.

Organized psychology has turned up its nose at "pop psychology" on the question of self-fulfillment, and in particular has entirely missed the point that extremely powerful drives exist in humans in relation to work. Certainly, work for many people is the source of highly reinforcing experiences and in that case can be the venue for unique personal exertions and for lasting personal change. In ignoring the extent to which people can invest themselves in their work, psychological formalisms seem to me to have demonstrated a colossal lack of imagination and insight.

The truth is that working life offers some of the most ecstatic experiences and relationships people encounter in their lives. Not only that: in the early stages of life, a working apprenticeship can be a microcosm of almost every affirming principle of human life, a conservatory for the solution of the great existential struggle of our teens and twenties: Who will I be? What will I accomplish?

The standardization of education, offered in many schools, often fails to answer individual need. You would never know that this is true unless you happened to spend some time with people whose ambition and originality has little to do with their "formal" education. There are actually quite a few people around who just don't make an issue of the fact that they love what they do. You would also never guess how powerfully people are motivated, and how refreshed and productive they become, when they stumble into who they really are; when you find out that there were hints about a secret love all along, you can see that people get delirious when life suddenly offers them exactly their cup of tea for the first time and they hear themselves saying, "Yes, I will." Why is the formal psychology of work (of employment) nothing but a manipulator's manual for employers? Why don't parents, doctors, all of us, *demand* that education proceed on the basis of spontaneous combustion—signaled by the explosion of a child's questions about something that has captured his or her interest?

Do we really have specific skill and work aptitudes, and if so, how do we identify them? In the course of preparing this book, I interviewed David Ransom of the Johnson O'Connor Research Foundation in San Francisco, and Richard Unger, founder of the International Institute of Hand Analysis in Sausalito. David presides over aptitude tests conducted in the "human engineering laboratory," and Richard is a modern and stunningly provocative hand reader (or "palmist"). I made myself a guinea pig for both of them and came away intrigued with what David said (I have "ideaphoria") and *floored* at what Richard said (my "life purpose" is to see to it that something I believe in, having to do with other people's creativity, comes to pass).

In talking with David Ransom I learned that scores obtained in repeat tests of individuals are extremely stable; many of Johnson O'Connor's clients are first tested at or near college age; when they return for retesting, sometimes many years later, they always perform the same as they did on the first testing. There is no way one can prepare for these tests, which are focused only on what are seen as biologically based correlates of work success (dexterity, hearing, vision, and so on, plus specific cognitive and psychological orientations). These were the principles that motivated Johnson O'Connor to establish aptitude testing in the first place—he wanted to be able to hire non-English-speaking immigrants and place them immediately in jobs where they had a high probability of success. His book *Born That Way* might not fare well with "sophisticated" industrial psychologists, but I could find nothing in the test, or in the conclusions, to suggest that it was exploring anything other than what Ransom said it was—a relatively simple set of reactions and skills whose "profiling" can be used to help match people physically, cognitively, and psychologically to particular demands and circumstances found in almost any kind of work.[2]

Richard Unger's work is a major enigma. I had never met a palmist before. I certainly did not expect to be knocked right out of my chair by what he told me, which is what happened.

> Every pair of hands has a set of fingerprints, which do not alter after five months prior to birth. Looking at a person's fingerprints, to use an analogy I like, is like looking at an acorn and being able to see the type of oak tree that could develop. When people move toward becoming the type of oak tree that is possible for them, I tell them they are pursuing their life purpose. And as they move toward their life purpose, life gets good.
>
> By the same token, there are all sorts of accidents of life and circumstances—even the fear of doing your own life purpose—that get in the way. When you're doing your own *real* thing, it's scary because then a failure would be a *real* failure. But these are always failures specific to a given type of life path. If you're going to be a writer, you face a writer's challenges; if you're going to be a philosopher or a cop, you face other kinds of challenges. The fact that the challenges are connected with your life purpose is what gives them their meaning. They're still scary, but you can face the frustrations because your interest in the goal is so strong: if you always wanted to climb Mt. Everest, you *expect* it to get cold on the way up there and you can *take* the cold because you're fighting your particular battle. It's

fighting battles that do *not* directly pertain to your essential nature that drives you bonkers. Even if you *win* that kind of battle, going through whatever it takes, you get no quench, no great "ah!" at the end.

From a palmistic sense, there are two psychologies marked in a person's hands. There's a personality, which is a combination of learned behaviors, innate talents, and things of that sort. The personality includes behavior patterns in relationships and certain kinds of repeatable behaviors that are predictable in relation to the personality structure.

There are also the fingerprints, which reflect an unalterable aspect of you that's built in before you are born. It's like a soul imprinting, something that comes to you from the "gene pool." The gene pool selected a certain number of people to be healers, others to be leaders, and so forth.[3]

I asked Richard to tell me a little about the history of palmistry.

Palmistry is very old; actually it predates recorded history. Aristotle wrote about palmistry as "that most ancient of arts." There are even some interesting quotes from Aristotle—assuming that they're accurate. He said, "It tells you who you are and to what vice you will addicted be." That's how I remember the quote. Supposedly palmistry came from India. The Greeks got theirs from Indian palmistry, and as far as anyone can tell, the practice goes back anywhere from four to ten thousand years in India.

It uses the same nomenclature as astrology. There are areas on the palm named the Venus, the Mars, and the Jupiter area. No one knows who it was in India who five thousand years ago came up with the system—*somebody* did it. There are reports that there are some pieces of wood with the first written rules of how to read hands—ten thousand years old—kept somewhere in India. But nobody's allowed to see them, so who knows if any of this is right?

But we do know that as soon as there is *any* written history, they're talking about palmistry as old stuff; Julius Caesar is supposed to have read hands, Alexander the Great is supposed to have read hands. In fact, the history of palmistry is in large measure the history of *individuals* who read hands, rather than that of an organized movement. Somewhere I've read that the third book printed by the Gutenberg press was a palmistry book.

Richard jokes about almost everything, but he himself is no joke. During the several years that have intervened since I first met Richard, I have had many opportunities to observe him and to learn from him, and as a result have come to appreciate how radically his work differs from what is generally expected from practitioners of this somewhat arcane art. Richard is not entirely a renegade from tradition: much of his work draws heavily upon classical teaching and its venerable nomenclature. The thumb has to do with self-will, discipline, and the personal drive to manifest one's own potential; the index finger is named Jupiter (sometimes Zeus) and is associated with leadership and the desire to influence others; the middle finger, Saturn, is associated with the tendency to seek or impose order or simply to attend to small detail; the ring finger, Apollo, is associated with artistic and creative drives and public persona; the small finger, Mercury, is associated with "wit" and the quality and character of internal communication, and to an extent with negotiating skill. Richard holds to this scheme, and in addition uses the classical interpretations of head, heart, and life lines, as well as a variety of special "gift" and other markings on the hand.

But what he has added to classic palmistry is original, theoretically provocative, and—I am no longer as surprised as I initially was—remarkably useful in a medical practice seeing any significant number of patients whose ailments have a significant psychological component or relationship to life stress. Based on a straightforward and reproducible system for scoring fingerprints, supplemented with a more conventional evaluation and interpretation of other aspects of the hands' appearance and their markings, Richard produces a report that would be completely familiar to anyone who ever consults (either personally or professionally) with psychological specialists. And since his conclusions come as readily from reviewing a printed impression of the hand as from a personal encounter, the risks of deliberate or unintended invention is really quite small.

Several years after we first met, and after I had begun to use him as a consultant in one of my own clinics, I enrolled in the year-long course he offers at his institute. This turned out to be quite comparable to a medical school course in radiology; here, though, we worked through volumes of hand prints arranged and presented on the basis of a very sophisticated theory of developmental and relational psychodynamics. What Richard could not, and did not, offer was a theory about *why* his system works. "You guys need to work that out," he explained.

Richard is gregarious, serious, curious, witty, self-effacing, highly intelligent, and as honest as anyone I've ever met; the implications of his skill, and the questions they raise, make me dizzy. He also has a warning for me: that I will press very hard to get people to change something I think needs to be changed. I admit it. I will try, in my own loopy way, to explain why we will never know how the brain works, and why you shouldn't cry about that. What does matter is that you realize that *you* are the only one who will ever know what you were cut out for. The same is true of your children, if you have any; you have no way of knowing what is best for them. Only they can find the answer to that secret, and they will do it if given half a chance. People find out or they don't, and whether or not they do find out in time for it to matter depends largely on how willingly society supports individual choice and broad definitions in the matter of education, from beginning to end.

Let me try to illustrate the difficulties raised by the attempt to base an understanding of behavior on a physiologic science of the brain. The example reflects my belief that one could as easily explain politics in this way.

November 17, 1993, was the day the American Congress voted to establish the North American Free Trade Agreement. The vote itself was a paragon of simplicity—by a simple majority, yes or no, the question would be decided. Let us imagine the following situation existing just before the vote on November 17. The president of the United States wished to telephone the president of Mexico and the prime minister of Canada as soon as he knew what the outcome would be. Prior to the day of the vote he had learned that two congressmen remained undecided, and each was known to be edgy about the political risks connected with their own votes on the NAFTA question. One of them had been judged by the president's staff as being a "soft yes," the other a "soft no" vote. Not surprisingly, the president's staff had been up all night trying to shore up the first and convert the latter. Wishing to be able to report good news to his counterparts in the Mexican and Canadian governments as soon as he was certain of the outcome, Mr.

Clinton had assigned a special telephone line representing each of the two wavering congressmen. A staff person assigned to watch the "soft yes" congressman would call on line 1 immediately if and only if he learned that this man had remained firm. The staff person assigned to the "soft no" congressman would call on line 2 if this man changed his mind and decided to vote for the treaty. Thus, if the first phone rings, the president can prepare to make his call; if the second phone rings, he *will* call. If neither phone rings, the president can leave it to the news media to carry the bad news over the borders.

The long chain of preliminaries did not stop with those who actually voted. All of the people who sent messages to a congressman had gone through something like the same process, weighing alternatives, considering the arguments, trying to imagine potential costs and rewards in their own lives, and talking to *their* friends, co-workers, relatives, and so on. It is truly a stupefying exercise to try even to imagine all the events that preceded the big event on November 17.

This brings us to our limitations. The president, metaphorically speaking, is just one nerve cell out of millions of similar cells having the power to activate just one muscle fiber located somewhere in the body. The simplest decision taken by a nerve cell within the brain—shall a single nerve cell in the motor cortex send a message to some other nerve cells living a great distance to the north or south?—almost always has an antecedent history as complex as that of the NAFTA example.

The brain represents the principle of nature that states that function arises from obligate, recurrent, finite synergies. There is a left hemisphere and a right hemisphere, a sensory and a motor nerve. Each functions independently and, in parallel, represents the existence of a channel along/through/within which information exists. When it interacts with another channel (another neuron, another nerve, another nucleus, another hemisphere, another person, and so on), the continuation of the relationship depends both on the immediate outcome and on distant effects that cannot be specifically anticipated. Many such relationships may exist simultaneously and nonexclusively, and they may be synergistic, neutral, or antagonistic. I, too, am like that. I am all at the same time a father, a husband, a physician, a producer of toxic waste, a taxpayer, a Democrat who sometimes votes for Republicans, a white guy who likes the blues, a spiritual atheist, an honest person with considerably less than a perfect record for telling the truth.

The brain has countless thousands of simultaneous functional entities integrated with other functional entities that exist and maintain themselves in shifting arrangements of neural networks, coordinative structures—past, present, and future. The brain gives rise to worries about the weather, about enemies and pals, and to uncontrollable urges for oxygen, coffee, gin, love, speed, and attention. Its anatomy and physiology support both confusion and coherence, continuity and change. The brain seems to be even more complicated than we thought it was when we realized that it had both sequential and simultaneous operations (which we gratefully assigned, one each, to a "dominant" and "nondominant" hemisphere). But the brain seems also to demonstrate a gift for what are now called intermaths, which are techniques for representing massively complex sets of minute data that are themselves in constant flux. As computer scientist James Baily explains, intermaths are what you need to predict the weather or to understand why it really *wasn't* the iceberg that sank the *Titanic:*

> As products of the culture of sequential thought, we prefer our big results to come from big causes. Since early in the century, we have been comfortable, for example, with the image of a giant iceberg taking down the equally giant *Titanic.* Here is a villain worthy of its deed; the huge ship and the huge iceberg go well together in a picture. Reports that the collision itself was minor and that rumblings worked their way through the hull after the impact have been discounted because they did not fit with our causal preconceptions. Also rarely mentioned is the fact that the *Titanic* was built from a cheap grade of steel known to be brittle in the cold. . . . It remains hard for our minds to let go of the symmetry of one iceberg, one *Titanic.*[4]

The human brain is essentially an open market of infinite flux and dynamic reassociations. If you imagine a great city at the intersection of numerous trade routes, you can see that information, goods, energy, and people all arrive by ship, car, air, rail, and they also arrive *virtually* by telephone, fax, modem, fiber optics, dish antennas, and so on. It is only rarely that a person, parcel, or message enters this complex without a specific destination. It is going there for a reason, and the reason has space-time coordinates within the city. Once there, many other things may happen, many events will happen (both expected and unexpected, to all of which one can at least theoretically assign probabilities). No one can be sure exactly what will happen after the rendezvous; there are so many other meetings going on, relationships being built, tested, and extinguished in the city that no one—not the mayor, not the police chief, not the governor of the state, not the president of the United States, not even the head of the mafia—can be totally aware of. Yet the city exists: controlled chaos in perpetual, constrained evolution.

So, too, with the brain and the body. A single human is an analogue of Alexandria, New York City, Beijing, and (of course) Galápagos. It is full of things old and new, familiar combinations and routines, unexpected delays. Experiments. Liaisons. Opportunity. Danger. Hope and despair. Death and reproduction. Messages, like taxis from JFK to Manhattan, flock to the brain along nerves, the familiar trade routes, and they arrive with addresses and appointments. How any of these messages will influence the brain can never be predicted except as a probability.

We have always wanted to understand function by scientific reductionism. We examine the brain, examine many brains, and notice similarities. We notice that functions are "specialized" in patterned ways. We find segregation and integration. We find stability and instability. We find that the loss of a small area in the left hemisphere interferes with speech, and we conclude that this is a "speech area." We find an optic nerve and an occipital cortex, and when we find people blinded by damage to these structures, we tell ourselves that we have discovered a brain "substrate for vision." But upon closer inspection we find that things are not so simple, and even if I take into account the dorsal and ventral visual systems and the history of development of hand-eye coordination, and the ability to interpret ASL, I will never understand why, exactly, I sat and stared at Picasso's *Guernica* for four hours the first time I saw it as a college student in 1955 or cried when I came to the end of the Matisse exhibit of cutouts at the National Gallery in

California artist Barry Kite has recently poked fun, as only he can, at what has become known as the "cubicle culture." As with all his work, insolent sentiment abounds. (Reprinted with permission.)

Washington twenty years after that, and yet another twenty years after *that* can't help laughing at the nutty collages of a California artist named Barry Kite.

The brain is changed by active involvement; in that sense, it remembers. It is not content to limit its sampling of the observable world or the intermingling of its impressions. It not only sees, it hears. It may keep sound and vision separate or fuse them. It associates external objects with their emanations: correlated patterns of both "sight" and "sound" energy. A tiger growls and the image of fangs flashes in front of the eyes, adrenaline is unleashed, and the whole body is instantly transformed into a maniacal tree-climbing machine. The word "tiger" is born somewhere in this maze of neurons and it (like the word "fire") can have the same effect on us as if we were face to face with the real thing. The brain is ready for experience, it learns from experience, and it uses whatever tricks it can to be prepared both for familiar troubles and for new ones.

The brain does not live inside the head, even though that is its formal habitat. It reaches out to the body, and with the body it reaches out to the world. We can say that the brain "ends" at the spinal cord, and that the spinal cord "ends" at the peripheral nerve, and the peripheral nerve "ends" at the neuromuscular junction, and on and on down to the quarks, but brain is hand and hand is brain, and their interdependence includes everything else right down to the quarks.

We lose ourselves in the details. We find the right and the left hemispheres respond to certain challenges in distinctive ways—one does a job faster than the other, or better, or more reliably—and we conclude that this hemisphere is *specialized* for that function. It is important and useful (maybe) that we make these discoveries, but they are true only as artifacts of the tests we devise. Once the left hemisphere and right hemisphere join forces, our attention, or perspective, must shift to take account of that unique combination and a new set of consequences. Language is not just words, semantics, syntax. It is also melody, and sometimes it is dance. Sometimes, as in the deaf, it is a silent dance of the hands. It is a voice, a face, and the words between the lines. It is a small carved object, a submarine, a chair, underarm odor, a hairdo, a chant, a puff of smoke from the Vatican chimney.

The strength of a biologic system cannot be judged independent of context. This is why there is no point getting upset over the meaning of the word "intelligence." It is just a word. Educators and many others have been grateful to Howard Gardner for explaining that there is not one intelligence but several. I think this is a move in the right direction, but the part of the theory of multiple intelligences that leaves me cold is its strong tendency to divide one giant preposterous elitism into six diminished duchies of elitism: "Who is the most brilliant genius of all time?" becomes "Who is the greatest mathematician, musician, programmer, astrophysicist?" I don't care. People at the very top of anything got there because they got the important things right and somehow managed to sidestep the killer banana peels. And they *never* do it alone.

One night several years ago, although (unlike Harlan Lane) I am allergic to graphs and diagrams, I woke up dreaming of a circle. Standing around the circle, actually forming it, were the people I had already interviewed and others who seemed to want to talk to me. It was somehow like a small tribe gathered around a fire enjoying a meal, stories, and music, and somehow they were also like a wagon train circled together to fend off an attack. It was a *mandala* of the consciousness of a number of people who had bound themselves and their fates to one another.

This mandala-like form suggested to me that although we are different, we almost certainly benefit by capitalizing on our differences. The knowledge to do this, if it exists, must exist collectively, as a distributed, heritable trait outside the individual but within the community itself. Any tribe whose gene pool consistently produces a *mix* of individuals with basic compatibility in physical, cognitive, and emotional makeup probably has a survival advantage. Might we detect this *community* trait by gene testing? I doubt it. Can we duplicate its effectiveness by aptitude testing, educational policy, and social engineering, or by stuffing our children with vitamins, surrounding them with interactive toys, and badgering them about getting into Ivy League colleges? What do *you* think?

Russia and the United States. Different, distinct, at odds, combative, independent and interdependent, temporary. A vague memory, almost certainly lost a thousand years from now, yet two countries, tribes, personify themselves in two men at any given moment: the president of the United States and the president of Russia. These men are flesh and blood, fallible individuals whose lives become obsessed with chiefdom; they are men (so far) who have given up a lesser existence to inhabit a plane of consciousness called "the state." Their lives are rife with contradiction. They eat food, make love, use toilet paper, love their grandchildren, and are sworn to protect the lives of the citizens of their country. Yet they send young men and women to war, knowing that they will die. They grieve that they do this, but they are not criminals and they do not repent. The "state" has its own life and its own existence. No biologic cohort, no tribe, no state can exist without such a figure empowered to think and act with authority that can be revoked only by expulsion, revolt, or death. Is this just a habit? If it is a heritable survival trait of societies, what chromosome might it be on?

When it is time to mate, any male able to do so will fight and may even kill another male to win access to a female. The brain makes war and love at the same time. The Egyptians, the Aztecs, the Greeks, the Asians, the Africans *all* knew this; every culture's mythology is the collective story of the brain trying to catch hold of its own reins. Wagner's opera cycle *The Ring of the Nibelungen* is both the music and the story of his own brain seeking to resolve the inherent vitality, turmoil, and ruthlessness born from the collision of great appetites and great opportunities.

If a bear catches its foot in a trap, it will chew its foot off to get out, and it will go on about its business after that as well as it can. Just like the president sacrificing troops to save the country. Wagner, like the bear, loves, eats, protects, and kills. Sometimes, inside his brain, he must bite off the foot of one of his gods or set it on fire, or the god will catch him and drive him mad.

No one knows what will happen to any of us or why we do what we do. Understanding or reengineering the brain will not save us; neither will sitting our children in front of computers when they are three years old so that they can skip the "pointless" experiences of childhood during which they find out what a baseball, or a puppet, or a toy car, or a swing can do to their body, and vice versa. We have no idea what will happen to the child who *watches* eye-catching imitations of juggling over the Internet if that child never gets around to trying a three-ball toss himself or herself. Since kids are turning out to be better and better computer users—and hackers!—at younger and younger ages, we must be prepared to accept that *their* ideas of baseball, excitement, and partnership will be something new, not at all like ours. We can thank Plotkin's secondary heuristic for that certainty.

The fully computerized kid may turn out to be just like us or strikingly different, as a consequence of having replaced haptics with vision as the primary arbiter of reality and having substituted virtual baseball for the old-fashioned kind at an age when the brain's sensorimotor system hasn't settled on the time constants it will use for its own perceptual-motor operations.

There really *is* something quite new about bonding very early in life with keyboard, mouse, and 3-D graphics, and it will be very interesting to see what it produces by way of new heuristics (problem-solving behaviors) in adult life. I am not surprised that we are so eager as a society to welcome the Internet into our public schools. I *am* a little surprised that we are so ready to say goodbye to the playground and the books in the school library. And I am actually *stunned* that we imagine that commercial sponsors of in-school computer networks will not take their lesson from the tobacco companies as they eagerly underwrite the development of more appealing ways to help children learn how to be happy and successful adults.

It is time we stop fretting so much about our brains and squeezing them so hard, as if by doing so we could improve our own lives. As individuals we have every reason to rejoice at the gifts bequeathed us by our ancestors 100,000 years ago and no good reason not to find out for ourselves what kind of game we're going to be good at. I have no better idea than anyone else what the new millennium might hold for us, but I am certain we still have a great deal to learn about increasing the independence of our own lives and about providing effective encouragement to children, young people, and *old* people to build and retain autonomy in theirs. In the final analysis, humans have the same chance at life that bears or sharks or even flatworms have. A specific bear may be shot by a hunter or killed by another bear or die of hepatitis. If he's lucky, he will live out his life in some leafy glade where there are plenty of berries and some seasonal companionship. When the time comes, after he has left the scent of his urine on a bush and the mark of his claws on a tree (and on other bears), and has sired his cubs, he will finally lie down peacefully and die, making room for the next generation, leaving them to their own fate.

I was recently asked to give a talk to a group of doctors about migraine headaches, and I was reminded of a story Oliver Sacks told about one of his own patients, a mathematician who suffered from migraines. This man was nearly disabled by his headaches, but finally a medicine was found that worked. As it happened, curiously enough, the man's headaches were somehow linked to days of imagining and euphoria and productivity, and, said Dr. Sacks, "I cured his headaches, but unfortunately I also cured his mathematics."

There is something odd about migraines. An extremely high percentage of healthy people have them, and they are caused by a reaction within the brain to a neurotransmitter called serotonin. Serotonin has a number of important chemical effects in the body: it is involved in both sleeping and dreaming, and LSD causes hallucinations by blocking its chemical breakdown.

When I was preparing the lecture it occurred to me, especially since migraines run in families, that they must be part of the price we pay for having the godlike power of dreaming. As we have discussed in this book, the ability to imagine—to dream—is now strongly linked to the unique human ability to create complexity both in our thoughts and in whatever we create with our own hands. The Latin word *rabere* means "to dream," and it is fascinating to me that it means both "to dream" and "to rage" (as a *rabid* animal does). The French word *rêver* is even better, meaning both "to dream" and (as in English) "to *rave*." It is the capacity to be possessed, to be driven wild, by an idea or a possibility, the sound of an engine, the appearance of a desired image in a collection of lines drawn by your own hand on a piece of paper. We humans are what we are not only because of our rationality but because of our capacity to harness our irrationality. So my final recommendation is addressed to the official species-namers: celebrate the hand and its central role in human life by reclaiming the heritage of Leakey's handyman, *Homo habilis.* Then, remind us that our rationality is nourished by the unconscious and animated with playful curiosity. This leaves us with a somewhat longer name—*Homo habilis rabens ludens sapiens*—but it is a name that comes far closer to capturing the essential (and colorful) human than what we have now.

Appendix

A BRIEF TRIBUTE
TO SIR CHARLES BELL
AND TO JOHN RUSSELL NAPIER

As it turns out, not only am I not the first person to have had the idea that guided this book, I am not the first person to have written such a book. Certainly there is a long and impressive literature (including a number of books) on the hand. But the list includes two previous books in the English language whose thesis is virtually the same as my own: that the hand figures critically in human cognitive, physical, and emotional development.

The physician who initiated this sequence, Sir Charles Bell, was a Scottish surgeon who wrote his book in 1833, a work specifically commissioned to employ the human hand as proof of the existence of God. This is made explicit in the book's full title, which is:

THE BRIDGEWATER TREATISES
ON THE POWER, WISDOM, AND GOODNESS OF GOD
AS MANIFESTED IN THE CREATION

TREATISE IV.
THE HAND, ITS MECHANISM AND VITAL ENDOWMENTS
AS EVINCING DESIGN
SIR CHARLES BELL, K.G.H.F.R.S.L.&E.

It is worth commenting briefly on Bell's illustrious career. A minister's son born in 1774, he had been taught as a young boy to draw by the painter David Allen. Although he eventually went to university and medical school, his description of his two years of high school has a familiar ring to it: "The education at the High School which I attended was a torture and

THE HAND,

ITS MECHANISM AND VITAL ENDOWMENTS,

AS EVINCING DESIGN.

BY

SIR CHARLES BELL, K. G. H.

F. R. S. L. & E.

NEW-YORK:

PUBLISHED BY HARPER & BROTHERS,

NO. 82 CLIFF-STREET.

1840.

The title page from the first American edition of Sir Charles Bell's *The Hand.*

humiliation." Bell's older brother John, a surgeon, had founded an anatomy school in Edinburgh, and when Bell was a university student he assisted in his brother's school. Another older brother, George, was his enthusiastic patron, and wrote this in his own memoirs: "Charles' natural clearness of head and neatness of hand, and the vigorous correctness of his conception, with hard labour, made him an admirable surgeon and one of the first anatomists of the day while he was yet a boy not entered upon life."

Encouraged by John to pursue his interest in anatomy and surgery, by his mid-twenties he had already published an important artist's book, *The Anatomy of Expression.* After moving to London to establish his practice, he continued to study and teach anatomy. He also developed a reputation as a skilled surgeon, at a time when speed counted. Since neither antiseptics nor anesthesia were available, one understands why the ability to remove a bladder stone surgically in under three minutes would have been a powerful boon to his reputation.

His reputation has lasted longest in neurology, however: he did important original work on the nerves of the face, and established a basic principle of the anatomy of motor and sensory nerves at the level of the spinal cord. Every sufferer of acute facial paralysis learns his name, since the condition is still called Bell's palsy, and all medical students learn that the "Bell-Magendie Law" describes the separation of motor and sensory pathways into two separate roots at their attachments to the spinal cord.

Bell's biographers, Sir Gordon Gordon-Taylor and E. W. Walls, have included in their book a transcript of a little-known manuscript written by Bell in July 1808, *Idea of a New Anatomy of the Brain.* The book was never published, nor intended to be published. It was what might now be called a "think piece," "submitted for the observations of his friends," wherein he took the liberty of expressing the conclusions he had reached on the basis of his anatomical studies of the brain. This brief work shows him to be not only an anatomist of the first rank, but such a complete modernist in his conceptualizations that no one yet seems to have grasped that he had the basic idea of brain function, at least in the sense that we understand it now, absolutely right. Imagine these lines being spoken by a Shakespearean actor, and let the ideas settle in. Bell understood the brain nearly two hundred years ago as well as we do now. He was the real father of modern neuroscience!

> The operations of the brain may be said to be three-fold: 1. The frame of the body is endowed with the characters of life, and the vital parts held together as one system through the operation of the brain and nerves; and the secret operations of the vital organs suffer the controul of the brain, though we are unconscious of the thousand delicate operations which are every instant going on in the body. 2. In the second place, the instinctive motions which precede the development of the intellectual faculties are performed through the brain and nerves. 3. In the last place, the operation of the senses in rouzing the faculties of the mind, and the exercise of the mind over the moving parts of the body, is through the brain and nerves. The first of these is perfect in nature, and independent of the mind. The second is a prescribed and limited operation of the instrument of thought and agency. The last begins by imperceptible degrees, and has no limit in extent and variety. It is that to which all the rest is subservient, the end being the calling into activity and the sustaining of an intellectual being.[1]

Bell's ideas about the hand were as remarkably advanced as were his ideas about the brain, and it was not until a century and a half later that another physician-anatomist would undertake to tell this story—neither the topic itself nor Bell's masterful treatment of it make for the sort of challenge one would lightly undertake.

But a worthy sequel did, in fact, appear when anthropologist-physician John Napier published *Hands* in 1980. Napier's book reflects the accomplishments of his own long career as a comparative anatomist of the hand and draws on a full century of work by anthropologists to give us the story of its evolutionary history. Napier's biography has not been written, but Russell H. Tuttle, in the foreword to his newly edited and revised 1993 edition of *Hands,* has this to say:

> Professor John Russell Napier (1917–1987) is preeminent among the founders of modern primatology. He is renowned for his descriptions and interpretations of the hand of a new species, *Homo habilis,* and for his

creative, functional, ecological, behavioral, and evolutionary overviews of the order Primates, which stimulated many of his contemporaries and younger students to develop and challenge his ideas and keen observations. *Hands* is Dr. Napier's last major single-authored work, written at the pinnacle of a distinguished research career. Here he shares his vast knowledge of human and nonhuman anatomy, evolutionary history, and broader anthropological and artistic perspectives on our hands in a highly accessible and entertaining manner.

The Hand has been written with deep respect and appreciation for these two exceptionally gifted physician-scientists and for their landmark contributions to humanistic science.

Notes

ACKNOWLEDGMENTS

1. D. H. Lawrence, *Fantasia of the Unconscious* (New York: Penguin Books, 1960; first published in 1922), p. 23.

PROLOGUE

1. The rare individuals born without hands, or with profound congenital deformities of the upper limbs, not uncommonly transfer this manual fluency to their feet and think nothing of it.
2. The neurologic side of the hand story was originally explored in *Tone Deaf and All Thumbs?* (1986), my small paean to amateur piano study (and an imperturbable piano teacher). To be honest, I knew nothing at all about the neurologic control of skilled movement until I began piano study, when watching my own fumbling fingers acquire some discipline impelled me to take a longer and much closer look at the science I thought I already knew something about.
3. John Russell Napier, the eminent British anthropologist and comparative anatomist, was almost singlehandedly responsible for the modern revival of interest in Bell's ideas about the hand; see the Appendix.

1. DAWN

1. Stephen Jay Gould, *The Panda's Thumb: More Reflections in Natural History* (New York: Penguin Books, 1980), p. 20.
2. Lucy is now thought to have descended from *Australopithecus anamensis,* so far the earliest known of the hominids and estimated to have lived approximately 4 million years ago.
3. Sherwood L. Washburn, "Tools and Human Evolution," *Scientific American* 203, no. 3 (1960): 63–75.
4. The Columbia University anthropologist Ralph Holloway argues that the brain did not evolve last but in company with other structures; that not only size but internal structure underwent modification and reorganization in this process. The increase

in size could not by itself account for the behavioral advances which separate *Homo* from earlier hominids; not only may the cellular arrangements of homologous structures differ, but proportional relations may not be comparable. Disproportions in the primary visual cortex (associated with surface markings which appear in fossil skulls) have in fact provided a critical tool in making species identifications of fossil brains. See Ralph Holloway, "Evolution of the Human Brain," in A. Lock and C. Peters, eds., *Handbook of Human Symbolic Evolution* (Oxford: Clarendon Press, 1996). See also endnote 6, below.

5. See the recently published biography of the Leakey family: Virginia Morell, *Ancestral Passions: The Leakey Family and the Quest for Humankind's Beginnings* (New York: Simon and Schuster, 1995).
6. "[In 1962] . . . despite the nearly one-hundred-year search for human origins, the genus *Homo* was at first only vaguely defined. Most authorities considered brain size the distinguishing feature, although they had never agreed on the boundaries of this 'mental Rubicon,' as Sir Arthur Keith called the dividing line between humans and apes. Keith himself thought the transition occurred when a hominid's brain measured 750 cubic centimeters, while Franz Weidenreich, an anthropologist who specialized in *Homo erectus,* thought a volume of only 700 cc would do, and Henri-Victor Vallois, who studied the Neanderthal specimens, boosted the figure to 800 cc. . . . Although the numbers were somewhat arbitrary, most anatomists accepted the idea . . . that humankind began when the brain attained a capacity between seven hundred and eight hundred cubic centimeters" (Morell, *Ancestral Passions,* pp. 225–26).

   Brain size among modern humans, it should be noted, can vary considerably: "The increase of brain size in these terms is responsible for the success of our highly adapted species, though, as has been pointed out . . . there are considerable variations from the mean (1350 cc) that indicate a large margin of tolerance. For example, of the ranges in brain volume, Jonathan Swift, 2000 cc, and Anatole France, 1100 cc, were cited" (Ragnar Granit, *The Purposive Brain* [Cambridge, Mass.: MIT Press, 1977; 1980 edition], pp. 52–56).
7. Lucy herself (*Australopithecus afarensis*) lived about 3.2 mya, and *A. afarensis* is believed to have been the sole extant hominid species from 3.9 to 2.9 mya. The first specimen of *Homo habilis* was a shin bone, or tibia, found in the Olduvai Gorge in Tanzania in 1959 by Jonathan, the ten-year-old son of Louis and Mary Leakey. The second specimen, a piece of a child's skull, was found at Olduvai by Louis himself on December 1, 1960. Although Leakey immediately reported finding the skull and speculated that it was from "Chellean Man"—the first makers of stone handaxes—it took nearly three years for anthropologist Phillip Tobias to calculate the cranial capacity, and the species was officially named in a paper published in the April 4, 1964, issue of the journal *Nature.* In that report, *Homo* was defined as having "an erect posture and an habitual bipedal gait, a precision grip, and a brain capacity of

only 600 cc" (Morell, *Ancestral Passions*, p. 235). A more complete *H. habilis* skeleton was recovered in the Olduvai in 1986 by a group led by Tim White and Donald Johanson. See Donald Johanson and James Shreeve, *Lucy's Child* (New York: Avon, 1990). *Homo erectus* is the name given to the most successful and widely dispersed of early human precursors. Fossils have been found in Africa, Europe, and Asia, and a recently re-dated specimen from Java has been reported to be 2 million years old. *H. erectus* developed an elaborate tool-based society, and the term "Acheulian Stone Culture" refers to the entire collection of stone tools associated with *erectus*. It has now been claimed that a fossil found in China of what anthropologists formally designate "archaic" *H. sapiens* is 200,000 years old. Once again in anthropology all bets are off. Are there parallel, isolated lines of *H. sapiens*, or did the global migration start much earlier than previously thought? Stay tuned.

8. Neurology is very much interested in how things got the way they are, but it has yet

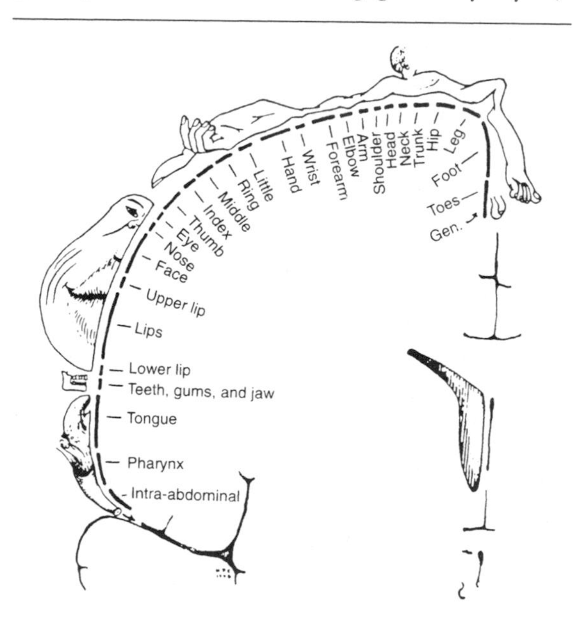

Above: The human hand and face receive disproportionate representation in the human brain, as depicted here in the widely recognized "homunculus" diagram first published by Canadian neurosurgeons Penfield and Rasmussen in 1950. The diagram suggests that refined control of hand and orofacial movements imposes heavy computational demands on the central nervous system. (Reprinted with the permission of Simon and Schuster from *The Cerebral Cortex of Man* by Wilder Penfield and Theodore Rasmussen. Copyright 1950 Macmillan Publishing Company; copyright renewed © 1978 Theodore Rasmussen.)

to explain how the hand came to occupy such a huge territory in the human brain. That circumstance alone cries out for an evolutionary branch of neurologic research on arm and hand control. Wilder Penfield first demonstrated this stunning disproportion in the brain's surface topography during operations on awake humans carried out at the Montreal Neurological Institute. The resulting "homunculus"—a hand/mouth-dominated human figure seen on the surface of the brain—became world famous with the publication of his findings. John C. Eccles's recent book, *Evolution of the Brain* (New York: Routledge, 1989), can be taken as the definitive starting point for the modern liaison of anthropology and neurology. In his chapter 3, "Evolution of the Hominid Brain," Eccles gives a detailed account of the basic components of the motor system and their role in the progressive refinement of the locomotor profile of humans. Obviously, any collaborative future involving neurology and anthropology will require thinking past the shards of ancient skulls. Those bone fragments were once connected to other parts of a skeleton which was itself changing, and neither cranial nor postcranial changes can be interpreted without reference to environmental and archaeological history. So, too, with what was *encased* in the skull. No credible theory of human brain evolution can ignore, or isolate from environmental context, the co-evolution of locomotor, manipulative, communicative, and social behaviors of human ancestors. Addressing the need to define these relationships is what neuroanthropology would be up to, if it existed.

9. General references to the discussion on the co-evolution of hand and brain:

John C. Eccles, *Evolution of the Brain: Creation of the Self* (New York: Routledge, 1989)

K. Gibson and T. Ingold, eds., *Tools, Language and Cognition in Human Evolution* (New York: Cambridge University Press, 1993)

R. Holloway, "Evolution of the Human Brain," in A. Lock and C. Peters, eds., *Handbook of Human Symbolic Evolution* (Oxford: Clarendon Press, 1996).

D. Johanson and M. Edey, *Lucy: The Beginnings of Humankind* (New York: Simon and Schuster, 1981)

F. Jones, *The Principles of Anatomy as Seen in the Hand* (Baltimore: Williams and Wilkins, 1942)

R. Klein, *The Human Career* (Chicago: University of Chicago Press, 1989)

O. Lewis, *Functional Morphology of the Evolving Hand and Foot* (Oxford: Clarendon Press, 1989)

M. Marzke, "Evolution of the Hand and Bipedality," in A. Lock and C. Peters, eds., *Handbook of Human Symbolic Evolution* (Oxford: Clarendon Press, 1996).

W. McGrew, *Chimpanzee Material Culture: Implications for Human Evolution* (New York: Cambridge University Press, 1992)

D. Morris, *The Naked Ape* (New York: Dell Publishing, 1967)

J. Napier, *Hands* (Princeton: Princeton University Press, 1980; revised by R. H. Tuttle, 1993)

K. Schick and N. Toth, *Making Silent Stones Speak: Human Evolution and the Dawn of Technology* (New York: Simon and Schuster, 1993)

10. Success in moving across a supporting surface depends on a number of factors, including height, weight, and the area available to support weight. For a small animal, walking along a narrow branch is no trick; for a large animal, walking along the same branch could be as difficult as a ballet dancer crossing a tightrope on her toes. See Napier, *Hands*, pp. 77–82.
11. McGrew points out that chimpanzees in the wild use a very sophisticated "tool kit" of plant materials for gathering of food; he argues that nothing in the celebrated Oldowan collection of stone tools exceeds in complexity what chimpanzees routinely use. "What chimpanzees lack is what may have been important in hominisation: Tools for obtaining vertebrate prey and a means of collecting and transporting food for exchange" (McGrew, *Chimpanzee Material Culture*, p. 119).
12. Lewis comments on these changes in the ape arm as follows: "Thus, hard morphological fact seems to support the hypothesis, which has formed the mainstream of informed opinion for a century, that man is especially closely related to the African apes, and that 'brachiation' was the significant factor that molded much of their shared postcranial morphology and provided the apprenticeship and many of the pre-adaptations needed for the assumption of a habitual erect bipedal posture" (Lewis, *Functional Morphology*, p. 86).
13. Although it is not a matter of great importance to our consideration of the hand in human evolution, it is of great interest that paleoanthropologists continue to debate Lucy's entitlement to the status of direct human ancestor. One of the most interesting objections to the family tree constructed by Donald Johanson is that of Dean Falk, who argues that Lucy is an ancestor of the robust australopithecines, who ultimately became extinct.

    Falk proposes that Lucy and the robust australopithecines inhabited both the ground and the trees, never venturing out into the savannah (and into the hot sun) long or far. The gracile australopithecines did move away from the thermal protection of the trees, into a far more diverse and challenging environment. Brain growth ("intelligence") would have provided a competitive advantage but could not have occurred without a solution to the problem of temperature regulation inside the cranial cavity. The upright posture itself had the advantage of reducing the body area exposed to direct sunlight, but shading the body with the head has major drawbacks. No increase in brain size could have occurred, Falk argues, without improving the brain's capacity to maintain its internal temperature within the close tolerances required for normal physiologic function. In effect, the brain could not grow without improving heat dispersal. The new venous drainage system (coupled with bipedalism, altered hair patterns, sweat glands, and altered skin pigmentation) solved that problem. Falk has called this the "radiator theory of brain evolution." See Dean Falk, *Braindance* (New York: Henry Holt and Co., 1992), chapters 4, 6, and 7.

14. See Mary Marzke, "Joint Functions and Grips of the *Australopithecus afarensis* Hand, with Special Reference to the Region of the Capitate," *Journal of Human Evolution* 12 (1983): 197–211; also see Marzke, "Evolution of the Hand and Bipedality," in A. Lock and C. Peters, eds., *Handbook of Human Symbolic Evolution* (Oxford: Clarendon Press, 1996).
15. Lewis (*Functional Morphology*), however, dissents from Marzke's view of the changes at the base of the index finger in *A. afarensis,* regarding the evidence as insufficient to make the claim of humanlike mobility in this finger (p. 178).
16. M. Marzke, J. Longhill, and S. Rasmussen, "Gluteus Maximus Muscle Function and the Origin of Hominid Bipedality," *American Journal of Physical Anthropology* 77 (1988): 519–28.
17. It is not so easy to see that the gluteus maximus would be useful to a biped intent on bringing down a moving target with a rock, but a quick trip to the gym after a visit to Marzke's laboratory in Tempe convinced me that this really *is* how things would work. The gluteus and several other muscles underneath it are external rotators of the femur—if you lift your leg off the ground, they will rotate the thigh at the hip joint and cause the foot to turn out to the side; you also use these muscles when you cross one leg over the other when you are sitting.

The situation changes, however, when the leg is solidly planted on the ground. If the foot is free to move, contraction of the gluteus maximus rotates the femur so that the toes point outward. If the foot is firmly planted on the ground, however, contraction of the gluteus will pull the sacrum toward the outside of the thigh, causing the upper body to rotate in a direction away from that toward which the planted foot is pointing. (If you can't visualize what happens, just try it.)

Now we see what Marzke is driving at. If you stand as a right-handed pitcher stands before the windup, with your left side facing the plate, standing only on your right foot, a *right* gluteus contraction initiates a torso rotation to the left by pulling the upper body toward the right hip. This movement will accelerate an arm swing, as you can prove to yourself by observing what happens by going through the pitching motion; if you are more a dancer than a hurler, let your arms hang loosely at your side and imitate Chubby Checker doing the twist. The next phase of the pitch begins as the weight comes off the right foot and is placed on the left foot, where a rapid contraction of the gluteus will have the opposite effect on torso rotation: it will rapidly decelerate. This braking effect adds a final kick to the acceleration of the *arm,* as you can also prove to yourself by watching what happens to your hands each time you reverse the direction of pelvic rotation. You get the same effect if you try to dislodge something stuck to a spoon, or get catsup out of a bottle, by slamming your wrist against your opposite hand. The sudden braking of forward motion uses inertia to propel in a forward direction anything *not* attached.

> Toyoshima and Hosikawa have shown that approximately 46.9% of the velocity of the overhand throw comes from the step and rotation of the trunk. Deceleration of trunk rotation allows the throwing forelimb to accelerate, thus increasing the speed and distance of the thrown object. Atwater found that skilled throwers tend to decelerate the trunk to zero velocity prior to ball release. (Marzke, "Gluteus Maximus Muscle Function," p. 525)

18. See M. Marzke, K. Wullstein, and S. Viegas, "Evolution of the Power ('Squeeze') Grip and Its Morphological Correlates in Hominids," *American Journal of Physical Anthropology* 89 (1992): 283–98. The authors note that chimpanzees can and do hold sticks in an oblique orientation, using fingers and the thumb, and do use sticks as clubs with this grip; however, since the stick cannot be seated against the palm, a resulting strike will lack "clout." Here a simple personal demonstration makes the point. Hold a stick in your hand with the same grip you would use holding a suitcase. Then hold it as you would hold a baseball bat. Raise the stick over your head and see which grip would work better for protecting yourself.
19. Most readers are very familiar with the famous dictum of Descartes: *Cogito, ergo sum* (I think, therefore I am). An eminent neurologist and cognitive neuroscientist, Antonio Damasio, has recently written a neurologic treatise on the emptiness of pure reason: *Descartes' Error: Emotion, Reason and the Human Brain* (New York: Grosset/Putnam, 1994). In his introduction, Damasio argues, as do I, that the human brain does not exist in isolation: "Surprising as it may sound, the mind exists in and for an integrated organism; our minds would not be the way they are if it were not for the interplay of body and brain during evolution, during individual development, and at the current moment" (p. xvi).

    Not a week before I read Damasio's book, a musician friend had sent me a T-shirt inscribed with the following motto: "I jam, therefore I am." To a musician, to "jam" means to engage musically in a form of improvisational fun, freely mixing ideas and emotion. Hence, for me, "I jam, therefore I am" fully (and sublimely) redresses Descartes' error.
20. As current techniques for the study of cognitive processing advance, it should be possible to say with much greater precision how specific neurologic operations explain both language and skilled tool use. Perhaps we will be able to confirm that these operations are anatomically or developmentally linked. But doing so will require a new tradition of generating and collating evidence simultaneously from neurologic and anthropological sources. This kind of research seems plausible but daunting in the extreme.

    Nor will it be sufficient merely to argue the point on theoretical grounds. We must establish as specifically as we can how the structure of the hand evolved, and relate this process to the archaeological record of tool industry and culture and to specific modifications in the brain itself. We will never obtain examples of intermediate stages of brain evolution, of course, and as Kathleen Gibson has ruefully noted, "speech and gesture do not fossilize" (Kathleen Gibson, "Animal Minds, Human Minds: General Introduction," in Gibson and Ingold, eds., *Tools, Language and Cognition in Human Evolution*, p. 9).

Indeed, the hand itself has proven nearly as ephemeral as behavior from that standpoint: hand and wrist bones are notoriously scarce in the fossil record. Still, the specimens that have been recovered from early humans, apes, and monkeys may now be sufficiently complete to permit the roughing out of a first-order chain of events underlying the hand/brain coevolution. Certainly the ingenuity of brain investigators—and the technology at their disposal—should give us a crack at reconstructing the neurologic changes that would have accompanied the anatomic changes we *can* document.

21. Peter C. Reynolds, "The Complementation Theory of Language and Tool Use," in Gibson and Ingold, eds., *Tools, Language and Cognition in Human Development*, pp. 407–28. (Another influential voice in this discussion is that of Patricia M. Greenfield, a psychologist at UCLA. We will look at her ideas concerning tool use and language evolution later in the book.)

There are *profoundly* important implications if this hypothesis has any truth to it. Specifically (as suggested in the logic of universal Darwinism) it links language and toolmaking behaviorally through the Darwinian engine of selection. In essence, it states that the long association of mutually reinforcing behaviors with critical survival value will be reflected in the genetic capture, storage, and transmission of neurologic control mechanisms that govern those behaviors. An extended exploration of this hypothesis is the subject of the book in which Reynolds' paper appears. Actually, what has already been written on the subject would fill a large library, since a sizable and prolific branch of neurobehavioral science is devoted to a study of those brain mechanisms supporting language and those supporting skilled hand use. For the most part, however, research concerning these two behaviors is segregated by discipline: language is studied by linguists, cognitive scientists, and neurologists; hand function by surgeons and movement scientists. This division, however, says more about the organization of research and medical practice than it does about the organization of the brain.

2. THE HAND-THOUGHT-LANGUAGE NEXUS

1. Tolstoy, from his *Pedagogical Writings* (1903), cited in L. Vygotsky, *Thought and Language* (Vygotsky's own translation [1934] revised and edited by Alex Kozulin) (Cambridge, Mass.: MIT Press, 1986), p. 151.
2. For a complete and authoritative comparison of animal and human tool use, see

Thomas G. Wynn, "The Evolution of Tools and Symbolic Behavior," in A. Lock and C. Peters, eds., *Handbook of Human Symbolic Evolution* (New York: Oxford University Press, 1996), pp. 263–87.

3. We have a very limited understanding of the meaning of most of these intraspecies communications, but field observations have shown that they are sometimes used for remarkably sophisticated purposes, including deception. For example, a juvenile chacma baboon, after watching an adult female laboriously unearth a large edible root, called out as if he were being attacked just as the root came unstuck. This call immediately brought the youngster's mother, who attacked the other adult in retribution for the assault she apparently believed had taken place. During this ruckus the young baboon "nonchalantly picked up the abandoned tuber and settled in to enjoy lunch" (R. Byrne and A. Whiten, "The Thinking Primate's Guide to Deception," *New Scientist* 116, no. 1589 [1987]: 54–57; cited in Robin Dunbar, *Grooming, Gossip, and the Evolution of Language* [Cambridge, Mass.: Harvard University Press, 1996], p. 23).
4. Body language can signal more than emotion or a brewing fight. A few animals have elaborate systems of signaling that contain quite specific information about matters of importance. The best-known example is the use of aerial acrobatics by honeybees to "spell" the precise location of pollen-bearing blossoms. And, although there are severe limits on these abilities and on their inclination to use them without extensive training and encouragement, a few apes (especially the celebrated bonobo, Kanzi) have the capacity to understand human speech and to communicate with hand signs, and they even respect elementary human linguistic conventions (syntax) when they do.
5. See also Steven Pinker's new book *How the Mind Works* (New York: W. W. Norton, 1997). Pinker acknowledges but does not explore in any detail the relationship between brain development and the "frightfully complicated trigonometry" of the human arm (p. 11). He also asserts that "Hands are levers of influence on the world that made intelligence worth having. Precision hands and precision intelligence coevolved in the human lineage, and the fossil record shows that hands led the way" (p. 194).
6. R. Dunbar, *Grooming, Gossip, and the Evolution of Language,* pp. 60–65.
7. Merlin Donald, *Origins of the Modern Mind: Three Stages in the Evolution of Culture and Cognition* (Cambridge, Mass.: Harvard University Press, 1991).
8. The justification for the use of this term and a careful exploration of the implications are the subject of an extended discussion in Donald's book (*Origins,* pp. 167–86).
9. Are the cognitive, or "computational," underpinnings of speech and gesture the same? See D. McNeill, "So You Think Gestures Are Nonverbal?" *Psychological Review* 92, no. 3 (1985): 350–71.

10. Henry Plotkin, *Darwin Machines and the Nature of Knowledge* (Cambridge, Mass.: Harvard University Press, 1993).
11. Noam Chomsky, "Language and Thought," Anshen Transdisciplinary Lectureships in Art, Science, and the Philosophy of Culture, Monograph Three (Wakefield, Rhode Island: Moyer Bell /The Frick Collection, 1993), pp. 26–27.
12. Terence Deacon, *The Symbolic Species: The Co-Evolution of Language and the Brain* (New York: W. W. Norton, 1997).
13. Robertson Davies, *What's Bred in the Bone* (New York: Penguin Books, 1986).

3. THE ARM WE BROUGHT DOWN FROM THE TREES

1. J. Wilke, *Wall Street Journal,* January 7, 1992, pp. A1, A4 (Western edition).
2. Richard's own story was actually quite interesting. He first became acquainted with cranes while working for an aerospace company, where his job was to maintain communications equipment for remote tracking stations. Since the job required that he occasionally clean or paint large parabolic dish antennas, he had to learn to operate a crane. He later worked for a company that maintains large cranes, where he made safety inspections of, and occasionally repaired, bridge cranes. "I found myself going back into the mechanical side of cranes," he told me, "tearing gearboxes apart, replacing wire ropes, doing inspections. I was either telling people how to fix their cranes or doing it myself, mostly troubleshooting electrical problems for them. I do have a pretty good mechanical background, and have done quite a bit of welding as well as design and drafting, so I had no problem doing mechanical and electrical work on these cranes." But since the repair work felt like a demotion to him, he made a deal with his boss: because he liked the idea of preventing crane accidents, he agreed to develop a training division for the company.
3. Although this is not the topic here, it is fair to remind you that, for any particular movement, weights, distances, and speeds can vary by many orders of magnitude, and *can do so independently of one another.* The existence of these variables, which demand extreme flexibility in the control of force and timing in muscle contraction, adds so much additional complexity to the computational load of coordinated movement that no one has the slightest idea how it is done. Humans also play with these complexities by employing tools whose size and action differ hugely in scale—the puppeteer moves a marionette weighing about four pounds at the end of strings about 10 feet long, while the operator of the world's tallest mobile crane lifts 33 tons a vertical distance of 525 feet. Another extreme example: a vascular microsurgeon holds a forceps, needle, and thread and with infinite delicacy moves them through a semicircular arc as he sews together blood vessels barely larger than a red blood cell; when his job is done he goes out and, with a practiced forehand stroke, swings a tennis racquet through the same semicircular arc, moving the racquet thousands

of times faster, farther, and more forcefully than the forceps, with no loss of accuracy.

4. Interestingly enough, there is another paired structure in the body whose movements mimic those of the shoulder in its geometry: the eye. Although there is no official status for this notion (as far as I know), the pairing of shoulder and eye rotation in the act of pointing seems to me deserving of closer official inspection.
5. You can test this end-to-end biomechanical linkage yourself very easily. Stand close to a wall, your side turned toward it, just far enough away so that when you hold your arm straight out to your side your fingertips nearly touch the wall. Now raise your hand the way a traffic policeman does in making the "Stop!" gesture. Holding this arm position, place the thumb of your other hand at the base of the collarbone, right beneath your Adam's apple. When you're all set, make your policeman's arm absolutely straight and stiff and then just let yourself fall toward the wall. As the heel of your hand hits the wall, you will feel a slightly delayed impact in the thumb of the other hand. The force of impact has been transmitted directly from the heel of the hand (the wrist) through the forearm, through the elbow, through the upper arm, through the shoulder joint, through the scapula, through the AC joint, through the clavicle, to the sternoclavicular joint (where your thumb is). A cushioned bump but an honest, palpable bump nonetheless. You lined up all those bones and joints absolutely straight, like boxcars on a track, so that the force of impact was transmitted from one end to the other, just as it is when a railroad engine backs into the first car and the guy in the caboose feels a jolt a couple of seconds later.
6. Constant fluctuations in muscle activity occur with changes in posture and arm use, and are not only essential to the fluency of hand movement but probably critical to the maintenance of physiological health of the shoulder and arms. Secretaries, data-entry clerks, and others who use computers for long periods often complain of wrist and hand pain. Although they may be told they have developed so-called carpal tunnel syndrome, they often have far greater problems in the shoulder, neck, and upper back than they do in the wrist. Part of the explanation must be that when people root themselves at keyboards in front of computer monitors (especially if they are under stress or extremely intent on their work), the shoulder muscles seem to enter a state of low-level continuous contraction. This is an abnormal state for these muscles, and even though the muscular *forces* may be small, unrelieved muscle contraction by itself is enough to cause neck and shoulder pain, and it almost certainly contributes to the development of numb arms and hands, and even headaches. A further contribution to the common musculoskeletal disorders associated with computer use is impaired circulation to the arm (and the nerves to the arm) induced by sustained muscle contraction and the resulting abnormal holding postures of the shoulder—the so-called thoracic outlet syndrome, or neurovascular compression syndrome.

7. An impressive example of this mutually dependent, bidirectional reactivity and control is the familiar gymnastic trick of the two-man tandem handstand. One gymnast stands, hand overhead, holding the hand of the second gymnast, who is doing a single handstand over his partner. Each member of the pair must not only control the movements of his own hand but adjust his own bodily movements to maintain the stability of a top-heavy column. Success depends upon the mutual strength of the gymnasts, as well as the extreme precision of a sensorimotor system that detects and reacts instantly to minute deviations in the position of the head of the humerus within the shoulder joint.

4. PUPPET LESSONS FROM ALEXANDRIA AND DÜSSELDORF

1. Maxine Kumin, "The Absent Ones" in *House, Bridge, Fountain, Gate* (New York: Viking Press, 1975).
2. H. Joseph, *A Book of Marionettes* (New York: Viking Press, 1929), chapter 1 ("Puppets of Antiquity").
3. Yale's Heinrich von Staden, who has translated many original medical texts of the ancient Greeks, quotes Rufus of Ephesus as having written:

   > "But according to Herophilus . . . the nerves that make voluntary things possible have their origin in the cerebrum and spinal marrow, and some grow from bone to bone, others from muscle to muscle, and some also bind together the joints."
   >
   > Von Staden himself concludes: "While the last phrase suggests that Herophilus was not always successful in differentiating nerves from ligaments and tendons, the rest of the passage indicates that he knew of two kinds of nerves, *aesthetic* (sensory) and *prohairetika* (motor; literally 'capable of choosing, purposive') nerves" (Von Staden, *Herophilus: The Art of Medicine in Early Alexandria* [New York, Cambridge University Press, 1989], pp. 250–51).

4. C. Goss, "*On Movement of Muscles* by Galen of Pergamon," *American Journal of Anatomy* 123 (1968): 1–26. (This is a complete translation of the original work, with commentary by Dr. Goss.)
5. In Greek, *agon* means a contest, and an *agonist* is a contestant for a prize. An *antagonist* is the adversary the agonist must challenge to win the prize. Any of three different kinds of terms can be used to refer to a specific muscle. Each muscle has a unique *anatomic* name: for example, the muscle that connects the scapula to the head of the radius is called the biceps. Second, a *functional* term can be applied to any named muscle to indicate what that muscle does with respect to the joints on which it acts: the biceps is a flexor of the upper arm and of the elbow, and a supinator of the forearm. Third, a *physiologic* term indicates a temporary state of the mus-

cle's activation: while the biceps is contracting and shortening, causing the elbow to flex, it is an agonist; while the biceps is lengthening but actively contracting in order to slow or brake the extension of the elbow, it is an antagonist.

6. This is not such an unreasonable idea, after all, since a fatter muscle is a *shorter* muscle—think of the flexed biceps. Descartes was not an experimentalist, however, and this idea was quickly punctured by the Dutch physiologist Swammerdam, who showed that the volume of a muscle is the same whether it is relaxed or contracted.
7. See K. Ciuffreda and L. Stark, "Descartes' Law of Reciprocal Innervation," *American Journal of Optometry and Physiological Optics* 52, vol. 10 (1974): 663–73. It is possible that Descartes's choice of eye muscles for his model may have inadvertently drawn Bell into this research. Bell was preeminently an anatomist of the cranial nerves, and it was his work on the fourth cranial nerve that led him to propose, and then demonstrate, in 1829, that a nerve can have the effect of *inhibiting* muscular contraction. In 1898 Charles Sherrington, citing Bell's work, showed that the active contraction of an agonist is *preceded* by the inhibition (or relaxation) of its antagonist.
8. Sherrington's monograph *The Integrative Action of the Nervous System* (1906) marks the beginning of the modern era of descriptive and quantitative study of the circuitry and functional relationships of brain structures in relation to muscle function. Sherrington's work not only provided the necessary foundation for contemporary neurophysiology but set both the agenda and the methodology for modern clinical research in neurologic disease. For their joint work, Sherrington and Adrian shared the Nobel Prize in 1932. For an informative and extremely readable account of this history, see chapter 8 ("Springs of Action") of Jonathan Miller's *The Body in Question* (New York: Random House, 1978). The core of Sherrington's work on reflex inhibition, including an essay on the "knee jerk," can be found in chapter VII, "On Reciprocal Innervation," in D. Denny-Brown, ed., *Selected Writings of Sir Charles Sherrington* (New York: Paul B. Hoeber, Inc., 1940).
9. Galen himself proposed doing this in his book on the movement of muscles: see Goss, "On Movement of Muscles," p. 12.
10. Guillaume Duchenne, *Physiology of Motion,* translated and edited by E. Kaplan (Philadelphia: J. B. Lippincott Co., 1949; original French edition, 1867).
11. See George Speaight, *The History of the English Puppet Theatre,* second edition, (Carbondale, Ill.: Southern Illinois University Press, 1990), especially chapters 1 and 9.
12. Sometimes both muscles of a functionally reciprocal pair will contract at the same time in order to keep the joint from moving. For example, it may be necessary to "lock" one joint while moving an adjacent joint. This happens, for example, when a weightlifter is doing wrist curls and biceps curls. In the former exercise, a weight is held in place by the flexed fingers while the wrist flexes and extends. During this movement the elbow is stabilized by the simultaneous and sustained contraction of the biceps and triceps, which is called "flexor-extensor cocontraction." In the biceps

curl, the wrist is locked while the elbow is flexed and extended. Myriad variations of this shifting back and forth of contraction and relaxation occur throughout the musculoskeletal system during all active movement in real life. Without this built-in physiologic flexibility our repertoire of voluntary movements would be extremely limited.

13. The muscles around the eye are ideally arranged to direct the eye to any target it needs to inspect, and represents the ideal case of obligate reciprocal muscle activity in the body. Four muscles (the rectus—or "straight"—muscles) run front to back, from attachments at the top, bottom, and both sides of the eye to a common point at the back of the orbit. The top muscle (superior rectus) moves the eye (and gaze) upward, the bottom (inferior rectus) moves eye and gaze down, the inside muscle (medial rectus) causes the eye to point toward the nose, and the outside (lateral rectus) causes the eye to point away from the nose. Two other muscles (the obliques) are oriented at right angles to the rectus muscles, and their main job is to rotate the eye either clockwise or counterclockwise. In real life, the functions of these muscles are not quite so straightforward. The medial and lateral rectus muscles do, in fact, pull only in a side-to-side direction, toward or away from the nose. But the obliques and the superior and inferior rectus muscles all produce combined vertical and rotary movement. See Fig. 4.3.
14. J. Macpherson, "How Flexible Are Muscle Synergies?" in D. Humphrey and H.-J. Freund, eds., *Motor Control: Concepts and Issues* (New York: John Wiley and Sons, 1989).
15. The main body (shaft) is elongated, and from it three smaller bars extend to the side like wings. The front bar holds the three leads to the ears and nose and is solidly fixed in the middle, so that any roll of the longitudinal bar will tilt the head to the side; dropping the head of the shaft drops the head of the puppet. In the center of the shaft, where the right hand grips the main body of the device, is a larger, hinged bar holding the strings that go to the knees. Manipulating this bar at its pivot causes the knees to raise without any change in the head or shoulders. At the back end of the shaft is a loosely hinged bar whose strings run to the shoulders. This bar tends to stay level no matter what else is being done with the shaft, so that the shoulders of the puppet tend to remain horizontal. The arms and hands are controlled by a very long bar connected to the front of the shaft by strings that run from the shaft to each end of the bar. Because the hand bar is separately connected and is manipulated with the left hand, arm and hand movements completely independent from those of the rest of the body are possible.
16. In other words, the puppet's forearm, like that of the human, can pronate and supinate. Remember that the biceps is a supinator of the forearm; pronation is controlled primarily by two muscles, one that crosses the elbow and one near the wrist. In the puppet, if the string pulls on the thumb side of the wrist, the hand will supinate. If the string pulls on the small-finger side of the wrist, the hand will

pronate. If *both* strings are pulled together, the hand will lift. This arrangement is strictly analogous to that which determines joint movement in a real arm, dependent on agonist-antagonist reciprocity and (as in humans) a different movement entirely when simultaneous cocontraction occurs. If the puppeteer happens to use his own biceps to supinate his own elbow in the gesture that supinates the *puppet's* arm, there is a close match not only in surface choreography but in the underlying mechanics of the movement.

5. HAND, EYE, AND SKY

1. From "Boy and Top," in Octavio Paz, *Early Poems* 1935–1955 (New York: New Directions, 1973), p. 45.
2. Nicolai Bernstein, *The Coordination and Regulation of Movements* (Oxford: Pergamon Press, 1967), p. 33.
3. S. Culin, *Games of the North American Indians* (New York: Dover, 1975).
4. Seven texts are recommended for the reader who wishes to pursue in greater detail the technical issues in motor control raised in this chapter. The Rosenbaum and Wing volumes were written for advanced undergraduate students; the others require a substantial reading background in neurophysiology or movement research. The latest Jeannerod monograph is very strongly oriented toward research in visuomotor control of reaching and grasping movements. Also well worth reading is S. Zeki's piece, "The Visual Image in Mind and Brain," *Scientific American,* Sept. 1992, pp. 69–76.

   M. Jeannerod, *The Neural and Behavioral Organization of Goal-Directed Movements* (Oxford: Clarendon Press, 1988)
   ———, *The Cognitive Neuroscience of Action* (Cambridge, Mass.: Blackwell Publishers, 1997)
   D. Humphrey and H.-J. Freund, *Motor Control: Concepts and Issues* (Chichester: John Wiley, 1991)
   C. Phillips, *Movements of the Hand* (Liverpool: Liverpool University Press, 1986)
   D. Rosenbaum, *Human Motor Control* (New York: Academic Press, 1991); especially chapters 5 and 6
   J. Rothwell, *Control of Human Voluntary Movement,* second edition (New York: Chapman and Hall, 1994)
   A. Wing, P. Haggard, and J. Flanagan, *Hand and Brain: The Neurophysiology and Psychology of Hand Movements* (San Diego: Academic Press, 1996)

5. Recently, the answer to this question has changed from a definite "no" to a definite "maybe." The rapid evolution of what are called functional brain imaging tech-

niques makes it possible under highly controlled and circumscribed conditions to investigate brain correlates of learning. One general trend emerging from this work seems to be that during the early stages of learning, the cortex is most metabolically active; once motor routines have been memorized, performing the activity tends to be associated with activation of a deep set of structures called the basal ganglia. See, for example, R. Seitz and P. Roland, "Learning of Sequential Finger Movements in Man: A Combined Kinematic and Positron Emission Tomography (PET) Study," *European Journal of Neuroscience* 4 (1992): 154–65. One wonders what would happen in such a study if the practicing were done on a performer who had learned, say, to juggle, using a virtual reality instruction course.

Neurologists have also been interested in patients who, following brain injury, are able to learn new skills but retain no recollection of practicing them. As their training progresses, improvements are noted; yet the patients still approach each training session declaring that it is their first attempt. This dissociation of skill from the conscious awareness of it has led to the proposal that there are two kinds of skill learning: one depends upon a postulated motor, or *procedural,* memory; the other, associated with conscious awareness and factual information that can be communicated verbally to others, depends upon what is called *declarative* memory.

Some sports psychologists doubt that these two kinds of learning, or memory, function independently. Careful studies of expert athletes confirms that superior performance demands both, and that they are *highly* interactive. See, for example, chapters 1 and 6 in J. Starkes and F. Allard, eds., *Cognitive Issues in Motor Expertise* (Amsterdam: Elsevier Science Publishers B.V., 1993).

6. Charles Bell, *The Hand, Its Mechanism and Vital Endowments, As Evincing Design: The Bridgewater Treatises on the Power, Wisdom, and Goodness of God as Manifested in the Creation* (Treatise IV; New York: Harper and Brothers, 1840), p. 188.
7. With advances in visualization of cellular structure at the beginning of this century, it was learned that a number of specialized cells called "mechanoreceptors" were located in skin, muscle, and joints, and that these are collectively responsible for both conscious and unconscious monitoring by the brain of head, limb, and body orientation and movement—in other words, they account for what Bell meant by the term "muscle sense." Among these receptors, the most complex (called the muscle spindle) has both sensory and motor roles, and is known to be essential to the critical tuning and correction of skilled movements as they are performed.
8. Recent investigations of this developmental process have confirmed Bell's basic contention about the important role active visual seeking plays in building a visual reference system for movements of the hand. See A. Hein and R. Diamond, "Contribution of Eye Movement to the Representation of Space," in Hein and Jeannerod, eds., *Spatially Oriented Behavior* (New York: Springer-Verlag, 1983), pp. 119–33.

   Information acquired though tactile exploration is called *haptic* information. For an extended discussion of this topic, see Phillips, *Movements of the Hand*, especially chapters 1, 7, and 8; and Wing et al., *Hand and Brain*, chapters 16–21.

9. Jeannerod, *Neural and Behavioral Organization*, p. 51.

10. "Active touch" (a term coined by Sherrington) means conscious guidance of the hand to examine and identify objects in the external world; it cannot be mastered until hand and eye movements have been coupled. And that cannot happen until *head* and eye movements have been coupled. So much research has been and is being done on this topic alone that it is generally considered to be a legitimate specialty in its own right.

    The developmental sequence of hand and finger skills is a matter of considerable interest because of the unique human ability to control finger movements individually; these are called "fractionation" movements. Young children can make crude approximations of thumb to index finger within the first six months of life, but they cannot make other controlled movements of individual digits or control small objects in the hand with a precision grip until the age of one year—the time they also begin to show left- or right-hand preference. According to a comprehensive study of the maturation of grip in children, it is not until much later—sometimes not until the age of five—that children gain refined control of the frictional forces required to manipulate objects without using excessive force (H. Forssberg et al., "Development of Human Precision Grip" [4. Tactile Adaptation of Isometric Finger Forces to the Frictional Condition], *Experimental Brain Research* 104 [1995]: 323–30).

    Refinement of motor skills of the hand has long been associated with the maturation of a special pathway—the pyramidal tract—between the primary motor area of the cortex and the spinal cord. The ability of children to move fingers rapidly has recently been shown to depend not on training but simply on maturation of this pathway (see K. Müller and V. Hömberg, "Development of Speed of Repetitive Movements in Children Is Determined by Structural Changes in Corticospinal Efferents," *Neuroscience Letters* 144 [1992]: 57–60). Damage to the primary cortical control area for the hand or the pyramidal tract has disastrous consequences for humans because it deprives the hand of its dexterity. Such damage is a common and tragic consequence of strokes and head injuries; it may also appear in children as part of the more generalized motor system disorder, cerebral palsy.

11. See "Serge Percelly Does Not Take Juggling Lightly" in *The New Yorker*, November 8, 1993, pp. 50–51.

12. Howard Austin, "A Computational Theory of Physical Skill" (doctoral dissertation, Department of Electrical Engineering and Computer Science, Massachusetts Institute of Technology, March 1976), pp. 88–91. Two recent articles are also worth consulting for those interested in juggling as a topic within which to explore neuroscience and the pragmatics of learning at the same time: P. Beek and A. Lewbel, "The Science of Juggling," *Scientific American* 273 (November 1995): 92–97; and S.

Kemper, "If It's Impossible, Michael Moschen Will Do It Anyway," *Smithsonian,* August 1995, pp. 38–47.

13. Austin, "A Computational Theory of Physical Skill," p. 363.
14. Austin, "A Computational Theory of Physical Skill," pp. 43–44.
15. Sports psychologists have investigated the interaction of declarative and procedural knowledge via the following proposition: "Were knowing and doing linked, the link should work both ways: if knowing facilitates doing, doing should facilitate knowing" (F. Allard et al., "Declarative Knowledge in Skilled Motor Performance: Byproduct or Constituent?" in Starkes and Allard, eds., *Cognitive Issues in Motor Expertise*). The authors of this paper examined the relationship between qualification by former athletes for international competition-level judging and their own earlier participation in that sport *at the same level,* and found them to be positively correlated: "If you are qualified to judge at the most prestigious competitions, it helps to have competed at the same level."

    Another study, attempting to discern what experts do that novices do *not,* returns us to the fascinating world of visuomotor performance. Eye movements of table tennis players tracked by infrared sensors show that experts "seek predictive information" and "reduce the amount of information that needs to be processed in the very short periods of time available to them" (F. Allard, "Cognition, Expertise, and Motor Performance," in Starkes and Allard, eds., *Cognitive Issues in Motor Expertise*).
16. Austin, "A Computational Theory of Physical Skill," pp. 367–71.
17. Jeannerod, *The Cognitive Neuroscience of Action,* pp. 28–31.
18. Jeannerod, *The Cognitive Neuroscience of Action,* p. 21.
19. Austin, "A Computational Theory of Physical Skill," p. 50.

6. THE GRIP OF THE PAST

1. D. H. Lawrence, *Fantasia of the Unconscious* (New York: Penguin Books, 1960), p. 49; first published in 1922.
2. Like father, like son. Several years ago when David was working as a carpenter, he suffered a serious injury to his hand: a power saw nearly severed his small finger, at the same time cutting his ring finger and fracturing one of the bones in it. Here is how he described what happened next:

> I picked up the pieces, wrapped them in a shirt, and went with a friend of mine to the hospital. My little finger was detached and my ring finger was mutilated, broken all the way down to the base of the finger. I walked up to the gal at the desk and said, "I severely cut my fingers," and she said, "Well, I need your card." My buddy who was there, he pulled my wallet out and gave her the card.
>
> Next she says, "Okay, that'll be five dollars." So I gave her the five dollars and

she said, "You can go take a seat." By now it was about forty-five minutes since the accident had happened and the pain was really intense, but I figured she knew what she was doing. After sitting there for about ten minutes, talking to my friend, keeping the pressure on, I said to him, "You know, I'd really like to see somebody." So I walked back up to the desk. "I'm sorry, ma'am," I said, "I don't think you realize how severe my injury is." So she said, "Well, let me see what it looks like." And I said, "You'll regret it. Pieces will fall out." She said, "Come on!" She thought I was exaggerating like most people would do in that kind of a situation, but when I unwrapped it and part of my finger fell on her desk, she was gone. The next thing I was lying on the gurney.

Then a doctor came over and talked to me and told me he would set my finger. So, he reaches over and he began manipulating and pulling and cracking and stuff, and I reached out my other hand and started biting on my index finger. I was sweatin' bullets. I was right on the very edge, and he looked at me and saw what I was doing, and he said, "Have you received any kind of medication—pain medication?" And I said, "No." He stuck his hand over his badge and ran out of the room. I was being tested, but I said to myself, "If he wants to do it this way, I can do it. I can handle this."

I was not covered by Workmen's Compensation on that particular job; I had just finished graduate school and couldn't afford to stay home. So, the next week I went back on the job with my hand in a cast and a sling. I found out right away that I couldn't use my left hand at all, because of the cast. So I went home and cut the cast between my middle finger and my ring finger, then cut it crosswise so that I could have access to my index finger and my middle finger and my thumb. That way I could hold things. My wife walked in on me when I had the saw out and was cutting my cast. She about had a fit.

I was able to go back on the job by holding the chisel or holding nails with my left hand—that kind of thing—being cautious with it. And the same thing with hunting. I went on a hunting trip about two weeks later, still in a cast and everything. Tromped through the hills of Pennsylvania—I think it was only two weeks after the accident—and hunted.

3. David's approach to horses resonates strongly with that described in two recent books: Tom Dorrance, *True Unity: Willing Communication Between Horse and Human* (Fresno, Calif.: Pioneer Publishing, 1987); and Monty Roberts, *The Man Who Listens to Horses* (New York: Random House, 1997).
4. J. Napier, "The Prehensile Movements of the Human Hand," *Journal of Bone and Joint Surgery* 38-B, no. 4 (1956): 902–13.
5. Thirty years after Napier proposed this rule, neurophysiologists were finding its counterpart in the behavior of single cells within the brain. Jeannerod (*The Cogni-*

tive Neuroscience of Action [Cambridge, Mass.: Blackwell Publishers, 1997]) reviews this startling discovery:

> It can be shown, for example, that a monkey corticospinal neuron which fires during a movement of a given type (for example, a precision grip) will not fire during another type (for example, a power grip), in spite of the fact that the same muscles are implied in both cases. This result indicates that a given motoneuron can be connected to several sets of cortical cells, and that each set becomes activated in relation to a certain type of movement, not to the contraction of a certain muscle. The logical consequence of these findings is that there are several cortical representations for the same muscle, each representation coming into play as a function of the type of movement to be performed (p. 41).

Jeannerod cites as his source for this paragraph R. Muir and R. Lemon, "Corticospinal Neurons with a Special Role in Precision Grip," *Brain Research* 261 (1983): 312–16; and R. Lemon et al., "Corticospinal Facilitation of Hand Muscles During Voluntary Movements in the Conscious Monkey," *Journal of Physiology* 381 (1986): 497–527.

6. Napier pointed out that whether it is bracing a small or a large object, the thumb is mechanically stable only in two positions: pushing against the side of the index finger (adducted) or fully rotated into opposition (abducted, pronated, and either flexed or extended, depending on the size of the object). He then noted that the thumb is almost always *adducted* when some degree of precision is required in the power grip (as in pointing with a flashlight, watering with a garden hose, or holding a rapier). In contrast, the thumb is moved as far as possible into opposition when force is the main requirement (swinging a club). Conceivably, the adducted thumb position (the "precision-power") grip simply marks this as a pointing or aiming gesture with the implement being held; in studies of open-hand reaching, the thumb normally aligns with the axis of the forearm, while the index finger extends (opens) by an amount sufficient to match the size of the target.
7. Jeannerod, *The Cognitive Neuroscience of Action,* pp. 39–40.
8. C. Long et al., "Intrinsic-Extrinsic Muscle Control of the Hand in Power Grip and Precision Handling," *Journal of Bone and Joint Surgery* 52-A, no. 5 (1970): 853–67. In the hammer lift, both the dorsal and the palmar interossei associated with the index, middle, and ring fingers became active. The tiny lumbrical muscles—with the exception of the lumbrical to the small finger—were never involved in any power grip. In commenting on the fourth-lumbrical exception, the authors suggested that this muscle was acting as an adductor and ulnar rotator of the fifth finger. This, of course, is the signature movement of ulnar deviation, upon which the obliquely oriented squeeze grip depends, and which is unique to humans. See Fig. 6.4.

9. H. Forssberg et al., "Development of Human Precision Grip" (4. Tactile Adaptation of Isometric Finger Forces to the Frictional Condition), *Experimental Brain Research* 104 (1995): 323–30.
10. According to the *1993 Guinness Book of Records,* Robert Chisnell was thirty years old in 1982 when he did set the world's record of 22 one-arm chins, 18 two-finger chins, and 12 one-finger chins. Lee Chin-yong set the record for consecutive two-arm chin-ups in 1988: 370 without stopping. He was sixty-three years old at the time. (Forget the testosterone surge at sixteen; life begins at thirty.)

7. THE TWENTY-FOUR-KARAT THUMB

1. Raoul Tubiana, "Architecture and Functions of the Hand," in R. Tubiana, ed., *The Hand,* vol. 1 (Philadelphia: W. B. Saunders, 1981), p. 19.
2. John Napier, *Hands,* revised ed. (Princeton: Princeton University Press, 1993), p. 55.
3. Some people believe—and Merlin Donald is one of them—that Darwin's theory fails either to account for the status of, or to provide an appropriate predictive evolutionary model for, *Homo sapiens sapiens.* Why? Because *we* have started to meddle in a big way with the formerly unorchestrated rearranging of life in the jungle. We don't just fight our own niche battles—we aggressively seek to remake niche structure and dynamics. First came *Homo habilis,* the first professional toolmaker. Then came *Homo erectus,* who apparently discovered that a natural force like fire could be used, just as stones could, as a tool. Then came *Homo sapiens,* who expanded this process of appropriation, modification, and elaboration (concocting gunpowder and harnessing electricity, for example) to such an extent that humans now enjoy an unprecedented degree of protection from the blind processes that shaped the evolution of our species. Having seen how easy it is to "improve on nature," we humans have, so to speak, taken the gloves off: assisted by computers, we are busily tinkering with the ozone layer and global warming. If, one day, we launch a nuclear warhead into outer space and, by exploding it, deflect an asteroid headed straight for earth, no one will deny that humans have become masterful cheaters in Darwin's natural selection game.

   Merlin Donald (*Origins of the Modern Mind: Three Stages in the Evolution of Culture and Cognition* [Cambridge, Mass.: Harvard University Press, 1991]) puts it this way:

   > One thing is certain: if we compare the complex representational architecture of the modern mind with that of the ape, we must conclude that the Darwinian universe is too small to contain humanity. . . . Nineteenth-century biology . . . had no adequate vocabulary for assessing the cognitive dimensions of human evolution. Our genes may be largely identical to those of a chimp or a gorilla, but our cognitive architecture is not. And having reached a critical point

> in our cognitive evolution, we are symbol-using, networked creatures, unlike any that went before us. (p. 382)

Tim Ingold (*Tools, Language and Cognition*) puts it this way:

> What is required is a much broader conception of evolution than the narrowly Darwinian one embraced by the majority of biologists. Central to this broader conception is the organism-person as an intentional and creative agent, coming into being and undergoing development within a context of environmental relations (including social relationships with conspecifics), and through its actions contributing to the context of development for others to which it relates. In this account, behaviour is generated not by innate, genetically coded programmes, nor by programmes that are culturally acquired, but by the agency of the whole organism in its environment. (p. 470)

Of course, this change in the rules may not ultimately prove to be to our advantage. It just means that we can now do *ourselves* in out of ignorant misuse of power as easily as Mother Nature can when she is merely being herself. Read, for example, "Los Angeles Against the Mountains" in John McPhee's *The Control of Nature* (New York: The Noonday Press, 1990).

4. Charles Bell, *The Hand, Its Mechanism and Vital Endowments, As Evincing Design* (New York: Harper and Brothers, 1840), p. 157.
5. Frederick Wood Jones, *The Principles of Anatomy as Seen in the Hand,* second edition (Baltimore: Williams and Wilkins, 1942), pp. 298–300.
6. O. J. Lewis, *Functional Morphology of the Evolving Hand and Foot* (Oxford: Clarendon Press, 1989) p. 89.
7. Napier, *Hands,* p. 55.
8. D. Johanson and M. Edey, *Lucy: The Beginnings of Humankind* (New York: Simon and Schuster, 1981), pp. 348–49.
9. Mary Marzke, "Evolution," in K. M. B. Bennett and U. Castiello, eds., *Insights into the Reach to Grasp Movement* (Amsterdam: Elsevier Science B.V., 1994), chapter 2.
10. Mary W. Marzke, "Precision Grips, Hand Morphology, and Tools," *American Journal of Physical Anthropology* (1997) 102:91–110, p. 99.
11. Mary W. Marzke, "Evolution of the Hand and Bipedality," in A. Lock and C. Peters, eds., *Handbook of Human Symbolic Evolution* (Oxford: Clarendon Press, 1996), p. 131; Mary W. Marzke and Kathryn L. Wullstein, "Chimpanzee and Human Grips: A New Classification with a Focus on Evolutionary Morphology," *International Journal of Primatology* (1996): 117–39, p. 135 (Table III).
12. Mary Marzke, "Evolutionary Development of the Human Thumb," *Hand Clinics* 8, no. 1 (February 1992).
13. Some controversy remains as to whether the changes described by Marzke permit

Lucy's second metacarpal to pronate (rotate toward the thumb) at the wrist, as it does in humans. Also, it is not possible from existing specimens to determine whether the long and extremely powerful flexor of the distal joint of the thumb is a separate muscle in *A. afarensis,* as it is in humans. In most apes this muscle originates in common with the deep flexors of the fingers, so that, for them, flexion and extension of the thumb is difficult to execute as an independent motion.

When Napier examined the hand of *H. habilis,* the specimen being dated to about 1.82 million years ago, there was a suggestion that the tendon of the flexor pollicis longus might now be a separate muscle, with the possibility of flexing independently of the deep finger flexors. The saddle joint at the base of the thumb was also wider and flatter, making possible improved contact between the thumb and the remaining fingers.

Evolution next brought several additional modifications that make all of the grips described by Long available to the human hand. To recap, the changes in the human hand and wrist consist of the following major changes:

- Further modification of the base of the second metacarpal and its attachments to the trapezium, trapezoid, and capitate, permitting it up to 10 degrees of flexion-extension, and slight pronation, at the wrist joint.
- Flattening and curvature of the articulating surface of the hamate (i.e., a modified saddle joint), so that approximately 25 degrees flexion-extension can occur at the wrist-metacarpal joint of the small finger, plus supination during flexion movements; and about 10 degrees of flexion at the same joint of the ring finger, as well as supination in flexion. In addition, the fourth metacarpal is shorter than the third, and the fifth shorter than the fourth, giving a slight slope to the obliquity of the hand in the squeeze grip, making it more effective. The fifth metacarpal is also extremely thick, so it can absorb and transmit large forces developed in the fifth finger during the squeeze grip.
- Configuration of the metacarpal-phalangeal (knuckle) joints permits considerable rotation in the direction of both pronation and supination. The ability to manipulate small objects is increased by this change (and it would be impossible to finger chords on the guitar without it). See Fig. 7.2.
- A widened volar tuft on the tip of the thumb with two distinct soft tissue compartments has been formed. The result is that there is an enlarged surface for tactile exploration with the tip of the thumb, and a gradient of low distal to high proximal "give" in the pad.

Recently, a number of refinements have been made to Napier's system for categorization of grip. Marzke has adopted Long's model, which divides the power grip into the *squeeze* grip (hammers, tennis rackets); the *disc* grip (jar lids and doorknobs); the *hook* grip (suitcases); and the *spherical* grip, for large ball-shaped objects.

Long divides precision grips, or *handling motions,* into the three given above by Marzke for *A. afarensis,* plus a *pad-to-pad* grip, in which the thumb pad can be brought into direct contact with any or all of the palmar pads of the distal digits of the remaining fingers. See Charles Long, "Electromyographic Studies of Hand Function" (fig. 4), in Tubiana, ed., *The Hand,* vol. 1, pp. 427–40; and Charles Long et al., "Intrinsic-Extrinsic Muscle Control of the Hand in Power Grip and Precision Handling," *Journal of Bone and Joint Surgery* 52-B (1970): 853–67.

14. Mary Marzke, "Evolutionary Development of the Human Thumb," *Hand Clinics* 8 (1992): 1–8.
15. James Shreeve, *The Neandertal Enigma: Solving the Mystery of Modern Human Origins* (New York: William Morrow and Company, 1995), pp. 20–21.
16. Sometime, long after its first appearance in primitive fish over 400 million years ago, the five-ray fin became adapted for land use. Then, at different times and in different places on the globe, the outside chain of bones on the radial side of the forelimb (arm) began to assume a new shape and to carve a new niche for itself on the limb. Somehow, the radial chain or "ray" (the thumb) lost one of the three bones that make up each digit. Why does the thumb have only two phalanges while the fingers have three? Where did the missing bone disappear to? One possibility is that it didn't disappear at all. Instead, it could be that the first metacarpal (the hand bone at the base of the thumb) is what is missing. What we *call* the first metacarpal has a growth plate located at the end closest to the wrist, as do all twelve phalangeal bones of the four fingers. The other metacarpals have their growth plates located at the end *away from* the wrist. Thus, it is possible that the original proximal phalanx of the thumb became elongated, then migrated proximally to a position alongside the other four metacarpals at some time in the very, very distant past. If that is so, where did the *real* first metacarpal disappear to? Maybe it just got squeezed out of existence. Maybe it turned into the trapezium. It is indeed fascinating to imagine that the tiny, biomechanically unique bone at the base of the first metacarpal, the trapezium, is the ancient *real* first metacarpal, living a new life as the liberator of the terminal three phalanges of the radial ray. No one knows what actually happened, and no one knows how to find out.
17. If you can't visualize this sequence, do it yourself. Lay your hand and thumb flat on your leg. Notice how far out on the index finger the tip of the thumb reaches. Press the tip of the thumb inward against that point. Now, slowly turn your hand about 90 degrees toward a palm-up position, and, while you are doing that, slide the tip of the thumb toward the tip of the index finger. Stop moving when the tips of the thumb and index finger touch. The thumb and index finger should form an "O," almost as though you are holding a teacup. Reverse the motion, do it again, and keep watching what happens to your thumb as its tip moves out toward the tip of the index finger. The rotation of the thumb as it swings around to face the palm is called pronation. Repeat the sequence again and watch what happens to the angle

between the forearm and the wrist, and between the hand and the first row of finger bones. This momentous, complex, magical move of the human hand—thumb opposition—cannot occur without simultaneous *extension* of the wrist and *projection* of the thumb.

18. A peek into George's "tool kit" produces an instant revelation about the meaning of expanded tool use. It is jammed full of hammers, files, pliers, hand saws, and clamps, all in various shapes and sizes. A short paragraph from the school's textbook (A. Revere, *Professional Goldsmithing: A Contemporary Guide to Traditional Jewelry Techniques* [New York: Van Nostrand Reinhold, 1991]) gives you something of the exotic flavor of the tools of this profession:

> Creation of a forged ring employs two different types of forging procedures: spread forging (multidirectional forging) and directional forging with a cross peen. Afterward, the form is filed to perfection and a pearl post is sweat soldered in place. Procedures include the use of ring-bending pliers, ring-holding pliers, and a circle template, as well as both the planishing and cross-peen faces of a goldsmithing hammer (p. 91).

19. See D. Sloane and L. Sosniak, "The Development of Accomplished Sculptors," in B. Bloom, *Developing Talent in Young People* (New York: Ballantine Books, 1985).

8. THE RIGHT HAND KNOWS WHAT THE LEFT HAND JUST DID

1. Gerald Young, "Changes, Constancies and Continuities in Lateralization Development," in G. Young et al., *Manual Specialization and the Developing Brain* (New York: Academic Press, 1983), p. 403.
2. For a comprehensive review of research on hand preference among nonhuman primates, see P. MacNeilage, M. Studdert-Kennedy, and B. Lindblom, "Primate Handedness Reconsidered," *Behavioral and Brain Sciences* 10 (1987): 247–303.
3. Michael Corballis, *The Lopsided Ape* (New York: Oxford University Press, 1991).
4. See, for example, Stanley Coren, *The Left-Hander Syndrome: The Causes and Consequences of Left-Handedness* (New York: Free Press, 1992), especially the chapter "Evolution and Handedness," pp. 74–83.

   Extensive reviews and references to recent work on the biology of handedness are available in:

   R. Davidson and K. Hugdahl, eds., *Brain Asymmetry* (Cambridge, Mass.: MIT Press, 1995)
   N. Geschwind and A. Galaburda, eds., *Cerebral Dominance: The Biological Foundations* (Cambridge, Mass.: Harvard University Press, 1984)
   S. Springer and G. Deutsche, *Left Brain, Right Brain,* fourth edition (New York: W. H. Freeman and Co., 1993)

Gerald Young et al., eds., *Manual Specialization and the Developing Brain* (New York: Academic Press, 1983)

5. William Calvin, *The Throwing Madonna* (New York: McGraw-Hill, 1983). See also Calvin, "The Unitary Hypothesis: A Common Neural Circuitry for Novel Manipulations, Language, Plan-ahead, and Throwing?" in K. Gibson and T. Ingold, eds., *Tools, Language and Cognition in Human Evolution* (New York: Cambridge University Press, 1993), pp. 230–50.
6. There is now a substantial body of research that helps explain how this happens. Michael Merzenich and his colleagues at the University of California, San Francisco, have shown that a region of the cerebral cortex that has a central role in controlling precision movements of hand and fingers reconfigures the assignment of cortical tissue related to specific digits in response to experimental manipulation of the activities of the hand and fingers. This cortical remodeling, or "remapping," takes place almost immediately under ideal learning conditions, specifically when rapid repetitive movements take place under conditions of high motivation. See, for example, R. Nuda et al., "Neurophysiological Correlates of Hand Preference in Primary Motor Cortex of Adult Squirrel Monkeys," *Journal of Neuroscience* 12, no. 8 (1992): 2818–947; and M. Merzenich and K. Sameshima, "Cortical Plasticity and Memory," *Current Opinion in Neurobiology* 3 (1993): 187–96.

   Although this is a technical point, it is worth mentioning that the sensorimotor cortex in primates is influenced by activity in the lateral cerebellum, an essentially new structure in the mammalian brain. Lateral cerebellum is particularly important in relation to control of skilled hand movement, although it is not yet known specifically how it is involved in learning. Most researchers now believe that it is most important in the later, or final, stages of automatizing of precise movements (as in musical instrument playing). See W. Thach, H. Goodkin, and J. Eating, "The Cerebellum and the Adaptive Coordination of Movement," *Annual Review of Neuroscience* 15 (1992): 403–42. For a dissenting opinion, see R. Lanais and J. Welsh, "On the Cerebellum and Motor Learning," *Current Opinion in Neurobiology* 3 (1993): 958–65.
7. It is important to understand the difference between these two scenarios, which we are inclined to gloss over in an era dazzled by the feats of genetic engineering. To refresh on a canon of Darwinism: traits can be passed on only through genetic change. Behavioral change in an individual does *not* alter the genetic information that can be passed to any offspring. Thus, the one-arm sharpshooters who gained an early "upper hand" in the biped niche could have passed on this advantage to succeeding generations only by *instruction.*

   How, then, could this clan that had discovered the value of one-arm sharpshooting transform its strategy into a heritable trait? The answer is, they could not, because we cannot alter our own genes (not *yet,* at any rate). However, they *could* over time increase the probability of such a trend by increasing the breeding rate of individuals who were quick learners and/or high achievers in sharpshooting school. And they might do this without even knowing they were doing it; it doesn't matter. Families with this particular bit of genetic good fortune would *select themselves,* because more of their members would survive into their reproductive years.

8. Subsequent evidence for the prevalence of right-handedness comes from the cave drawings left by the Cro-Magnons, who colonized Ice Age Europe about 40,000 years ago. Not only did they leave the earliest human paintings, but they left behind tracings of their own hands. (See R. Hughes, "Behold the Stone Age," *Time,* February 13, 1995, pp. 52–62.) Eighty percent of the tracings found have been of the left hand (which means they were executed by the right hand), giving a right-to-left dominance ratio of 8:2. On the basis of Egyptian drawings in which people were depicted using tools, the ratio had reached its present state, 9:1. (See Springer and Deutsche, *Left Brain, Right Brain,* chapter 5.)
9. See the chapter "Fast Fingers and Hot Chops" in Frank R. Wilson, *Tone Deaf and All Thumbs?* (New York: Viking-Penguin, 1986), in which I attempted to explain this process in a musical context. Neurophysiology has advanced a good deal since then, but the basic ideas are reasonably intact. For a current, highly condensed review of the fascinating topic of motor learning and motor memory, see U. Halsband and H.-J. Freund, "Motor Learning," *Current Opinion in Neurobiology* 3 (1993): 940–49.
10. The *apparent* speed with which a novel complex motor skill can appear to gain genetic support can be breathtaking, or utterly mystifying. For example, musical keyboards are quite recent additions to the human warehouse of tools, gadgets, and toys, and they are among the most demanding to operate skillfully. Yet examples abound of high-level pianistic skill among young children (whose prodigious playing can only be explained by "inborn" talent), and by multiple members and generations of single families (among which the Bachs and Mozarts reign as the supreme examples).

    How could the code for a piano-playing "chip" have been introduced into human DNA so quickly? It *wasn't,* of course. It must already have been there, waiting for the invention of pianos for us to discover it. Once pianos were introduced into the environment and skillful piano playing proved to be an effective strategy for earning money and attracting members of the opposite sex, the genetic code (and the "pianist's chip" it produced) prospered mightily. We take up this subject in considerable detail, in chapter 11, "In Tune and Evolving *Prestissimo.*"

    Recently, it has been established that a mechanism of this kind governs the body's defenses to infectious disease. The appearance of a new antigen (akin to pianos in our story) does not require the creation of a new gene for the appropriate antibody but simply the "unpacking" of the dormant one that is already there, or the rearranging of existing components into whatever new assembly or configuration will suffice to solve the problem. This is simply another version of the strategy we encountered, in chapter 5 (Percelly's gradual construction and refinement of a finished performance), where an *ad hoc* performance assembly (a *virtual* "chip" stored in RAM, so to speak) appears to gain integrative control of an array of subtasks which must be meshed during the performance of a complex, learned skill. Nicolai Bernstein followers conceive of this "chip" as a hierarchical control theme (the "dominanta") and its more familiar name in Western psychology is the "coordinative structure."

11. R. C. Oldfield, "The Assessment and Analysis of Handedness: The Edinburgh Inventory," *Neuropsychologia* 9 (1971): 97–113. The survey comprised ten questions designed to represent a range of unimanual and bimanual activities with minimal redundancy. Respondents were asked to indicate which hand they would use in writing, drawing, throwing, or using scissors, toothbrush, knife, spoon, broom (upper hand), striking a match (hand holding the match) or removing a lid (hand holding the lid).
12. The distribution is strangely like that of the celebrated Kinsey study which concluded that human sexual affinities are arranged along a continuum of preferences from *strongly* for the opposite sex, through bisexuality, to *strongly* for the same sex, in an approximate ratio of 9:1 at the poles. The idea that there might actually be a real connection between left-handedness and homosexuality (or ambidexterity and bisexuality) seems to have originated with Sigmund Freud's close friend, Wilhelm Fleiss, who discussed this idea in his book *The Course of Life,* published in 1906. For a fascinating discussion of continuing reverberations, see chapter 12 ("On the Other Hand") in Marjorie Garber's *Vice Versa* (New York: Simon and Schuster, 1995).
13. M. Annett, *Left, Right, Hand and Brain: The Right Shift Theory* (Hillsdale, N.J.: Lawrence Erlbaum Associates, 1985). For an update, with suggested revisions of the right shift theory, see I. McManus and P. Breton, "The Neurobiology of Handedness, Language, and Cerebral Dominance: A Model for the Molecular Genetics of Handedness," in M. Johnson, ed., *Brain Development and Cognition* (Oxford: Oxford University Press, 1993).
14. See L. Harris, "Laterality of Function in the Infant: Historical and Contemporary Trends in Theory and Research," in Young et al., eds., *Manual Specialization and the Developing Brain.*
15. "Apraxia . . . was for many years almost entirely neglected and even today some authors regard it as rare. The fact is that apraxia is a frequent phenomenon that can be identified in many patients with cerebral lesions, especially those caused by stroke. It can be a useful sign for a clinical neurologist, but its greatest importance lies in what it tells us about cerebral organization in relation to movement" (N. Geschwind and A. Damasio, "Apraxia," in J. A. M. Frederiks, ed., *Handbook of Clinical Neurology,* vol. 1: *Clinical Neuropsychology* [Amsterdam: Elsevier Science Publishers, 1985], p. 430).

For a comprehensive current review, see H.-J. Freund, "The Apraxias," in A. Asbury, G. McKhann and W. McDonald, eds., *Diseases of the Nervous System: Clinical Neurobiology II* (Chichester: Wiley, 1992); or P. Pramstaller and C. Marsden, "The Basal Ganglia and Apraxia," *Brain* 119 (1996): 319–40. See also the chapter "Praxis and the Left Brain," in Corballis, *The Lopsided Ape;* and E. Ray and P. Square-Scorer, "Evidence for Common Expressions of Apraxia," in G. E. Hammond, ed., *Cerebral Control of Speech and Limb Movements* (Amsterdam: Elsevier Science Publishers B.V., 1990).

16. "It is doubtful that one will ever be able to understand this system's behavior solely on the basis of detailed analysis of its individual components; they appear to be too complex and too probabilistic in their behavior" (M. Wiesendanger et al., "Two Hands, One Action," in A. Wing, P. Haggard, and J. Flanagan, eds., *Hand and Brain: The Neurophysiology and Psychology of Hand Movements* (San Diego: Academic Press, 1996), p. 299.

It should also be mentioned that endocrine influences on brain development have entered the picture, remain in it, and require integration into theories of motor development. See N. Geschwind and A. Galaburda, *Cerebral Lateralization: Biological Mechanisms, Associations and Pathology* (Cambridge, Mass.: MIT Press, 1987). A highly condensed synopsis is contained in I. McManus and M. Breton, "The Neurobiology of Handedness, Language and Cerebral Dominance":

> The core of the GAG theory is that levels of testosterone in the fetus affect the relative growth of the right and left hemispheres, so that high testosterone levels, which of course will more typically be found in males than females, will result in the phenomenon of "anomalous dominance," which is defined as left-handedness, or right-hemisphere language, or left-hemisphere visuospatial ability, or a reduced degree of dominance for any of those characteristics, whichever hemisphere they may be in. Raised testosterone levels are also said to result in delayed growth of the posterior part of the left hemisphere, and thereby cause developmental learning disorders, such as autism, dyslexia, stuttering, hyperactivity, and poor artistic, musical, or mathematical ability. Finally, and most importantly for testing the theory, raised testosterone levels are postulated to result in modifications of immune functioning, so that there is an increased incidence of a wide range of pathological conditions, particularly those of an auto-immune nature, such as myasthenia gravis, ulcerative colitis, systemic lupus erythematosus (SLE), asthma, hay fever, and atopy, or of conditions that have a putative auto-immune status, such as migraine (p. 683).

17. J. Healey, J. Liederman, and N. Geschwind, "Handedness Is Not a Unidimensional Trait," *Cortex* 22 (1986): 33–53; M. Peters and M. Pang, "Do 'Right-Armed' Left-handers Have Different Lateralization of Motor Control for the Proximal and Distal Musculature?" *Cortex* 28 (1992): 391–99; R. Steenhuis and M. Breton, "Different

Dimensions of Hand Preference That Relate to Skilled and Unskilled Activities," *Cortex* 25 (1989): 289–304; M. Peters, "Subclassification of Non-pathological Left-handers Poses Problems for Theories of Handedness," *Neuropsychologia* 28, no. 3 (1990): 279–89; G. Hammond, "Manual Performance Asymmetries," in Hammond, ed., *Cerebral Control of Speech and Limb Movements;* M. Corballis, "Human Handedness," chapter 4 in *The Lopsided Ape.*

18. Yves Guiard, "Asymmetric Division of Labor in Human Skilled Bimanual Action: The Kinematic Chain as a Model," *Journal of Motor Behavior* 19, no. 4 (1987): 488.
19. Guiard, "Asymmetric Division of Labor," p. 493.
20. Guiard, "Asymmetric Division of Labor," p. 502.
21. Mary Marzke, "Precision Grips, Hand Morphology and Tools," *American Journal Physical Anthropology* 102 (1997): 91–110.
22. Doreen Kimura originally suggested that these functions were comparably dependent on sequencing operations. See D. Kimura, "Neuromotor Mechanisms in the Evolution of Human Communications," in D. Steklis and M. Raleigh, eds., *Neurobiology of Social Communication in Primates: An Evolutionary Perspective* (New York: Academic Press, 1979), and her recent book, *Neuromotor Mechanism in Human Communication* (New York: Oxford University Press, 1993).

9. BAD BOYS, POLYLITHS, AND THE HETEROTECHNIC REVOLUTION

1. Patricia Greenfield, "Language, Tools and Brain: The Ontogeny and Phylogeny of Hierarchically Organized Sequential Behavior," *Behavioral and Brain Sciences* 14 (1991): 531–95.
2. "The sequences of motions that we practice and internalize in the process of carrying out familiar activities—most particularly sequences of actions that we internalize in learning to perform a piece of music on an instrument, sequences that we both *make and follow* with each new performance—these action-paths become our most intimate way of knowing the piece. I call these internalized action paths 'felt paths.' . . ." (Jeanne Bamberger, *The Mind Behind the Musical Ear: How Children Develop Musical Intelligence* [Cambridge, Mass.: Harvard University Press, 1991], pp. 9–10.)
3. Ralph Holloway, "Paleoneurological Evidence for Language Origins," *Annals of the New York Academy of Sciences* 280 (1976): 330–48.
4. Stephen Jay Gould discusses the influence of various developmental rates (heterochrony) and the impact of protracted early development (neoteny) in his book *Ontogeny and Phylogeny* (Cambridge, Mass.: Harvard University Press, 1977). As Andrew Lock points out (in "Language Development and Object Manipulation," in K. Gibson and T. Ingold, eds., *Tools, Language and Cognition in Human Evolution* [New York: Cambridge University Press, 1993]):

> Neoteny is claimed as the major process responsible for human evolution. However, while such an approach rehabilitates comparisons between ontogeny and phylogeny in the morphological field, it is not directly applicable to the cultural elaboration of human action and "knowledge." Neoteny clearly plays little role in this sphere, for human knowledge does not remain infantile during ontogeny (p. 293).

5. Peter C. Reynolds, "The Complementation Theory of Language and Tool Use," in K. Gibson and T. Ingold, eds., *Tools, Language and Cognition in Human Evolution*, p. 411.
6. Reynolds, *Tools, Language and Cognition,* p. 412.
7. Reynolds, *Tools, Language and Cognition,* p. 423.
8. The sixteen-inch Cirkut camera, the largest model of a panoramic camera first produced in Rochester, New York, in 1905, could produce a twenty-foot negative. See Stephen J. Fletcher, "Cirkut Photography in Indiana Since 1906," *Traces* 3, no. 1 (Winter 1991): 18–31 (published by the Indiana Historical Society).
9. His driver was apparently born for the job. Jack describes him:

> Strangest guy. He was originally from Fall River Mill up by Mt. Shasta, and he was with the Bank of America. He got transferred to Red Bluff to be the loan manager there. I would see him at the drags all the time, always interested in the car. So I just asked him one day if he was interested in driving, because he didn't weigh anything—he was just a little guy. He jumped right on my offer. He was totally fearless, did stuff with a car that would have scared me to death. His timing was great and he had a lot of guts. He was just *smart!*
>
> I have this image of him and the car goin' off the track, can't see nothing but dust and wheels flying off through the air, kind of sparkling as they go past the trees. Just *ruin* a car. He'd walk away from it, come over and say, "Gee, I'm sorry about your car." The faster they went, the more he loved it. Wild.

10. Kathleen Gibson, "Beyond Neoteny and Recapitulation: New Approaches to the Evolution of Cognitive Development," in K. Gibson and T. Ingold, eds., *Tools, Language and Cognition in Human Evolution*, p. 275.
11. *Is* it art? The cover story of the July 1993 *Smithsonian,* "Sculpture on Wheels," by Richard Wolkomir, features a stunning collection of renovations of the kind Jack is now devoting his time to. Jack's career is the subject of an article by Mike Bishop in *American Rodder* (March 1998), pp. 26–31, and his business can be contacted through his web site at www.flatheadjack.com

10. THE ARTICULATE HAND

1. See A. Damasio and H. Damasio, "Brain and Language," *Scientific American* 267, no. 3 (1992): 89–95. For a more extended review, see Frank Benson, *Aphasia, Alexia and Agraphia* (New York: Churchill Livingston, 1979), pp. 12–17.
2. Steven Pinker, *The Language Instinct* (New York: William Morrow and Co., 1994), pp. 313–14. With this book, Pinker, director of the Center for Cognitive Neuroscience at the Massachusetts Institute of Technology, has set a new high-water mark in discussions of language. At the outset he explains his choice of the term "instinct": "It conveys the idea that people know how to talk in more or less the sense that spiders know how to spin webs. Web-spinning was not invented by some unsung spider genius and does not depend on having had the right education or on having an aptitude for architecture or the construction trades. Rather, spiders spin spider webs because they have spider brains, which give them the urge to spin and the competence to succeed" (p. 18). The citation quoted is from pp. 313-14.
3. Attempts are occasionally made to construct a time line for language emergence based on fossil evidence. Much is made of markings on the inside of the cranium that might establish asymmetries in the parietal, occipital, and temporal lobes of the brain. The pronouncement of the man who has done most of the original work, anthropologist Ralph L. Holloway at Columbia University, is: "Bluntly, the ugly fact is that there is not one single well-documented instance of paleoneurological evidence that unambiguously demonstrates a relative expansion of the parietal-occipital-temporal (POT) junction in early *Homo.*" Quoted in W. Wilkins and J. Wakefield, "Brain Evolution and Neurolinguistic Preconditions," *Behavioral and Brain Sciences* 18 (1995): 191. For a general discussion of the problems associated with interpretation of cranial endocasts—can there be a science of *paleoneurology?*—see Holloway's thorough and carefully argued presentations, "Paleoneurological Evidence for Language Origins," *Annals of the New York Academy of Sciences* 280 (1976): 330–48, and his more recent discussion of that topic in "Evolution of the Human Brain," in A. Lock and C. Peters, eds., *Handbook of Human Symbolic Evolution* (Oxford: Clarendon Press, 1996), pp. 90–98.
4. David Armstrong, William Stokoe, and Sherman Wilcox, "Signs of the Origin of Syntax," *Current Anthropology* 35 (1994): 349–68, p. 350.
5. Harlan Lane, in personal interview with author, February, 1993. The differences between gesture and sign are explored in detail in H. Poizner, E. Klima, and U. Bellugi, *What the Hands Reveal about the Brain* (Cambridge, Mass.: MIT Press, 1987); see chapter 1 ("Preliminaries: Language in a Visual Modality," pp. 1–30). Another source is a recent article about this research by Peter Radetsky: "Silence, Signs, and Wonder," *Discovery,* August, 1994, pp. 62–68.
6. Pinker, *The Language Instinct*, pp. 73–78.

7. Lev Vygotsky, *Thought and Language* (Cambridge, Mass.: MIT Press, 1986), p. 65. Vygotsky's reference to Stern is to C. and W. Stern, *Die Kindersprache* (Leipzig: Barth, 1928), pp. 160, 166.
8. Vygotsky, *Thought and Language,* p. 62.
9. Although we now know that the timing and general character of this process is largely under genetic guidance, when watched closely, its improvisational character can be striking:

> On the 251st day of his life, a child applies the word *bow wow* to a china figurine of a girl that usually stands on the sideboard and that he likes to play with. On the 307th day, he applies *bow wow* to a dog barking in the yard, to the pictures of his grandparents, to a toy dog, and to a clock. On the 331st day, he applies it to a fur piece with an animal's head, noticing particularly the glass eyes, and to another fur stole without a head. On the 334th day, he applies it to a rubber doll that squeaks when pressed, and on the 396th, to his father's cufflinks. On the 433rd day, he utters the same word at the sight of pearl buttons on a dress and of a bath thermometer. Although we now know that the timing and general character of this process is largely under genetic control, neither is it automatic, nor is the outcome predetermined. (Vygotsky, *Thought and Language,* p. 127)

Pinker (*The Language Instinct,* pp. 269–71) charts the exploding sentence structure of a little boy named Adam, beginning at 2 years, 3 months through 3 years, 2 months. Excerpts at three-month intervals:

> 2/3: Play checkers. Big drum. I got horn. A bunny rabbit walk.
>
> 2/6: Write a piece of paper. What that egg doing? I lost a shoe. No, I don't want to sit seat.
>
> 2/9: Where mommy keep her pocket book? Show you something funny. Just like turtle make mud pie.
>
> 3/0: I going come in fourteen minutes. I going wear that to wedding. I see what happens. I have to same them now. Those are not strong mens.
>
> 3/2: So it can't be cleaned? I broke my racing car. Do you know the lights went off? What happened to the bridge? When it's got a flat tire it's need a go to the station. I'm going to mail this to the letter can't come off.

10. Vygotsky, *Thought and Language,* p. 107.
11. Karl Bühler, *The Mental Development of the Child* (New York: Harcourt, Brace, 1930), p. 30.
12. Vygotsky, *Thought and Language,* pp. 91–94.
13. U. Halsband et al., "The Role of Premotor Cortex and the Supplementary Motor Area in the Temporal Control of Movement in Man," *Brain* 116 (1993): 243–66.
14. Harlan Lane, ed., Franklin Philip, trans., *The Deaf Experience: Classics in Language and Education* (Cambridge, Mass.: Harvard University Press, 1984).

15. Sicard, an advocate of training the deaf in signed spoken language, remained convinced that the alternative (i.e., a pure signed language without reference to speech) meant life devoid of intellectual meaning or possibility:

> All ideas come to us either directly from the senses, or mediated through our different combinations of them (giving us all our ideas of nonsensory things). We express these ideas with spoken sounds and evoke them in the midst of others by impressions on their hearing; we combine ideas and fix them in our minds by means of words. Now because no sound can affect the deaf person's hearing (he has none) and because he consequently has no symbols for fixing and combining his ideas, *it is evident that no original idea can remain in his mind and that no unfamiliar idea can reach him.* Hence the total communication gap between him and other people—there his is alone in nature with no possible use for his intellectual faculties, which remain inactive and lifeless unless some kindly hand happens to pluck him from this deathlike sleep. (Italics added)
>
> As the deaf person knows no speech signals and hence has no means of communication, his sensory impressions must all be transitory and his mental impressions fleeting. Nothing remains in his mind to which he can relate what is happening in him and which he can use as a basis for comparison. And because he can never combine two ideas at a time (he has no signs for retaining them) even the simplest sort of reasoning is impossible for him. (Roch-Ambroise Sicard, "Course of Instruction for a Congenitally Deaf Person," in Lane, ed., *The Deaf Experience,* p. 85).

Contrast this with Bébian:

> We know that the abbé Sicard brought these signs to a quite satisfactory level of perfection. This kind of translation is a genuine, comprehensive grammatical analysis that reveals the categories of words as parts of speech, their composition, inflections, and interconnections forming the sentence.
>
> But, one senses, the more profoundly these signs decompose the sentence—thus revealing the structure of French—the further they get away from the language of the deaf, from their intellectual capacities and style of thinking. That is why the deaf never make use of these signs among themselves; they use them in taking word-for-word dictation, but to explain the meaning of the text dictated, they go back to their familiar language. (Roch-Ambroise Bébian, "Essay on the Deaf and Natural Language, or Introduction to a Natural Classification of Ideas with Their Proper Signs," in Lane, ed., *The Deaf Experience,* p. 148)

Bébian regarded sign as a full-fledged language in its own right, and spent virtually his entire working life in France and in his native Guadeloupe trying to bring the deaf and their language into full flower. Lane, and a good many others presently engaged in a civil rights movement for ASL, consider the situation for the deaf to be somewhat worse now that it was when Bébian died in 1834. See Harlan Lane, *The Mask of Benevolence: Disabling the Deaf Community* (New York: Alfred A. Knopf, 1992). In our first interview, he said:

> I'm extremely frustrated because the people that I'm criticizing are really estimable people. They're bright, they're trying to do the right thing. I think they are ethical people, yet I haven't succeeded in conveying to them how wrong-headed what they stand for is, and how much better off they would be—deaf kids, and hearing parents of deaf kids—if they would only see it my way, or see it *part* of my way. In any case, what I'm trying to get across to them is that there is something *unique* about not having your hearing, which is unlike not seeing, or not walking.
>
> Being blind doesn't entail having a different language; but being deaf, as a variation in human capacity, leads you to have a different language. And the result of having a different language for this group is that they're a true community, and I mean the word in a way beyond what is intended when important differences lead to social labeling—for example, the "blind community," or the "gay community."
>
> The deaf are a community in the same sense that Polish Americans or Navajo Indians are—with a history, a language, a culture, art forms and mores, values and beliefs. If you want to understand deaf people, just think about the Navajos. They have all the problems that Navajos have. People think they're not with it, they're not capable; they're socially oppressed, they have trouble with English. Of course the deaf are not the same as Navajos, but neither are they biologic victims, as the normal modelists who think of them as "hearing impaired" would have us believe. What I'm trying to get across is that this is a *cultural and linguistic minority.* If we really understood that, we would behave very differently as a society, and we might also learn a lesson about human nature that I think anybody with a reflective mind would be interested in. *The lesson is that biology and culture irrevocably shape one another into special varieties of humankind.*
>
> Look—in Burundi there is a tribe of forest dwellers who are very short. We now have growth hormones. Suppose the Burundi government decided to inject all these short people with growth hormones. I would say, "What are you doing? What's the point of this medical approach? Would you destroy them, destroy their culture? What if you started with the premise that theirs is a legitimate way of being, even if it is based on their being shorter? But *you* think short is bad news." "Come on, Harlan, it's really bad to be short in Burundi. There's not one legislator or professional person who is short," and so on. This is exactly what the audiologists and otologists say: "Any amount of hearing is worth it."
>
> But they're quite wrong. Giving a healthy, happy deaf child a little hearing is *wrong* if it means rearranging the life of that child, turning it into an object of

pity who is labeled "hearing impaired" and forced into speech therapy, oralism in the schools, and an incredible struggle to escape a life of failed communication *in English.* It's a disaster. It's a big mistake to give a deaf child a *little* hearing in this way, when there is already a language and a culture tailor-made for that child.

16. Bébian, "Essay on the Deaf and Natural Language," p. 150. Contemporary research on perception increasingly supports the hypothesis that the brain does in fact assemble and organize sensory data ("features") into higher-level categories or entities that have significance *as* entities, just as Bébian suggests in his explanation of how the "idea of a peach" is composed. But the brain is unlikely to maintain fixed cubbyholes for peaches (The Center for Peaches and Peachiness) or any of the countless other things we will find ourselves involved with or interested in during the course of our lives. Rather, peaches seem to arise as an *idea* when certain structures or systems in the brain become active at the same time. To see how this actually works in the visual system (and how complicated the details turn out to be), see S. Zeki, "The Visual Image in Mind and Brain," *Scientific American,* September 1992, pp. 69–76.

    Echoing Bébian, Lane commented to me: "Remember that ASL exists in a visual and spatial mode. Its poetry and its humor are so different from what spoken language creates that it's quite startling and quite wonderful when you get to know it."
17. H. Poizner, E. Klima, and U. Bellugi, *What the Hands Reveal about the Brain;* see especially chapters 3–6.
18. For a review of this work, done at McGill University by Dr. Laura Petitto, see the following articles.

Laura Petitto, "On the Autonomy of Language and Gesture: Evidence from the Acquisition of Personal Pronouns in American Sign Language," *Cognition* 27, no. 1 (1987): 1–52.

———, "In the Beginning: On the Genetic and Environmental Factors That Make Early Language Acquisition Possible," in M. Gopnick and S. Davis, eds., *The Biological Basis of Language* (Oxford: Oxford University Press, 1996).

———, "Modularity and Constraints in Early Lexical Acquisition: Evidence from Children's Early Language and Gesture," in M. Gunnar and M. Maratsos, eds., *The Minnesota Symposia on Child Psychology,* vol. 25 (Hillsdale, N.J.: Lawrence Erlbaum Associates, 1992). One particular aspect of this report struck me as being of remarkable interest. In looking at the signing behavior of children between the ages of 7 and 24 months, a set of specific gestures was used to collect information about hand use in both linguistic and nonlinguistic gestures (pp. 42–43). The list includes motor activity (banging and scratching); pointing; social gestures (waving and nodding); actions with object in hand (brushing); instrumental gestures (raising arms to signal desire to be picked up); and symbolic gestures (imitating combing hair with hand at side of head). None of these movements involved fractionation (individually executed or articulated) movements of the fingers. Neither in speech nor in sign do extended sequences or complex grammatical constructions occur before individual finger movements are developed! A simultaneous explosion of grammar and of finger movements begins in the third year.

——— and P. Marentette, "Babbling in the Manual Mode: Evidence for the Ontogeny of Language," *Science* 251 (1991): 1493–96.

For an intriguing manual on communicating manually with babies, see L. Acredolo and S. Goodwyn, *Baby Signs: How to Talk with Your Baby Before Your Baby Can Talk* (Chicago: Contemporary Books, 1996).

19. University of Colorado anthropologist Gordon Hewes has reviewed that history in considerable detail and restated his own longstanding conviction that early hominid tool use and the evolution of hemispheric specialization associated with hand use provide both the behavioral and neurologic context to account for the evolution of language. See Gordon W. Hewes, "A History of the Study of Language Origins and the Gestural Primacy Hypothesis," in Lock and Peters, eds., *Handbook of Human Symbolic Evolution.*
20. David F. Armstrong, William C. Stokoe, and Sherman E. Wilcox, *Gesture and the Nature of Language* (Cambridge/New York: Cambridge University Press, 1995). Stokoe's original papers were: William Stokoe, "Sign Language Structure: An Outline of the Communication Systems of the American Deaf," *Studies in Linguistics,* Occasional Papers 8 (1960); William Stokoe, "Sign Language Autonomy," *Annals of the New York Academy of Sciences* 280 (1976): 505-13; and William Stokoe, "Sign Language Structure," *Annual Review of Anthropology* 9 (1980): 365-90.
21. Armstrong, Stokoe, and Wilcox, *Gesture,* p. 161.
22. Armstrong, Stokoe, and Wilcox, *Gesture,* p. 197.
23. P. Pramstaller and C. Marsden, "The Basal Ganglia and Apraxia," *Brain* 119 (1996): 319–40.
24. N. Geschwind and A. Damasio, "Apraxia," in J. A. M. Frederiks, ed., *Handbook of Clinical Neurology,* vol. 1: *Clinical Neuropsychology* (Amsterdam: Elsevier Science Publishers, 1985), p. 423.
25. Apraxia, it turns out, is a sublimely tricky issue for some oth-r reasons. Occasionally a patient with a Broca's aphasia (and remember, comprehension is "normal" in these patients) plus paralysis of the right arm and right side of the face will be entirely unable to carry out skilled movements of the *left* hand, even though it is not weak, they understand what they are being asked to do, and they "know" how to do what

they are being asked to do. This is called a "sympathetic apraxia," and Geschwind and Damasio consider it to be caused by interruption of a pathway from the left hemisphere to premotor cortex on the right side. Apraxia, usually of both limbs, can also be caused by damage limited to a specific region of the *sensory* cortex: the lower parietal region near the junction of parietal, temporal, and frontal cortex.

For more current reviews, see H.-J. Freund, "The Apraxias," in A. Asbury, G. McKhann, and W. McDonald, eds., *Diseases of the Nervous System: Clinical Neurobiology II* (Chichester: Wiley, 1992); or Pramstaller and Marsden, "The Basal Ganglia and Apraxia." See also "Praxis and the Left Brain," in M. Corballis, *The Lopsided Ape* (New York: Oxford University Press, 1991); and E. Roy and P. Square-Storer, "Evidence for Common Expressions of Apraxia," in G. Hammond, ed., *Cerebral Control of Speech and Limb Movements* (Amsterdam: Elsevier Science Publishers, 1990).

Apraxia is also seen in patients who sign, and becomes a particularly interesting problem because of the opportunity to study the division of hand movement into gesture and *coded* sign. Oliver Sacks, in his book *Seeing Voices: A Journey into the World of the Deaf* (Berkeley: University of California Press, 1989), commented on the striking difference in performance of two tasks requiring visual-spatial skill in one of Ursula Bellugi's deaf patients (Brenda) who had suffered a large right-hemisphere stroke. In general, Brenda ignored everything presented visually to her left side; and when she constructed a map of her own room, she included in it everything that was in the actual room, but she placed everything on the *right* side of the room, as if its left half had been sealed off, or was otherwise invisible or unavailable to her. This is the sort of altered perception and behavior one typically sees in such patients.

However, Brenda continued using signs requiring movement of her hand into the left visual field in order to retain their full meaning, making only minor errors (e.g., omitting the left side of an in-the-air figure of a square). In other words, her access to the left side of space—her awareness of it—remained linguistically intact even while it had simply vanished from her nonlinguistic, visually based perceptual and motor behavior. (See Poizner, Klima, and Bellugi, *What the Hands Reveal about the Brain,* pp. 141–44, 209.)

Sacks comments:

> Thus there develops in signers a new and extraordinarily sophisticated way of representing space, a formal space, which has no analog in those of us who do not sign. This reflects a wholly novel neurological development. It is as if the left hemisphere in signers "takes over" a realm of visual-spatial perception, modifies it, sharpens it, in an unprecedented way, giving it a new, highly analytical and abstract character, making a visual language and visual conception possible. . . .
>
> One must wonder whether this linguistic spatial power is the only special

> development in signers. Do they develop other, non-linguistic, visual-spatial powers? Does a new form of visual *intelligence* become possible? (pp. 95–97)

After reviewing some of the evidence that indicates that signers do, indeed, demonstrate "enhancement" of spatial cognition, Sacks goes on to discuss the implications from the point of view of cognitive science:

> Such a finding also raises fundamental questions as to the extent to which the nervous system, or at least the cerebral cortex, is fixed by inborn genetic constraints (with fixed centers and fixed localization—areas "hardwired," "preprogrammed," or "prededicated" for specific functions) and to what extent it is plastic and may be modified by the particularities of sensory experience. (p. 103)

26. In my own work with musicians I have encountered pianists who had consulted competent neurologists because of difficulties with control of their hands and who had been told that there was nothing wrong, or that they were just tired and overworked, when in fact a careful evaluation of their keyboard performance confirmed the existence of a disabling loss of motor control specific to piano playing! This cannot be called a diagnostic error on the part of the neurologist, by the way. Anything less than a full-blown loss of control will fall off the established diagnostic charts. And unless the neurologist is a trained pianist or an experienced observer of piano technique, how could he judge the keyboard performance of a professional?
27. Perhaps because so many neurologists are interested in music, the *loss* of musical skill has been given its own apraxic syndrome name: amusia. The first serious look at the relationship between brain function and musical skill was prompted by the case of Maurice Ravel, whose symptoms were dominated by a profound impairment of artistic skill (both compositional and pianistic). The following papers are germane:

T. Alajouanine, "Aphasia and Artistic Realization," *Brain* 71 (1948): 229–41.
J. Brust, "Music and Language: Musical Alexia and Agraphia," *Brain* 103 (1980): 367–92.
A. Gates and J. Bradshaw, "The Role of the Cerebral Hemispheres in Music," *Brain and Language* 4 (1977): 403–31.
N. Wertheim and M. Botez, "Receptive Amusia: A Clinical Analysis," *Brain* 84 (1961): 19–30.

A new branch of neurologic interest, if not really a science, was launched in 1977 by two British neurologists, M. Critchley and R. A. Henson, with the publication of *Music and the Brain: Studies in the Neurology of Music* (London: Heinemann, 1977). Passionate jugglers and puppeteers are still waiting in the wings for similar recognition.

28. Oliver Sacks, "Neurology and the Soul," *New York Review,* November 22, 1990, pp. 44–50.

11. IN TUNE AND EVOLVING *PRESTISSIMO*

1. Justine Sergeant, "Distributed Neural Network Underlying Musical Sight-Reading and Keyboard Performance," *Science* 257 (1992): 106–9.
2. Justine Sergeant, "Mapping the Musical Brain," *Human Brain Mapping* 1 (1993).
3. Wanda Landowska, *Landowska on Music,* collected, edited, and translated by Denise Restout, assisted by Robert Hawkins (New York: Stein and Day, 1964), p. 44.
4. Dorothy Taubman and Pat O'Brien were the first music teachers to call my attention to the importance of "distal flexion" as a source of problems for instrumentalists. Flexion of the terminal phalanx—the tip joint—of the finger is powered by a muscle called the *flexor digitorum profundus* (FDP), or deep flexor. Ideally, for a musician, in the sequence of joint movements that produce folding of the fingers into the palm, the first joint to flex is the knuckle joint, or *metacarpal-phalangeal* (MP) joint. Next is the *proximal interphalangeal* (PIP) joint, and lastly the *distal interphalangeal* (DIP) or tip joint. As the fingers increasingly flex, or fold, into the palm, the wrist joint begins to extend. If you grasp a hammer, or a tennis racket, firmly, you will see that the strength of the grip is dependent upon the combination of finger flexion and wrist extension. This *synergistic* movement is organized automatically by the motor system, and probably has its roots in the overhand grasp of a branch during brachiation.

   The problem with flexion of the DIP or tip joint for musicians is twofold: first, the contraction of the deep flexor is always accompanied by contraction of the muscles that extend the wrist, preparing the hand for a power grip. And the power grip, unfortunately, is not an appropriate postural base for rapidly repeated flexions and extensions of the fingers. Second, as demonstrated in a recent study of mechanical and physiological control of independent finger flexion, the FDP is not really a single muscle, either mechanically or electrically; consequently, flexion of the DIP joint of a single digit recruits flexor movement in other digits, which must be neutralized by active *extension* of any not intended to be moved. (See S. Kilbreath and S. Gandevia, "Limited Independent Flexion of the Thumb and Fingers in Human Subjects," *Journal of Physiology* 479, no. 3 [1994]: 487–97.)

   An additional complexity in the control of finger movements is the combination of ligamentous and muscular attachments not only to bone but *to each other.* Beginning with Duchenne and his classic study, *Physiology of Motion,* anatomists and surgeons have struggled to understand the biomechanical and physiologic mechanisms in the hand and brain needed to solve two related problems in finger control, both of which have major implications for successful instrumental technique. The first concerns the means by which combinations of joint flexion and extension, abduction and adduction, and rotation can be controlled, independently varied at each of the joints of the wrist and hand, and also *integrated* so that the hand and fingers function biomechanically as a unit. The second concerns the rapid adjustment of forces at the fingertip based on mechanical resistances encountered at the fingertip from the moment of contact (with string, key, or valve) until the moment of release. It is almost certainly the stupefying complexity needed for simultaneous control of these variables that explains the extreme length of training required for high-level musical performance, as well as the need for continuing rehearsal to sustain what is lightly referred to as "touch."

You will recall from chapter 6 that I was curious about rock climbing because of the "forbidden" use of the fingers that seemed to be required. Now you can see what I was driving at. Picture what Dave Hall's hand looked like as he was chinning himself on a door jamb (or a ledge). Put your own hand and fingers into that same position. Now, try to wiggle your fingers up and down as fast as you can, or try to flop your wrist up and down, keeping the tip joint flexed. This movement doesn't feel very loose, nor does it go very fast, since doing so requires very high forces to overcome the extensor pull on the hand. Active flexion of the DIP joint significantly impedes free repetitive movement of the entire finger at the MP joint. It is exactly what the rock climber needs to keep from falling, but ruinous for musicians playing with rapidly moving fingers.

To a large extent, we remain ignorant of the fine details of muscular control of rapid hand and finger movement. There are *thirty-nine* muscles located just in the forearm and hand, and at execution rates observed in finished performance of technically advanced instrumental play, most of those muscles are intermittently active many times each second. When one considers the calculation problem facing the nervous system, it becomes clear why motor-system psychologists continue to search for a theory of hierarchical control that can account for the precision and stability of performance in trained musicians. A severe restriction on research exists because there is no way directly to observe the fine details of muscle activity in the human arm and hand during such performances, just as there is no way directly to observe the brain as it controls skilled movement.

One of the most recent, and interesting, studies of this problem, which must be extrapolated to the human arm, is one carried out on monkeys by physiologist Mark Schieber at Washington University, in St. Louis. Schieber's imaginative efforts to unlock the sublime mystery of human digital dexterity have established, in the form of a detailed mapping (in monkeys) of patterns of muscle contraction in the forearm flexors and extensors, that individual fingers' movements always occur in response to the activity of an array of separate muscles, each contributing something in the way of agonist, antagonist, or stabilizing function. The nearly unimaginable complexity of the control strategy, incidentally, was established with the monkey's arm fixed in a restraining device, and the "instructed movements" were nothing but isolated movements of a single digit in one direction or the other. Were one to attempt to open the gates of unrestricted movement, and make the task playing a piano sonata, no one would even know where, or how, to begin to find out how it is done. (See M. Schieber, "Muscular Production of Individuated Finger Movements: The Roles of Extrinsic Finger Muscles," *Journal of Neuroscience* 15, no. 1 [1995]: 284–97; see also M. Schieber, "How Might the Motor Cortex Individuate Movements?" *Trends in Neuroscience* 13, no. 11 [1990]: 440–45.)

Despite these observations, the notion that "this muscle wiggles this finger" is so far from the truth as to be hilariously naïve. Out of this work, even at this preliminary stage, however, there are intimations of a physiologic principle that has immediate application to musical performance, and to practicing at an instrument.

First, Schieber, in the second of the two papers cited above, hypothesizes that the nervous system may be reducing the complexity of the control problem by taking advantage of an existing repertoire of grasping synergies in the hand. Refined movements of the fingers, in other words, would not have to be detailed in all their complexity, but only in relation to deviation from a smaller set of common patterns. This would imply that musical practice of a specific piece could be planned to take advantage of opportunities to use postural "resting places" for the hand wherever they appear in a long sequence of keystrokes. Doing so would permit rest even while the hand is in motion.

Second, as Marc Jeannerod comments on Schieber's work, "this model would require more neurons to produce the movement of a single finger than those of several fingers." In other words, it is easier to move the hand than to move one finger. This means that musical technique, where possible, should seek to base as much of a technical strategy on the hand and arm as possible: use them to get the finger into position, and keep finger movements confined to simple press and release wherever possible. (See M. Jeannerod, *The Cognitive Neuroscience of Action* [Cambridge, Mass.: Blackwell Publishers, 1997], pp. 42–43.)

Both Taubman and O'Brien agree on several basic principles as being critical to injury prevention in musicians. The small muscles located within the hand itself are physiologically geared for very rapid cycles of contraction and relaxation; they do not respond well to any sustained contraction. These muscles are also *sensors!* Together with the skin of the fingers, they are front-line scouts and spotters used by the central nervous system to report all the small details of tension and movement from the site of body-instrument interaction, and this information is critical to the adjustment and fine-tuning of timing and force needed to correct preplanned automatic movement sequences. When one couples these physiological preconditions with the extremely complex biomechanics of the hand itself, it becomes obvious why training must proceed slowly and deliberately and must be rehearsed (stabilized) only after the requirements of the music, the responsiveness of the instrument, and the sensations arising within the hand and arm have been taken into account and accommodated. Failure to respect this principle of individual variation through patient and thoughtful analysis and careful preparation accounts for almost all of the performance injuries suffered by instrumentalists.

Parenthetically, it appears that many of the principles that apply to healthy practices by instrumental musicians are directly applicable to other keyboard users, most of whom spend many hours a day operating an instrument for which they have had virtually no physiologically oriented training.

5. For nearly twenty years, Christoph Wagner, professor of physiology of music at the Hochschule for Music and Theater in Hannover, Germany, has documented both the normality and the variability of upper-limb biomechanics in musicians. While his work has shown that the distribution of biomechanical measures among musicians overlaps that of the general population, he has also shown that comparatively minor limitations in movement range or flexibility can impact profoundly on success with specific instruments. Ergonomics, in other words, is critically important for instrumentalists. See C. Wagner, "Success and Failure in Musical Performance: Biomechanics of the Hand," in F. Roehmann and F. Wilson, eds., *The Biology of Music Making: Proceedings of the 1984 Denver Conference* (St. Louis: MMB Music, 1988).

6. Not so obviously, the body has as much an impact on musical composition as on the design of musical instruments. The distance that can be spanned by the thumb and fifth finger of one hand rarely exceeds twelve inches, so it is pointless for a composer to ask a pianist to play with one hand a chord whose top and bottom notes are separated by more than that distance. Music for the clarinet never demands that the thumb reach *over* the third and fifth fingers to depress a key. Even if the idea is *musically* appropriate, the physical limitations of human musicians supervene.

   Even when music is composed with the physical constraints of the musician in mind, physical limitations will make the music unplayable for a certain percentage of competent musicians. Not every pianist is built to play every piece written for the piano, nor is every guitarist able to play any guitar piece he might wish to. Human biomechanics are extremely variable and are a critical factor in expert performance. The general failure to recognize this fact, or to address ergonomic factors except in the most primitive way, during a musician's training probably accounts for a very high proportion of injuries sustained by musicians early in their careers. For the musician whose hand or arm is not physically adapted to the instrument, or whose individual technique fails to accommodate limitations or distinctive characteristics in the movement characteristics of his own arms and hands, there is virtually no prospect of a pain-free career, and usually no prospect of a musical career of any kind.

7. John Blacking, *How Musical Is Man?* (London: Faber and Faber, 1976), p. 7 (originally published by the University of Washington Press, 1973, in the John Danz Lecture Series). Blacking, now deceased, was professor of social anthropology at Queen's University, Belfast, Northern Ireland.

8. Plotkin actually suggests, agreeing with Gerald Edelman, that human intelligence evolves "instinctively" under the control of genetically determined selective attention. Its operations are therefore analogous to human immune responsiveness, which depends for its specificity not on instruction but selection mechanisms. See Henry Plotkin, *Darwin Machines and the Nature of Knowledge* (Cambridge, Mass.: Harvard University Press, 1993), pp. 161–78.
9. Howard Gardner, *Frames of Mind: The Theory of Multiple Intelligences* (10th-anniv. ed.; New York: Basic Books, 1993), pp. xi–xii.
10. Merlin Donald, *Origins of the Modern Mind: Three Stages in the Evolution of Culture and Cognition* (Cambridge, Mass.: Harvard University Press, 1991), p. 382.

12. LUCY TO LULU TO ROSE

1. Laura Esquivel, *Like Water for Chocolate* (New York: Doubleday, 1992), p. 51.
2. Misia Landau, in *Narratives of Human Evolution* (New Haven: Yale University Press, 1991), with or without intending to do so, injects a new candidate for the moral of the Book of Genesis by offering another justification for the expulsion of Adam and Eve from the Garden of Eden. Another Eden existed—not the terrestrial habitat of Judaeo-Christian theology, but an arboreal one. In Landau's version of the Fall, we find another origin for the timeless theme of the vulnerability of human fortunes:

    - tree-dwelling pre-biped human ancestors, not humans, were expelled from the garden;
    - the reason for expulsion was neither sin nor the forbidden thirst for divine knowledge—the habitat had changed and survival in the trees was no longer an option.

    Landau's innovative use of narrative form, which mandates a search in scientific discourse for specific and important lessons hidden like Easter eggs, does not rob this defining stage in the story of human evolution of its great *moral;* one could hardly wish for a more trenchant lesson for a new species. The habitat in Eden, as I have suggested, must have encouraged the emergence of an aesthetic sense, and that is *good!* Landau indicates, though, that the habitat of arboreal life was characterized by "abundant food, little predation, refuge, and *complacency.*" Consequently, the habits of the primates were: "frugivorous, peaceable, *lazy,* and *arrested*" (p. 46). (Italics added)

    Landau, it seems to me, is offering an intriguing twist to the story of the Fall. Aesthetics, an unexpected gift from the situation, became a defining and *dignifying* trait of man; but man is by nature a creature of excess, and the blind pursuit of aesthetics becomes unfettered hedonism and therefore always gets us into trouble. So it had to stop. This is a pleasingly Puritanical message, even though it does tamper with the approved take-home message for exiles from the Garden of Eden: it was not lack of respect and violation of the taboo against knowledge that drove us out of Eden, but nearly the opposite: lack of ambition—*sloth!*

I like Landau's version, and we see immediately how intractable the problem is: the archetypical pampered contemporary teenager, cozy in the nest provided by its parents, will stay there indefinitely, watching television, playing video games, and eating fast food until forcibly ejected. Even after that the worthless creature will aggressively and repeatedly seek to reclaim its previous refuge, as any Parent of Adult Teenagers can attest.

3. M. A. Mariner, *San Francisco Chronicle,* February 3, 1993, p. 4 (food section).
4. Cafe Marimba opened on schedule, and in 1995 the *second* Cafe Marimba opened, this one in Burlingame. The chef was on a roll!
5. "Best in the Business," *San Francisco Focus Magazine,* August 1996, pp. 77–78.

13. TOUGH, TENDER, AND TENACIOUS

1. Moshe Feldenkrais, *Awareness Through Movement* (New York: Harper and Row, 1977), p. 46.
2. Todd Oppenheimer, "The Computer Delusion," *Atlantic Monthly,* July 1997, pp. 45–62.

14. HIDDEN IN THE HAND

1. Quoted in A. Schueneman and J. Pickleman, "Neuropsychological Analysis of Surgical Skill," in J. L. Starkes and F. Allard, eds., *Cognitive Issues in Motor Expertise* (Amsterdam: Elsevier Science Publishers B.V., 1993), p. 189.
2. See Ira M. Rutkow, *An Illustrated History of Surgery* (St. Louis: Mosby-Year Book, 1993). The barber pole stands for the red and white cloths that traditionally hung outside the barber shop. Early surgical procedures, mainly lancing, amputations, and "cutting for stone," were largely taken over by barbers from monastic brothers when the church decided that monks should be spending their time doing other things. In England, barber-surgeons were very influential by the seventeenth century, and it was not until shortly before Charles Bell's time that *physician*-surgeons separated their professional association from the older one that had been established by the barbers.
3. Schueneman and Pickleman, "Neuropsychological Analysis of Surgical Skill," pp. 189, 193, 197.
4. His books include a nine-volume series, *Classic Magic with Apparatus*, published in private edition beginning in 1977. The bulk of his collection is now owned by the magician David Copperfield.
5. Some months ago, my wife and I hosted a dinner for friends from Russia. The

guests included a Ukrainian surgeon (Val), his wife (Mila, who is a linguist and professional interpreter) and Misha, an educational psychologist from the Institute for Lifelong Learning in Moscow. We were discussing differences in Russian and American medical practice, and Misha told us about a new television personality in Russia, a man who "charms" his viewers' bottled water via television. Millions of people, he said, believe that by holding bottled water in front of the television while this man is offering his spell, the water will acquire healing powers.

We all had a good laugh over this, and then Val, smiling, said, "You know, the Russian word for physician is "Vrach" (Врач). It's related to the word for 'lying.' A doctor has to be a good liar!" Misha, a true Russian intellectual, is a broadly educated man, and he immediately corrected Val. "Actually," he said, "the word comes from the old Slavic word for 'witch'! " Mila, the professional linguist at the table, was unsure. We looked the words up in a Russian dictionary but could not find an answer. So Mila said she would do some research in an etymological dictionary and give us a verdict. The next day a fax arrived with her findings.

1. Врач (pronounced *vrach,* or *vrat'*), the contemporary word meaning "doctor," is the old Slavic word meaning "witch." It also means "prophet," and its source is from a verb which means "to mutter," as in performing an exorcism. The connection between this word and the Greek word for orator is strong, and the inference is that this is a person who "does not say what he thinks."
2. Ворожить (pronounced *vorozhit'*) means to do witchcraft, and is derived from an older word (Връшти, pronounced *vryatshti*) meaning "to cast," as in casting lots or to cast a spell.
3. Ведун (pronounced *v'edoon*) means magician or sorcerer, and comes from the ancient Slavic word Вѣдь (pronounced *vyatd'*), meaning "knowledge" or "spell." This word, in turn, exists in Sanskrit as two related words: "vidatham," meaning "knowledge," or "wisdom of priests," and "vidya," meaning "sorcerer" or "magic."

So the ancient links, indeed the original equivalence of medicine and magic described to us by Robert Albo, are authenticated in one of the oldest language families known to human history. In these early terms one finds that utterances not only represent knowledge but can act on their own as a primary medium through which special knowledge can exert its power, and that words can be used to cast either a favorable or an unfavorable spell—to exorcise or to harm.

The Greek and Latin roots of our own words *physician* and *doctor* are important markers for the passage of civilization into a more informed and less subordinate relationship with the physical and biologic (and spiritual) world. *Physics* and *physician* share the Greek root words meaning nature, and the knowledge of nature. *Doctor* comes from the Latin word meaning "to teach." Consciously, therefore, the medical practitioner bases his work rationally on what is known of the physical world—rather than on superstition. And he uses his special knowledge not to control, but to teach. This, surely, is as it should be. A final word of buried meanings: Val was not entirely mistaken in claiming a linguistic origin for his assertion that "a doctor has to be a good liar," since our own expression "to doctor the truth," means to deliberately falsify.

Physicians, however, are not immune from the seductive influence of the power which ancient tradition has placed in their hands. The long white coat of the hospital physician is not simply an apron worn as an antiseptic precaution. It is a vestment of authority, just as a priest's robe is, and both doctor and patient know it to be a ritual affirmation of knowledge and power.

6. Although the deeper levels of symbolism involving medicine and magic are not our topic, this is a fascinating topic in itself, and in fact is exploited by one of the best-known stage magicians of modern times. David Copperfield, who is a close friend of Albo's, is a sublime practitioner of this kind of magic, and his shows are full of illusions using boxes inside of which very unusual things happen to the body. Explains Albo:

> There is actually an interesting background history to illusions involving the body. From about 1915 on, it became common for magicians to have female assistants—the prettier, the better. The assistants were referred to as "box jumpers," because they were being produced and vanished. All kinds of things were done with these female assistants, and it was the beauty of the female assistant which made the show more attractive to the people watching.
>
> When Selbit invented the trick of sawing a woman in half, it was not nearly as spectacular as we see it today. The trick originally involved tying two ropes around the ankles and two ropes around the wrists, and bringing them through holes; you couldn't actually see the woman. They would pull the ropes taut and then saw the box in half. That was the way the trick was originally staged.
>
> The trick was modified in the United States by Thurston and Dante and Keller so that the head and feet were showing. Other modifications then made the box shorter and thinner, and so forth. But in the original concept you didn't even see that it was a woman. It could have been anybody because nothing showed.

Perhaps nowhere are the symbolic connections between medicine, magic, and ancient priesthoods closer to the surface than in the famous trick of sawing the woman in half. It is really a three-in-one spectacle: on its surface a dramatic and magical illusion of bodily separation and reattachment; by analogy, a real-life surgical event of trauma and repair; in our dreams, an ancient religious rite of human sacrifice.

7. A recent *New Yorker* profile of Ricky Jay, perhaps the world's best-known living sleight-of-hand artist, is nothing less than reverent about what Jay can do with his

hands. Its author describes Jay's hand as being small—"large enough that a playing card fits within the plane of his palm, with a slightly raised pad of flesh on the underside." Jay's own description is that his is "not the hand of a man who has done a lot of hard labor." See Mark Singer, "Profile: Secrets of the Magus," *The New Yorker,* April 5, 1993, pp. 54–73.

15. HEAD FOR THE HANDS

1. W. Gustin, "The Development of Exceptional Research Mathematicians," in Benjamin Bloom, ed., *Developing Talent in Young People* (New York: Ballantine Books, 1985), p. 287. Gustin's chapter reviews the findings of in-depth interviews of twenty mathematicians, all of whom were recipients of the Sloan Foundation Fellowship and all of whom, because of the frequency of references to their work in the publications of other mathematicians, are also listed in the *Science Citation Index.*
2. Seymour Sarason, *The Challenge of Art to Psychology* (New Haven: Yale University Press, 1990), pp. 76–77.
3. Kieran Egan, *The Educated Mind: How Cognitive Tools Shape Our Understanding* (Chicago: University of Chicago Press, 1997), p. 3.
4. Egan, *The Educated Mind,* p. 20.
5. Sarason, *Challenge,* p. 75.
6. John Dewey, *Art as Experience* (New York: Minton and Balch, 1934), p. 3. See Sarason, *Challenge* (chapter 5) for a full discussion.
7. Sarason, *Challenge,* pp. 68–69.
8. This, of course, is also what Dewey had in mind. Dewey's was an education of self-discipline: the power to control the means necessary to achieve, test, and assign value to whatever ends the individual chooses to seek. Discipline in the conventional sense, a punitive enforcement device, was really the opposite of what Dewey considered as the engine for learning. As he put this point: "All genuine education *terminates* in discipline, but it *proceeds* by engaging the mind in activities worthwhile for their own sake." (R. Archambault, ed., *John Dewey on Education: Selected Writings* [Chicago: University of Chicago Press, 1964].)
9. Lauren Sosniak, "Learning to Be a Concert Pianist," in Bloom, ed., *Developing Talent in Young People.*
10. Lauren Sosniak, "From Tyro to Virtuoso: A Long-term Commitment to Learning," in F. Wilson and F. Roehmann, eds., *Music and Child Development* (St. Louis: MMB Music, Inc., 1990), pp. 286–87.
11. Jeanne Bamberger, "The Laboratory for Making Things," in D. Schön, ed., *The Reflective Turn: Case Studies in and on Educational Practice* (New York: Teachers College Press, 1991), p. 38.
12. Bamberger, *The Reflective Turn,* p. 44.
13. Egan, *The Educated Mind,* p. 166.

14. Egan, *The Educated Mind,* pp. 84–85.
15. Charles Bell, *The Hand, Its Mechanism and Vital Endowments, As Evincing Design* (New York: Harper and Brothers, 1840), p. 157. Bell's choice of the word "evincing" is intentional, as it was his goal (in common with other biologic encyclopedists of the early nineteenth century) to establish that biologic forms are the living catalogue, and proof, of divine purpose.
16. Mark Jeannerod says this:

> Grasping cannot be reduced to its visuomotor aspects. It is the motor counterpart of a broader function. During handling and manipulation, for which grasping is a precondition, signals for object identification arising from sight and touch are co-processed. The fingerpads have been considered by some authors (for example, Sherrington) as the somatosensory "macula." Thus, the hand brings objects to be manipulated within the central field of vision, so that "the finest movements of the fingers must be under simultaneous control from the very centers of the visual and tactual maculae." Berkeley in his famous (1734) treatise on "A new theory of vision" emphasized the fact that objects can only be known by touch, which he considered as the ultimate means of exploration and knowledge of the world. Vision is subject to illusion, which arises from the distance-size problem (size must be extracted from apparent distance) or the 3-D reconstruction problem (the visual third dimension is extracted from a two-dimensional map using indirect cues from perspective). Touch, and particularly active touch, is not subject to these constraints, as it involves direct assessment of size and volume. In addition, touch is critical for perceiving object properties like hardness, compliance, texture, temperature, weight, etc., which can hardly be accessed by sight alone. (Jeannerod, *The Cognitive Neuroscience of Action* [Cambridge, Mass.: Blackwell 1997], pp. 38–39.)

17. Seymour Sarason has recently been working with Dennis Littky and Elliot Washor, co-principals of The Met, a small, innovative public high school in Providence, Rhode Island, where mentoring based on each student's self-defined interests (and community-involved teaching) is the core teaching strategy. ("The Met" is the nickname for the Metropolitan Regional Career and Technical Center.)
18. Jon-Roar Bjørkvold, *The Muse Within: Creativity and Communication, Song and Play from Childhood Through Maturity* (translated by William H. Halverson; New York: HarperCollins, 1992).
19. John Blacking has described another version of this process among the Venda of South Africa:

> The Venda theory of personality and cognitive development begins with the assumption that the child is an active rather than a passive participant in his or

> her own development and that one can only become a person through social interaction with others. . . . If an infant survived the first physically dangerous years of life, the innate characteristics of the deceased's ancestral spirit lodged in the child's body would gradually be modified by the different social experiences of the new person, who would eventually develop a strong and independent personality. Much of this discovery of self, discovery of "other," discovery of "real self," that is, ancestral self . . . was achieved through quite systematic musical training. (Blacking, "Music in Children's Cognitive and Affective Development: Problems Posed by Ethnomusicologic Research," in Wilson and Roehmann, eds., *Music and Child Development*, p. 76.)

20. Tim Ingold, "Tool Use, Sociality and Intelligence," in K. Gibson and T. Ingold, eds., *Tools, Language and Cognition in Human Evolution* (New York: Cambridge University Press, 1993), pp. 430–31.

EPILOGUE

1. Ernst Mayr, *This Is Biology: The Science of the Living World* (Cambridge, Mass.: Harvard University Press, 1997), chapter 3, "How Does Science Explain the Natural World?"
2. For information, write to Johnson O'Conner Research Foundation, The Monadnock, Suite 840, 685 Market Street, San Francisco, CA 94105.
3. Chapter 6 of Napier's book *Hands* (Princeton: Princeton University Press, 1993, revised edition) is about dermatoglyphics (fingerprints): "The first examples of fingerprints used as seals date from the third century B.C., when they were quite commonly used by Chinese businessmen" (p. 133). He does not mention palmistry. Richard Unger recommends *The Book of the Hand: An Illustrated History of Palmistry,* by Fred Gettings (London: Paul Hamlyn Ltd., 1965). If you are interested in further information or the newsletter published by his institute, write to: International Institute of Hand Analysis, Box 151313, San Rafael, CA 94915-1313.
4. James Bailey, *After Thought: The Computer Challenge to Human Intelligence* (New York: Basic Books, 1996), pp. 156–57.

APPENDIX

1. Quoted in Gordon-Taylor and Walls, *Sir Charles Bell: His Life and Times* (Edinburgh: E. & S. Livingstone Ltd., 1958), p. 223.

Bibliography

Acredolo, L., and S. Goodwyn. *Baby Signs: How to Talk with Your Baby Before Your Baby Can Talk.* Chicago: Contemporary Books, 1996.

Alajouanine, T. "Aphasia and Artistic Realization." *Brain* 71 (1948), 229–41.

Allard, F. "Cognition, Expertise, and Motor Performance." In Starkes and Allard, eds., *Cognitive Issues in Motor Expertise.*

———, et al. "Declarative Knowledge in Skilled Motor Performance: Byproduct or Constituent?" In Starkes and Allard, eds., *Cognitive Issues in Motor Expertise.*

Annett, M. *Left, Right, Hand and Brain.* Hillsdale, N.J.: Lawrence Erlbaum Associates, 1985.

Archambault, R., ed. *John Dewey on Education: Selected Writings.* Chicago: University of Chicago Press, 1964.

Armstrong, D., W. Stokoe, and S. Wilcox. *Gesture and the Nature of Language.* New York: Cambridge University Press, 1995.

———. "Signs of the Origin of Syntax." *Current Anthropology* 35 (1994): 349–68.

Asberg, A., G. McKhann, and W. McDonald, eds. *Diseases of the Nervous System: Clinical Neurobiology II.* Chichester: Wiley, 1992.

Austin, Howard. "A Computational Theory of Physical Skill." Doctoral dissertation, Department of Electrical Engineering and Computer Science, Massachusetts Institute of Technology, March 1976.

Bailey, James. *After Thought: The Computer Challenge to Human Intelligence.* New York: Basic Books, 1996.

Bamberger, J. "The Laboratory for Making Things." In D. Schön, ed., *The Reflective Turn: Case Studies in and on Educational Practice.* New York: Teachers College Press, 1991.

———. *The Mind Behind the Musical Ear: How Children Develop Musical Intelligence.* Cambridge, Mass.: Harvard University Press, 1991.

Bébian, Roch-Ambroise. "Essay on the Deaf and Natural Language, or Introduction to a Natural Classification of Ideas with their Proper Signs." In Lane, ed., *The Deaf Experience.*

Beek, P., and A. Lewbel. "The Science of Juggling." *Scientific American* 273 (November 1995): 92–97.

Bell, Charles. *The Hand, Its Mechanism and Vital Endowments, As Evincing Design: The Bridgewater Treatises on the Power, Wisdom, and Goodness of God as Manifested in the Creation.* (Treatise IV.) New York: Harper and Brothers, 1840.

Bennett, K. M. B., and U. Castiello, eds. *Insights into the Reach to Grasp Movement.* Amsterdam: Elsevier Science B.V., 1994.

Benson, Frank. *Aphasia, Alexia and Agraphia.* New York: Churchill Livingston, 1979.

Bernstein, Nicolai. *The Coordination and Regulation of Movements.* Oxford: Pergamon Press, 1967.

Bjørkvold, Jon-Roar. *The Muse Within: Creativity and Communication, Song and Play from Childhood Through Maturity.* Translated by William H. Halverson. New York: HarperCollins, 1992.

Blacking, John. *How Musical Is Man?* London: Faber and Faber, 1976.

Bloom, B., ed. *Developing Talent in Young People.* New York: Ballantine Books, 1985.

Brust, J. "Music and Language: Musical Alexia and Agraphia." *Brain* 103 (1980): 367–92.

Bühler, Karl. *The Mental Development of the Child.* New York: Harcourt Brace, 1930.

Byrne, R., and A. Whiten. "The Thinking Primate's Guide to Deception." *New Scientist* 116, no. 1589 (1987): 54–57.

Calvin, William. *The Throwing Madonna.* New York: McGraw-Hill, 1983.

———. "The Unitary Hypothesis: A Common Neural Circuitry for Novel Manipulations, Language, Plan-ahead, and Throwing?" In Gibson and Ingold, eds., *Tools, Language and Cognition in Human Evolution.*

Chomsky, Noam. "Language and Thought." Anshen Transdisciplinary Lectureships in Art, Science, and the Philosophy of Culture, Monograph Three. Wakefield, R.I.: Moyer Bell (The Frick Collection), 1993.

Ciuffreda, K., and L. Stark. "Descartes' Law of Reciprocal Innervation." *American Journal of Optometry and Physiological Optics* 52, vol. 10 (1974): 663–73.

Corballis, Michael. *The Lopsided Ape.* New York: Oxford University Press, 1991.

Coren, Stanley. *The Left-Hander Syndrome: The Courses and Consequences of Left-Handedness.* New York: Free Press, 1992.

Critchley, M., and R. A. Henson, eds. *Music and the Brain: Studies in the Neurology of Music.* London: Heinemann, 1977.

Culin, S. *Games of the North American Indians.* New York: Dover, 1975.

Damasio, Antonio. *Descartes' Error: Emotions, Reason and the Human Brain.* New York: G. P. Putnam's Sons, 1994.

———, and H. Damasio. "Brain and Language." *Scientific American* 267, no. 3 (1992): 89–95.

Davidson, R., and K. Hugdahl, eds. *Brain Asymmetry.* Cambridge, Mass.: MIT Press, 1995.

Davies, Robertson. *Fifth Business.* New York: Penguin Books, 1970.

———. *What's Bred in the Bone.* New York: Penguin Books, 1986.

Dawkins, Richard. *Climbing Mount Improbable.* New York: W. W. Norton & Co., Inc., 1996.

———. *The Blind Watchmaker: Why the Evidence of Evolution Reveals a Universe Without Design.* New York: W. W. Norton & Co., Inc., 1986.

Deacon, Terrence. *The Symbolic Species: The Co-evolution of Language and the Brain.* New York: W. W. Norton & Company, 1997.

Dehaene, Stanislaus. *The Number Sense: How the Mind Creates Mathematics.* New York/Oxford: Oxford University Press, 1997.

Dennett, Daniel. *Darwin's Dangerous Idea: Evolution and the Meanings of Life.* New York: Simon and Schuster, 1995.

Denny-Brown, D., ed. *Selected Writings of Sir Charles Sherrington.* New York: Paul B. Hoeber, Inc., 1940.

Dewey, John. *Art as Experience.* New York: Minton and Balch, 1934.

Donald, Merlin. *Origins of the Modern Mind: Three Stages in the Evolution of Culture and Cognition.* Cambridge, Mass.: Harvard University Press, 1991.

Duchenne, Guillaume. *Physiology of Motion.* Translated and edited by E. Kaplan. Philadelphia: J. B. Lippincott Co., 1948; original French edition, 1867.

Dunbar, Robin. *Grooming, Gossip, and the Evolution of Language.* Cambridge, Mass.: Harvard University Press, 1996.

Eccles, John C. *Evolution of the Brain.* New York: Routledge, 1989.

Egan, Kieran. *The Educated Mind: How Cognitive Tools Shape Our Understanding.* Chicago: University of Chicago Press, 1997.

Esquivel, Laura. *Like Water for Chocolate.* New York: Doubleday, 1992.

Falk, Dean. *Braindance.* New York: Henry Holt and Co., 1992.

Feldenkrais, Moshe. *Awareness Through Movement.* New York: Harper and Row, 1977.

Forssberg, H., et al. "Development of Human Precision Grip." (4. Tactile Adaptation of Isometric Finger Forces to the Frictional Condition.) *Experimental Brain Research* 104 (1995): 323–30.

Frederiks, J. A. M., ed. *Handbook of Clinical Neurology,* vol. 1: *Clinical Neuropsychology.* Amsterdam: Elsevier Science Publishers, 1985.

Freund, H.-J. "The Apraxias." In Asburg, McKhann, and McDonald, eds., *Diseases of the Nervous System: Clinical Neurobiology II.*

Garber, Marjorie. *Vice Versa.* New York: Simon and Schuster, 1995.

Gardner, Howard. *Frames of Mind: The Theory of Multiple Intelligences.* 10th-anniv. ed. New York: Basic Books, 1993.

Gates, A., and J. Bradshaw. "The Role of the Cerebral Hemispheres in Music." *Brain and Language* 4 (1977): 403–31.

Geschwind, N., and A. Damasio. "Apraxia." In Frederiks, ed., *Handbook of Clinical Neurology,* vol. 1: *Clinical Neuropsychology.*

———, and A. Galaburda. *Cerebral Lateralization: Biological Mechanisms, Associations and Pathology.* Cambridge, Mass.: MIT Press, 1987.

———, and A. Galaburda, eds. *Cerebral Dominance: The Biological Foundations.* Cambridge, Mass.: Harvard University Press, 1984.

Gettings, Fred. *The Book of the Hand: An Illustrated History of Palmistry.* London: Paul Hamlyn, Ltd., 1965.

Gibson, Kathleen. "Animal Minds, Human Minds: General Introduction." In Gibson and Ingold, eds., *Tools, Language and Cognition in Human Evolution.*

———. "Beyond Neoteny and Recapitulation: New Approaches to the Evolution of Cognitive Development." In Gibson and Ingold, eds., *Tools, Language and Cognition in Human Evolution.*

———, and T. Ingold, eds., *Tools, Language and Cognition in Human Evolution.* New York: Cambridge University Press, 1993.

Gopnick, M., and S. Davis, eds. *The Biological Basis of Language.* Oxford: Oxford University Press, 1996.

Goss, C. "*On Movement of Muscles* by Galen of Pergamon." *Journal of Anatomy* 123 (1968): 1–26.

Gould, Stephen Jay. *The Panda's Thumb: More Reflections in Natural History.* New York: Penguin Books, 1980.

———. *Ontogeny and Phylogeny.* Cambridge, Mass.: Harvard University Press, 1977.

Granit, Ragnar. *The Purposive Brain.* Cambridge, Mass.: MIT Press, 1977; 1980 ed.

Greenfield, Patricia. "Language, Tools and Brain: The Ontogeny and Phylogeny of Hierarchically Organized Sequential Behavior." *Behavioral and Brain Sciences* 14 (1991): 531–95.

———, and L. Schneider. "Building a Tree Structure: The Development of Hierarchical Complexity and Interrupted Strategies in Children's Construction Activity." *Developmental Psychology* 13 (1977): 299–313.

Guiard, Yves. "Asymmetric Division of Labor in Human Skilled Bimanual Action: The Kinematic Chain as Model." *Journal of Motor Behavior* 19, no. 4 (1987): 486–517.

Gunnar, M., and M. Maratsos, eds. *The Minnesota Symposia on Child Psychology,* vol. 25. Hillsdale, N.J.: Lawrence Erlbaum Associates, 1992.

Gustin, W. "The Development of Exceptional Research Mathematicians." In Bloom, ed., *Developing Talent in Young People.*

Halsband, U., and H.-J. Freund. "Motor Learning." *Current Opinion in Neurobiology* 3 (1993): 940–49.

———, et al. "The Role of Premotor Cortex and the Supplementary Motor Area in the Temporal Control of Movement in Man." *Brain* 116 (1993): 243–66.

Hammond, G. E., ed. *Cerebral Control of Speech and Limb Movements.* Amsterdam: Elsevier Science Publishers, 1990.

Harnad, Steven, H. Steklis, and J. Lancaster, eds. *Origins and Evolution of Language and Speech.* Annals of the New York Academy of Sciences, vol. 280. New York: New York Academy of Sciences, 1976.

Harris, L. "Laterality of Function in the Infant: Historical and Contemporary Trends in Theory and Research." In Young, et al., *Manual Specialization and the Developing Brain.*

Healed, J., J. Liederman, and N. Geschwind. "Handedness Is Not a Unidimensional Trait." *Cortex* 22 (1986): 33–53.

Hein, A., and R. Diamond. "Contribution of Eye Movement to the Representation of Space." In Hein and Jeannerod, eds., *Spatially Oriented Behavior.*

Hein, A., and Marc Jeannerod, eds. *Spatially Oriented Behavior.* New York: Springer-Verlag, 1983.

Hewes, Gordon. "The Current Status of the Gestural Theory of Language Origin." In Harnad, Steklis, and Lancaster, eds., *Origins and Evolution of Language and Speech.*

———. "A History of the Study of Language Origins and the Gestural Primacy Hypothesis." In Lock and Peters, eds., *Handbook of Human Symbolic Evolution.*

Holloway, Ralph. "Evolution of the Human Brain." In Lock and Peters, eds., *Handbook of Symbolic Evolution.*

———. "Paleoneurological Evidence for Language Origins." In Harnad, Steklis, and Lancaster, eds., *Origins and Evolution of Language and Speech.*

Humphrey, D., and H.-J. Freund. *Motor Control: Concepts and Issues.* New York: John Wiley and Sons, 1989.

Ingold, T. "Tool Use, Sociality and Intelligence." In Gibson and Ingold, eds., *Tools, Language and Cognition in Human Evolution.*

Jeannerod, Marc. *The Cognitive Neuroscience of Action.* Cambridge, Mass.: Blackwell Publishers, 1997.

———. *The Neural and Behavioral Organization of Goal-Directed Movements.* Oxford: Clarendon Press, 1988.

Johanson, D., and M. Edey. *Lucy: The Beginnings of Humankind.* New York: Simon and Schuster, 1981.

Johnson, M., ed. *Brain Development and Cognition.* Oxford: Oxford University Press, 1993.

Jones, Frederick Wood. *The Principles of Anatomy as Seen in the Hand.* 2d ed. Baltimore: Williams and Wilkins, 1942.

Joseph, H. *A Book of Marionettes.* New York: Viking Press, 1929.

Kemper, S. "If It's Impossible, Michael Moschen Will Do It Anyway." *Smithsonian,* August 1995, pp. 38–47.

Kilbreath S., and S. Gandevia. "Limited Independent Flexion of the Thumb and Fingers in Human Subjects." *Journal of Physiology* 479, no. 3 (1994): 487–97.

Kimura, Doreen. *Neuromotor Mechanisms in Human Communication.* New York: Oxford University Press, 1993.

Klein, R. *The Human Career.* Chicago: University of Chicago Press, 1989.

Kumin, Maxine. *House, Bridge, Fountain, Gate.* New York: Viking Press, 1975.

Lanais, R., and J. Welsh. "On the Cerebellum and Motor Learning." *Current Opinion in Neurobiology* 3 (1993): 958–65.

Landau, Misia. *Narratives of Human Evolution.* New Haven: Yale University Press, 1991.

Landowska, Wanda. *Landowska on Music.* Collected, edited and translated by Denise Restout, assisted by Robert Hawkins. New York: Stein and Day, 1964.

Lane, Harlan. *The Mask of Benevolence: Disabling the Deaf Community.* New York: Alfred A. Knopf, 1992.

———, ed. Translated by Franklin Philip. *The Deaf Experience: Classics in Language and Education.* Cambridge, Mass.: Harvard University Press, 1984.

Lawrence, D. H. *Fantasia of the Unconscious.* New York: Penguin Books, 1960; first published in 1922.

Lewis, O. J. *Functional Morphology of the Evolving Hand and Foot.* Oxford: Clarendon Press, 1989.

Lieberman, Philip. *Uniquely Human: The Evolution of Speech, Thought, and Selfless Behavior.* Cambridge, Mass.: Harvard University Press, 1991.

Lock, Andrew. "Language Development and Object Manipulation." In Gibson and Ingold, eds., *Tools, Language and Cognition in Human Evolution.*

———, and C. Peters, eds. *Handbook of Human Symbolic Evolution.* Oxford: Clarendon Press, 1996.

Long, Charles. "Electromyographic Studies of Hand Function" (fig. 4). In R. Tubiana, ed., *The Hand,* vol. 1.

———, et al. "Intrinsic-Extrinsic Muscle Control of the Hand in Power Grip and Precision Handling." *Journal of Bone and Joint Surgery* 52-A, no. 5 (1970): 853–67.

McGrew, W. *Chimpanzee Material Culture: Implications for Human Evolution.* New York: Cambridge University Press, 1992.

McManus, I., and P. Breton. "The Neurobiology of Handedness, Language, and Cerebral Dominance: A Model for the Molecular Genetics of Handedness." In Johnson, ed., *Brain Development and Cognition.*

MacNeilage, P., M. Studdert-Kennedy, and B. Lindlom. "Primate Handedness Reconsidered." *Behavioral and Brain Sciences* 10 (1987): 247–303.

McNeill, D. "So You Think Gestures Are Nonverbal?" *Psychological Review* 92, no. 3 (1985): 350–71.

Macpherson, J. "How Flexible Are Muscle Synergies?" In D. Humphrey and H.-J. Freund, eds., *Motor Control: Concepts and Issues.*

Marzke, Mary. "Evolution." In Bennett and Castiello, eds., *Insights into the Reach to Grasp Movement.*

———. "Evolution of the Hand and Bipedality." In Lock and Peters, eds., *Handbook of Symbolic Evolution.*

———. "Evolutionary Development of the Human Thumb." *Hand Clinics* 8, no. 1 (February 1992).

———. "Joint Functions and Grips of the *Australopithecus afarensis* Hand, with Special Reference to the Region of the Capitate." *Journal of Human Evolution* 12 (1983): 197–211.

———. "Precision Grips, Hand Morphology, and Tools." *American Journal of Physical Anthropology* 102 (1997): 91–110.

———, J. Longhill, and S. Rasmussen. "Gluteus Maximus Muscle Function and the Origin of Hominid Bipedality." *American Journal of Physical Anthropology* 77 (1988): 519–28.

——— and K. Wullstein. "Chimpanzee and Human Grips: A New Classification with a Focus on Evolutionary Morphology." *International Journal of Primatology* 17 (1996): 117–39.

———, K. Wullstein, and S. Viegas. "Evolution of the Power ('Squeeze') Grip and Its Morphological Correlates in Hominids." *American Journal of Physical Anthropology* 89 (1992): 283–98.

Mayer, Ernst. *This Is Biology: The Science of the Living World.* Cambridge, Mass.: Harvard University Press, 1997.

Merzenich, M., and K. Sameshima. "Cortical Plasticity and Memory." *Current Opinions in Neurobiology* 3 (1993): 187–96.

Miller, Jonathan. *The Body in Question.* New York: Random House, 1978.

Morell, Virginia. *Ancestral Passions: The Leakey Family and the Quest for Humankind's Beginnings.* New York: Simon and Schuster, 1995.

Morris, Desmond. *The Naked Ape.* New York: Dell, 1967.

Müller, K., and V. Hömberg. "Development of Speed of Repetitive Movements in Children Is Determined by Structural Changes in Corticospinal Efferents." *Neuroscience Letters* 144 (1992): 57–60.

Napier, John. "The Prehensile Movements of the Human Hand." *Journal of Bone and Joint Surgery* 38-B, no. 4 (1956): 902–13.

———. *Hands.* Rev. ed. Princeton, N.J.: Princeton University Press, 1993.

Nuda, R., et al. "Neurophysiological Correlates of Hand Preference in Primary Motor Cortex of Adult Squirrel Monkeys." *Journal of Neuroscience* 12, no. 8. (1992): 2818–947.

O'Connor, Johnson. *Born That Way.* Baltimore: Williams and Wilkins, 1923.

Oldfield, R. C. "The Assessment and Analysis of Handedness: The Edinburgh Inventory." *Neuropsychologia* 9 (1971): 97–113.

Oppenheimer, Todd. "The Computer Delusion." *Atlantic Monthly,* July 1997, pp. 45–62.
Paz, Octavio. *Early Poems* 1935–1955. New York: New Directions Publishing Corporation, 1973.
Penfield, W., and T. Rasmussen. *The Cerebral Cortex of Man.* New York: Macmillan, 1950.
Peters, M. "Subclassification of Non-pathological Left-handers Poses Problems for Theories of Handedness." *Neuropsychologia* 28, no. 3 (1990): 279–89.
———, and M. Lang. "Do 'Right-Armed' Lefthanders Have Different Lateralization of Motor Control for the Proximal and Distal Musculature?" *Cortex* 28 (1992): 391–99.
Petitto, L. "In the Beginning: On the Genetic and Environmental Factors That Make Early Language Acquisition Possible." In Gopnick and Davis, eds., *The Biological Basis of Language.*
———. "Modularity and Constraints in Early Lexical Acquisition: Evidence from Children's Early Language and Gesture." In Gunnar and Maratsos, eds. *The Minnesota Symposia on Child Psychology,* vol. 25.
———. "On the Autonomy of Language and Gesture: Evidence from the Acquisition of Personal Pronouns in American Sign Language." *Cognition* 27, no. 1 (1987): 1–62.
———, and P. Marentette. "Babbling in the Manual Mode: Evidence for the Ontogeny of Language." *Science* 251 (1991): 1493–96.
Phillips, C. *Movements of the Hand.* Liverpool: Liverpool University Press, 1986.
Pinker, Steven. *How the Mind Works.* New York: W. W. Norton, 1997.
———. *The Language Instinct.* New York: William Morrow and Co., 1994.
Plotkin, Henry. *Darwin Machines and the Nature of Knowledge.* Cambridge, Mass.: Harvard University Press, 1993.
———. *Evolution in Mind.* London: Penguin Books, Ltd., 1997.
Poizner, H., E. Klima, and U. Bellugi. *What the Hands Reveal about the Brain.* Cambridge, Mass.: MIT Press, 1987.
Pramstaller, P., and C. D. Marsden. "The Basal Ganglia and Apraxia." *Brain* 119 (1996): 319–40.
Radetsky, Peter. "Silence, Signs and Wonder." *Discovery,* August 1994, pp. 62–68.
Reynolds, Peter C. "The Complementation Theory of Language and Tool Use." In Gibson and Ingold, eds., *Tools, Language and Cognition in Human Evolution.*
Roehmann, F., and F. Wilson, eds. *The Biology of Music Making: Proceedings of the 1984 Denver Conference.* St. Louis: MMB Music, 1988.
Rosenbaum, D. *Human Motor Control.* New York: Academic Press, 1991.
Rothwell, J. *Control of Human Voluntary Movement.* New York: Chapman and Hall, 1994.
Roy, E., and P. Square-Storer. "Evidence for Common Expressions of Apraxia." In Hammond, ed., *Cerebral Control of Speech and Limb Movements.*

Sacks, Oliver. *Seeing Voices: A Journey into the World of the Deaf.* Berkeley: University of California Press, 1989.

———. "Neurology and the Soul." *New York Review of Books,* November 22, 1990, pp. 44–50.

Sarason, Seymour. *The Challenge of Art to Psychology.* New Haven: Yale University Press, 1990.

Schick, K., and N. Toth. *Making Silent Stones Speak: Human Evolution and the Dawn of Technology.* New York: Simon and Schuster, 1993.

Schieber, M. "Muscular Production of Individuated Finger Movements: The Roles of Extrinsic Finger Muscles." *Journal of Neuroscience* 15, no. 1 (1995): 284–97.

———. "How Might the Motor Cortex Individuate Movements?" *Trends in Neuroscience* 13, no. 11 (1990): 440–45.

Schueneman, A., and J. Pickleman. "Neuropsychological Analysis of Surgical Skill." In Starkes and Allard, eds., *Cognitive Issues in Motor Expertise.*

Seitz, R., and P. Roland. "Learning of Sequential Finger Movements in Man: A Combined Kinematic and Positron Emission Tomography (PET) Study." *European Journal of Neuroscience* 4 (1992): 154–65.

"Serge Percelly Does Not Take Juggling Lightly." *The New Yorker,* November 8, 1993, pp. 50–51.

Sergeant, J. "Mapping the Musical Brain." *Human Brain Mapping* 1 (1993): 20–38.

———. "Distributed Neural Network Underlying Musical Sight-Reading and Keyboard Performance." *Science* 257 (1992): 106–9.

Sherrington, Charles. *The Integrative Action of the Nervous System.* New York: Scribner, 1906.

Shreeve, James. *The Neandertal Enigma: Solving the Mystery of Modern Human Origins.* New York: William Morrow and Co., 1995.

Sicard, Roch-Ambroise. "Course of Instruction for a Congenitally Deaf Person." In Lane, ed., *The Deaf Experience.*

Sloane, D., and L. Sosniak. "The Development of Accomplished Sculptors." In Bloom, *Developing Talent in Young People.*

Sosniak, Lauren. "From Tyro to Virtuoso: A Long-term Commitment to Learning." In Wilson and Roehmann, eds., *Music and Child Development.*

———. "Learning to Be a Concert Pianist." In Bloom, ed., *Developing Talent in Young People.*

Spalteholz, Werner. *Hand Atlas of Human Anatomy.* Philadelphia/London: J.B. Lippincott Company, 1923.

Speaight, George. *The History of the English Puppet Theatre.* 2d ed. Carbondale, Ill.: Southern Illinois University Press, 1990.

Springer, S., and G. Deutsche. *Left Brain, Right Brain.* 4th ed. New York: W. H. Freeman and Co., 1993.

Staden, Heinrich von. *Herophilus: The Art of Medicine in Early Alexandria.* New York: Cambridge University Press, 1989.

Starkes, J., and F. Allard, eds. *Cognitive Issues in Motor Expertise.* Amsterdam: Elsevier Science Publishers B.V., 1993.

Steenhuis, R., and M. Breton. "Different Dimensions of Hand Preference That Relate to Skilled and Unskilled Activities." *Cortex* 25 (1989): 289–304.

Steklis, D., and M. Raleigh, eds. *Neurobiology of Social Communication in Primates: An Evolutionary Perspective.* New York: Academic Press, 1979.

Stern, C., and W. Stern. *Die Kindersprache.* Leipzig: Barth, 1928.

Tattersall, Ian. *Becoming Human: Evolution and Human Uniqueness.* New York: Harcourt Brace & Co., 1998.

———. *The Fossil Trail: How We Know What We Think We Know about Human Evolution.* New York/Oxford: Oxford University Press, 1995.

Thach, W., H. Goodkin, and J. Eating. "The Cerebellum and the Adaptive Coordination of Movement." *Annual Review of Neuroscience* 15 (1992): 403–42.

Tubiana, Raoul. "Architecture and Functions of the Hand." In Tubiana, ed., *The Hand,* vol. 1.

———, ed. *The Hand,* vol. 1. Philadelphia: W. B. Saunders, 1981.

Tufte, Edward. *Visual Explanations: Images and Quantities, Evidence and Narrative.* Cheshire, Conn.: Graphics Press, 1997.

Vygotsky, Lev. Translation revised and edited by A. Kozulin. *Thought and Language.* Cambridge, Mass.: MIT Press, 1986; originally published in Russian, 1934.

Wagner, C. "Success and Failure in Musical Performance: Biomechanics of the Hand." In Roehmann and Wilson, eds., *The Biology of Music Making: Proceedings of the 1984 Denver Conference.*

Washburn, Sherwood L. "Tools and Human Evolution." *Scientific American* 203, no. 3 (1960): 63–75.

Wertheim, N., and M. Botez. "Receptive Amusia: A Clinical Analysis." *Brain* 84 (1961): 19–30.

Wiesendanger, M., et al. "Two Hands, One Action." In Wing, Haggard, and Flanagan, eds., *Hand and Brain: The Neurophysiology and Psychology of Hand Movements.*

Wilkins, W., and J. Wakefield. "Brain Evolution and Neurolinguistic Preconditions." *Behavioral and Brain Sciences* 18 (1995): 161–226.

Wilson, Frank R. *Tone Deaf and All Thumbs?* New York: Viking-Penguin, 1986.

——— and F. Roehmann, eds., *Music and Child Development.* St. Louis: MMB Music, Inc., 1990.

Wing, A., P. Haggard, and J. Flanagan, eds. *Hand and Brain: The Neurophysiology and Psychology of Hand Movements.* San Diego: Academic Press, 1996.

Wynn, Thomas G. "The Evolution of Tools and Symbolic Behavior." In Lock and Peters, eds. *Handbook of Human Symbolic Evolution.*

Young, Gerald. "Changes, Constancies and Continuities in Lateralization Development." In Young, et al., eds., *Manual Specialization and the Developing Brain.*

———, et al., eds. *Manual Specialization and the Developing Brain.* New York: Academic Press, 1983.

Zeki, S. "The Visual Image in Mind and Brain." *Scientific American,* September 1992, pp. 69–76.

Permissions Acknowledgments

Grateful acknowledgment is made to the following for permission to reprint previously published material:

Blackwell Publishers, Inc.: Excerpts from *The Cognitive Neuroscience of Action* by Mark Jeannerod (Blackwell Publishers, Inc., Oxford, 1997). Excerpt from "The Neurobiology of Handedness, Language, and Cerebral Dominance, a Model for the Molecular Genetics of Behavior" by I. McManus and M. Breton (*Brain Development and Cognition,* edited by M. Johnson, Blackwell Publishers, Inc., Oxford, 1993, p. 683). Reprinted by permission of Blackwell Publishers, Inc.

Cambridge University Press: Excerpts from *Tools, Language, and Cognition in Human Evolution* by Kathleen R. Gibson and Tim Ingold (Cambridge University Press, New York, 1993, pp. 411, 430–431). Reprinted by permission of Cambridge University Press.

Elsevier Science: Excerpt from "Neuropsychological Analysis of Surgical Skill" by A. Shueneman and J. Pickleman (*Cognitive Issues in Motor Expertise,* edited by J. L. Starkes and F. Allard, Elsevier Science, Oxford, 1993, p. 189). Reprinted by permission of Elsevier Science.

Harvard University Press: Excerpts from *The Origins of the Modern Mind* by Merlin Donald (Harvard University Press, Cambridge, Mass.). Copyright © 1991 by the President and Fellows of Harvard College. Excerpts from *The Deaf Experience,* edited by Harlan Lane, translated by Franklin Philip. Copyright © 1984 by the President and Fellows of Harvard College. Reprinted by permission of Harvard University Press.

Maxine Kumin: "The Absent Ones" from *House, Bridge, Fountain, Gate* by Maxine Kumin (Viking Press, New York, 1975). Reprinted by permission of the author.

Massachusetts Institute of Technology: Excerpts from *A Computational Theory of Physical Skill* by H. A. Austin (MIT, Cambridge, Mass., 1976). Reprinted by permission of the Massachusetts Institute of Technology.

The MIT Press: Excerpts from *Thought and Language* by Lev S. Vygotsky, translated and edited by Alex Kozulin (The MIT Press, Cambridge, Mass., 1986, pp. 62, 65, 91–94,

107, 127, 222). Copyright © 1986 by The Massachusetts Institute of Technology. Reprinted by permission of The MIT Press.

MMB Music, Inc.: Excerpts from *Music and Child Development* by Frank Wilson and F. Roehmann, editors. Copyright © 1990 by MMB Music, Inc., St. Louis. All rights reserved. Reprinted by permission of MMB Music, Inc.

William Morrow & Company and Penguin Books Ltd.: Excerpts from *The Language Instinct: The New Science of Language and Mind* by Steven Pinker (William Morrow & Company, New York, and Allen Lane, The Penguin Press, London, 1994, pp. 18, 269–271, 313–314, 352). Copyright © 1994 by Steven Pinker. Reprinted by permission of William Morrow & Company and Penguin Books Ltd.

New Directions Publishing Corp.: "Object Lesson" from *Early Poems of Octavio Paz* by Octavio Paz, translated by Muriel Rukeyser. Copyright © 1973 by Octavio Paz and Muriel Rukeyser. Reprinted by permission of New Directions Publishing Corp.

The New York Review of Books: Excerpt from "Neurology and the Soul" by Oliver Sacks (*The New York Review of Books,* Nov. 22, 1990). Copyright © 1990 by NYREV, Inc. Reprinted by permission of *The New York Review of Books.*

Todd Oppenheimer: Excerpt from "The Computer Delusion" by Todd Oppenheimer (*The Atlantic Monthly,* July 1997, pp. 45–62). Copyright © 1997 by Todd Oppenheimer. Reprinted by permission of the author.

Penguin Books Ltd.: Approximately 750 words from *Darwin Machines and the Nature of Knowledge* by Henry Plotkin (Allen Lane, The Penguin Press, London, 1994, pp. 84, 139, 144, 149, 201–202, 203, 206). Copyright © 1994 by Henry C. Plotkin. Reprinted by permission of Penguin Books Ltd.

Laurence Pollinger Limited and the Estate of Frieda Lawrence Ravagli: Excerpt from *Fantasia of the Unconscious* by D. H. Lawrence. Copyright © 1922 by Thomas Seltzer, Inc. Copyright renewed 1950 by Frieda Lawrence. Reprinted by permission of Laurence Pollinger Limited and the Estate of Frieda Lawrence Ravagli.

W. B. Saunders Company: Excerpt from "Evolutionary Development of the Human Thumb" by Mary Marzke (*Hand Clinics,* vol. 8, no. 1, Feb. 1992). Reprinted by permission of W. B. Saunders Company.

Science: Excerpt from "Distributed Neural Network Underlying Musical Sight-Reading and Keyboard Performance" by Justine Sergeant (*Science* 257, 1992, pp. 106–109). Copyright © 1992 by the American Association for the Advancement of Science. Reprinted by permission of *Science.*

Scientific American: Excerpt from "Tools and Human Evolution" by Sherwood L. Washburn (*Scientific American* 203, no. 3, 1960, pp. 63–75). Reprinted by permission of *Scientific American.*

Teachers College Press: Excerpt from *The Reflective Turn: Case Studies in and on Educational Practice,* edited by D. A. Schön (Teachers College Press, New York, pp. 38, 44). Copyright © 1991 by Teachers College, Columbia University. All rights reserved. Reprinted by permission of Teachers College Press.

Index

(*Italicized* page numbers indicate illustrations)